With the compliments of the editor
John Stewart

# Scaling up in Hydrology using Remote Sensing

# Scaling up in Hydrology using Remote Sensing

*Edited by*

**J.B. STEWART**
*Institute of Hydrology, UK*

**E.T. ENGMAN**
*NASA Goddard Space Flight Center, USA*

**R.A. FEDDES**
*Wageningen Agricultural University, The Netherlands*

**Y. KERR**
*Centre d'Etudes Spatiales de la Biosphere, France*

BAHC

Institute of Hydrology

JOHN WILEY & SONS
Chichester • New York • Brisbane • Toronto • Singapore

Published in 1996 by John Wiley & Sons Ltd,
Baffins Lane, Chichester,
West Sussex PO19 1UD, England

National 01243 779777
International (+44) 1243 779777
e-mail (for orders and customer service enquiries): cs-books@wiley.co.uk
Visit our Home Page on http://www.wiley.co.uk
or http://www.wiley.com



*Other Wiley Editorial Offices*

John Wiley & Sons, Inc., 605 Third Avenue,
New York, NY 10158-0012, USA

Jacaranda Wiley Ltd, 33 Park Road, Milton,
Queensland 4064, Australia

John Wiley & Sons (Canada) Ltd, 22 Worcester Road,
Rexdale, Ontario M9W 1L1, Canada

John Wiley & Sons (Asia) Pte Ltd, 2 Clementi Loop #02-01,
Jin Xing Distripark, Singapore 051

***British Library Cataloguing in Publication Data***

A catalogue record for this book is available from the British Library

ISBN 0-471-96829-3

Produced from camera-ready copy supplied by the Institute of Hydrology
Printed and bound in Great Britain by Biddles Ltd, Guildford and King's Lynn
This book is printed on acid-free paper responsibly manufactured from sustainable forestation, for which at least two trees are planted for each one used for paper production.

# CONTENTS

# List of Contributors

**L.R. Ahuja**
*USDA-Agricultural Research Service-GPSR, 301 South Howes, PO Box E, Fort Collins, CO 80522, United States of America*

**W.G.M. Bastiaanssen**
*DLO-The Winand Staring Centre, Wageningen, The Netherlands*

**K.J. Beven**
*Centre for Research on Environmental Systems and Statistics, Institute of Environmental and Biological Sciences, Lancaster University, Lancaster LA1 4YQ, United Kingdom*

**A.L. Cognard**
*CEMAGREF-Hydrologie, Parc de Tourvoie, 92185 Antony, France*

**A. Chehbouni**
*ORSTOM, Laboratoire d'Hydrologie, 911 Avenue Agropolis, B.P. 5045, 34032 Montpelier, France*

**R.D. Crago**
*Department of Civil and Materials Engineering, 2095 Engineering Research Facility, University of Chicago, 842 West Taylor Street, Chicago, IL 60607-7023, United States of America*

**E.T. Engman**
*NASA Goddard Space Flight Center, Laboratory for Hydrospheric Processes, Hydrological Sciences Branch, Code 974, Greenbelt, MD 20771, United States of America*

**R.A. Feddes**
*Department of Water Resources, Wageningen Agricultural University, Nieuwe Kanaal 11, 6709 PA Wageningen, The Netherlands*

**J.W. Finch**
*Institute of Hydrology, Maclean Building, Crowmarsh Gifford, Wallingford, Oxon OX10 8BB, United Kingdom*

**J. Fisher**
*Centre for Research on Environmental Systems and Statistics, Institute of Environmental and Biological Sciences, Lancaster University, Lancaster LA1 4YQ, United Kingdom*

**J.R. Garratt**
*CSIRO Division of Atmospheric Research, Station Street, Aspendale, Vic. 3195, Australia*

**D. Gimenez**
*USDA-Agricultural Research Service Hydrology Lab, Beltsville, MD 20705, United States of America*

**A.E. Green**
*Hort Research, Kerikeri Research Centre, PO Box 23, Kerikeri, New Zealand*

**R.H. Harding**
*Institute of Hydrology, Maclean Building, Crowmarsh Gifford, Wallingford, Oxon OX10 8BB, United Kingdom*

**L.E. Hipps**
*Department of Plants, Soils and Biometeorology, Utah State University, Logan, UT 84322-4820, United States of America*

**D.H. Hoekman**
*Department of Water Resources, Wageningen Agricultural University, Nieuwe Kanaal 11, 6709 PA Wageningen, The Netherlands*

**S.J. Hook**
*Jet Propulsion Laboratory, California Institute of Technology, 4800 Oak Grove Drive, Pasadena, California 91109, United States of America*

**K.S. Humes**
*Department of Geography, University of Oklahoma, 100 E Boyd Street, Norman, Oklahoma 73019, United States of America*

**T.J. Jackson**
*USDA-Agricultural Research Service Hydrology Lab, Beltsville, MD 20705, United States of America*

**K. Johnsen**
*USDA-Agricultural Research Service-GPSR, 301 South Howes, PO Box E, Fort Collins, CO 80522, United States of America*

**W.P. Kustas**
*USDA-Agricultural Research Service Hydrology Lab, Beltsville 20705, MD, United States of America*

**J-P. Lagouarde**
*INRA Bioclimatologie, BP 81, 33883 Villenave d'Ornon, France*

**S. Leguizamon**
*University of Mendoza, 5500 Mendoza, Argentina*

**J-P. Lhomme**
*ORSTOM, Laboratoire d'Hydrologie, B.P. 5045, 34032 Montpellier, France*

**C. Loumagne**
*CEMAGREF-Hydrologie, Parc de Tourvoie, 92185 Antony, France*

**N.M. Mattikalli**
*NASA Goddard Space Flight Center, Laboratory for Hydrospheric Processes, Hydrological Sciences Branch, Code 974, Greenbelt, MD 20771, United States of America*

**K.J. McAneney**
*Hort Research, Kerikeri Research Centre, PO Box 23, Kerikeri, New Zealand*

**K.G. McNaughton**
*Hort Research, Kerikeri Research Centre, PO Box 23, Kerikeri, New Zealand*

**M. Menenti**
*DLO-The Winand Staring Centre, Wageningen, The Netherlands*

**B. Monteny**
*ORSTOM, Laboratoire d'Hydrologie, B.P. 5045, 34032 Montpellier, France*

**C.M.U. Neale**
*Department of Plants, Soils and Biometeorology, Utah State University, Logan, UT 84322-4820, United States of America*

**E.G. Njoku**
*Jet Propulsion Laboratory, California Institute of Technology, 4800 Oak Grove Drive, Pasadena, California 91109, United States of America*

**M. Normand**
*CEMAGREF-Hydrologie, Parc de Tourvoie, 92185 Antony, France*

**D. Or**
*Department of Plants, Soils and Biometeorology, Utah State University, Logan, UT 84322-4820, United States of America*

**C. Ottlé**
*CETP/CNRS, 10-12 Avenue de l'Europe, 78140 Vélizy, France*

**Y. Pachepsky**
*USDA-Agricultural Research Service Remote Sensing and Modeling Lab, Beltsville, MD 20705, United States of America*

**R. Parry**
*USDA-Agricultural Research Service Hydrology Lab, Beltsville, MD 20705, United States of America*

**H. Pelgrum**
*DLO-The Winand Staring Centre, Wageningen, The Netherlands*

**A.J. Prata**
*CSIRO Division of Atmospheric Research, Station Street, Aspendale, Vic. 3195, Australia*

**M.R. Raupach**
*CSIRO Centre for Environmental Mechanics, GPO Box 821, Canberra, ACT 2601, Australia*

**J.C. Ritchie**
*USDA-Agricultural Research Service Hydrology Lab, Beltsville, MD 20705, United States of America*

**T.J. Schmugge**
*USDA-Agricultural Research Service Hydrology Lab, Beltsville, MD 20705, United States of America*

**J.B. Stewart**
*Institute of Hydrology, Maclean Building, Crowmarsh Gifford, Wallingford, Oxon OX10 8BB, United Kingdom*

**C.M. Taylor**
*Institute of Hydrology, Maclean Building, Crowmarsh Gifford, Wallingford, Oxon OX10 8BB, United Kingdom*

**P.J. van Oevelen**
*Department of Water Resources, Wageningen Agricultural University, Nieuwe Kanaal 11, 6709 PA Wageningen, The Netherlands*

**D. Vidal-Madjar**
*CETP/CNRS, 10-12 Avenue de l'Europe, 78140 Vélizy, France*

# Preface

The increasing awareness that the hydrological cycle responds to feedbacks at local, regional, continental and global scales has led to the necessity for modelling atmospheric and hydrological processes at all of these scales. However, a major problem exists in obtaining quantitative data on key hydrological variables which affect land–atmosphere interactions, particularly at the larger scales. The ability of remote sensing techniques to measure radiances from areas of metres up to the complete globe, give it the potential to provide information which cannot easily be obtained by conventional ground-based techniques.

Research is currently under way to develop the use of remote sensing to determine a range of hydrological variables such as soil moisture, vegetation type, surface temperature, aerodynamic and surface conductance, for water resource, vegetation growth and land–atmosphere interaction models. Remote sensing can also be used to calibrate and test large-scale models by determining some of their outputs, such as sensible heat flux and radiation emitted at the top of the atmosphere. The use of satellite data for hydrological purposes has been limited so far by the availability of reliable algorithms to convert the measured radiances into the required hydrological variables.

Calibration and testing of remote sensing algorithms requires data at a variety of scales, ranging from tens to hundreds of metres (from gound-based measurements) through one to tens of kilometres (from planetary boundary layer measurements) up to hundreds of kilometres (from mesoscale models). Successful development of the necessary algorithms has been greatly impeded by the difference in spatial and temporal scales between remotely-sensed data and measured and modelled data. There is a need to scale up ground-based measurements, for example of rainfall and soil moisture, to that of the areal averages of the related satellite data. Similarly, there is a need to scale-up the instantaneous satellite measurements to the temporal averages obtained from conventional data and models.

Against this background, it was felt that the time was right to organise a Workshop of international experts to consider the issues involved in scaling-up of hydrological variables using remote sensing. The objectives of the Workshop were to assess the current status of the science, to identify major gaps in our knowledge and to bring together scientists with an interest in scaling-up issues; these will then be used to focus future research and collaboration in this field. The Workshop ran from 10-12 June 1996 and was held at and sponsored by the Institute of Hydrology, a component organisation of the UK's Natural Environment Research Council's Centre for Ecology and Hydrology. Co-sponsorship and assistance with the organisation of the meeting was also provided by the International Geosphere-Biosphere Programme via its Biospheric Aspects of the Hydrological Cycle (BAHC) Programme and the World Climate Research Programme's International Satellite Land Surface Climatology Project (ISLSCP).

This publication contains the 17 papers presented at the Workshop. All have been externally reviewed and then revised. There are nine papers which investigate the effect of scale on a range of hydrological variables using either field measurements or models. A further eight papers assess the relative merits of current methods of scaling up hydrological variables. Collectively, they report international state-of-the-art opinion on the scientific problems and progress in these important topics. The major findings and recommendations of the Workshop are summarised in a paper published after the meeting in the International Journal of Remote Sensing.

The Organising Committee would like to acknowledge the support and encouragement of Jim Wallace in establishing the Workshop and the efforts by Celia Kirby and her colleagues in preparing the camera-ready material for this book and helping with the domestic arrangements for the Workshop.

J.B. Stewart (Institute of Hydrology)
E.T. Engman (NASA Goddard Space Flight Center)
R.A. Feddes (Wageningen Agricultural University)
Y. Kerr (Centre d'Etudes Spatiales de la Biosphere)

# 1 Remote sensing and scaling in hydrology

**K.J. BEVEN and J. FISHER**
*Centre for Research on Environmental Systems and Statistics, Lancaster University, UK*

## PROBLEMS AND PRACTICE

The primary task of the hydrologist is the proper assessment of the water balance of a catchment area. This involves two main tasks. The first is the quantification of the inputs of water to the catchment. The second is the partitioning of those inputs into time series of discharges (both surface and subsurface as necessary) and evapotranspiration. At large (annual) time scales and large ($10^4$ km$^2$) space scales it has been suggested that these tasks can be accomplished with a reasonable degree of accuracy by very simple empirical relationships (e.g. Dooge, 1994). At smaller time and space scales, factors such as seasonality of climate and vegetation in time and heterogeneity of rainfalls, other meteorological variables, and catchment characteristics in space make both tasks very much more difficult; sufficiently difficult, in fact, that there is no adequately developed theory of scale dependence or scaling in hydrology. It will be argued in what follows that the classical approaches to the scaling problem in hydrology are fundamentally flawed and the potential for remote sensing data to aid alternative approaches will be assessed.

We should immediately recognise that there are two distinct problems involved in scaling. One is the requirement for a set of concepts that will allow the correct (or at least acceptable) partitioning of the water balance at any given scale. This is the *scale problem*. Hydrology lacks a theory for the scale problem but the problem has been solved in a pragmatic way by a variety of models, at least for cases where data are available for model calibration at the appropriate scale. The second problem is the requirement for a set of concepts that would allow information gathered at one scale to be used in making predictions at other scales. This is the *scaling* or *aggregation/disaggregation* problem. Hydrology lacks a theory for the scaling problem as well and in this case the pragmatic approaches that have been used have been simplistic in the extreme: so simplistic indeed as to be indefensible on any scientific grounds. Examples are the use of a single set of land surface parameters to represent the heterogeneity of the landscape in the hydrological components of large scale grid elements of General Circulation Models, or at smaller scales, and the use of 'effective' soil characteristic parameters at the grid scale in distributed

*Scaling up in hydrology using remote sensing*. Edited by J.B. Stewart, E.T. Engman, R.A. Feddes and Y. Kerr

'physically-based' hydrological models. In both cases there is an implicit assumption that the use of "effective" parameters in the small scale model is adequate to represent the heterogeneity of responses at the large scale. With the notable exception of developments in the theory of groundwater flow and transport (see for example Dagan, 1986), the *ad hoc* concept of 'effective' parameter values is the implicit scaling theory used in hydrology (including, still, in models of groundwater flow and transport). It has been criticised as inadequate elsewhere (e.g. Beven, 1989; 1995), and works, except for some simple analytical techniques for aggregating fluxes (e.g. Lhomme *et al.*, 1994; Raupach and Finnigan, 1995), only if data are available to calibrate those 'effective' parameter values at the desired scale of application.

Thus both the scale problem and the scaling problem depend, at present, on calibration data. Where such data are available the models used will have enough degrees of freedom to be able to simulate the calibration data to some acceptable level of accuracy. That is not the problem. This is true whether the models use small scale physical theory applied at much larger scales (e.g. Refsgaard *et al.*, 1992) or whether they are more explicitly conceptual in nature and designed to be applied only at larger scales (e.g. Jolley and Wheater, 1993). In fact, recent work has suggested that the problem is not that calibration will not provide adequate simulations but more that there are many different parameter sets, scattered throughout the parameter space, that will provide adequate simulations for whatever quantitative measure of performance is being used (e.g. Duan *et al.*, 1992; Beven, 1993; Beven and Quinn, 1994; Freer *et al.*, 1996). The implications of this are that parameter values derived in this way have meaning only within the context of the model structure and are conditional on both the values of the other parameters and the performance measure used in calibration. This must cast doubt on the predictions made for periods and conditions other than the calibration period, which are then best performed within an uncertainty estimation framework (e.g. Beven and Binley, 1992).

This situation has arisen in part because of the complexity of hydrological processes in space and time and the consequent difficulty of formulating an appropriate predictive theory, and partly because of the lack of suitable measurement techniques to study hydrological processes at different scales, particularly the scales at which predictions are required. Most measurement techniques used in hydrology are at the 'point' scale. This is useful in the case of discharge gaugings, since the discharge measurement integrates all of the complex flow processes of catchment response. It is less useful for most other hydrological variables, from rainfall to soil moisture, since it is not normally possible to have an adequate number of measurements to allow other than the very crudest integration up to the catchment scale at which predictions are normally required. This is true even of research catchments and underlies the lack of a proper scaling theory in hydrology. It is difficult to develop theory without examples, while the use of hypothetical examples (e.g. Binley *et al.*, 1989), which must be based on one particular (and not necessarily correct) theory, gives only an indication of what may be involved without providing much that is usable in practical situations where the structural parameters cannot be inferred from the small number of point measurements that might be available (if any).

In this context, a great deal of effort has been put into remote sensing as a source of distributed hydrologically useful information. Remote sensing has the unique potential to indicate the extent and nature of hydrological variability at the catchment and landscape scale. It has the potential to provide information on spatial and temporal patterns of catchment characteristics and hydrological variables instantaneously and synoptically in a way that no other technique can realistically achieve at large scales within reasonable financial constraints. Summaries of the potential of remote sensing in hydrology are given for example by van de Griend and Engman (1985) and Engman and Gurney (1991). This potential has led to considerable optimism.

Todini (1988) in an article on the future of hydrological modelling, for example, wrote: "*new information systems based on satellite and radar technologies will provide means for acquisition of distributed data and the size and speed of computers will allow for the solution of large distributed differential systems*".

Later, Engman and Gurney (1991) added: "*the ability to measure the state of the surface over large areas, whether soil moisture or snow or the surface energy balance, really distinguishes remote sensing from conventional data collection, and allows different types of models to be developed and different scientific questions to be addressed*".

Such optimism has not (yet) been borne out in practice. One of the reasons is precisely the problem of scale. Remote sensing provides a measure of the average electromagnetic emissions, as affected by surface roughness and modified by the atmosphere, from a pixel. The pixel size varies with the sensor but is very often much larger than the 'point' scale of ground measurements. Averaging of emissions by the sensor may not be a hydrologically appropriate averaging. Comparison of pixel scale responses with point scale measurements may be affected by significant sub-pixel scale variability (e.g. Charpentier and Goffman, 1992). Thus it has been difficult in practice to demonstrate the general utility of remote sensing information. Some of the evidence will be considered in the following sections.

## REMOTE SENSING AND SCALING OF SOIL MOISTURE AND RUNOFF PRODUCTION

There have been few studies that have incorporated remote sensing information into rainfall-runoff modelling. Most of those have used remotely-sensed vegetation classification as an input into the definition of different hydrological response units (e.g. Drayton *et al.*, 1992; Nemani *et al.*, 1993; Kite, 1995 and others). There have, however, been just a small number of studies that have attempted to use remotely-sensed soil moisture data as a direct input to a rainfall-runoff model, or as verification data. Engman and Gurney (1991) point out that not all hydrological models can, in any case, make use of remotely-sensed estimates of surface soil moisture since they are not formulated with surface soil moisture as a variable. In general, those models that can make use of such information are the more complex distributed 'physically-based' models. Engman *et al.* (1989) used passive microwave data to develop a time sequence of surface soil moisture maps for the FIFE area in Kansas. They compared these data with the simplified subsurface flow model of Sloan and Moore (1984). The observations provided a check on model discretisation and performance.

A further study using passive microwave data was described by Goodrich *et al.* (1994) for the Monsoon '90 experimental area in the Walnut Gulch catchment in Arizona. Surface soil moisture data were used directly as an input for pre-storm moisture conditions in a version of the KINEROS rainfall-runoff model. This is based on a Hortonian infiltration excess runoff production mechanism. The surface soil moisture initialisation allows the suction term in the infiltration equation to be temporally and spatially variable. Results were compared with average initial soil moisture estimates and model derived estimates using the CREAMS model. It was found that the representation of rainfall variability was far more important than variability in the initial soil moisture, to the extent that fine resolution rainfall intensity information is required even at small scales. The model-derived initial soil moisture estimates provided runoff simulations were as accurate as the remotely-sensed estimates despite the fact that the two patterns were very different.

Recently, Wood (1995) has examined scaling of surface soil moisture in the 525 $km^2$ Little Washita catchment in Oklahoma based on both passive microwave data (the ESTAR thin array radiometer sensor) with a model with a resolution of 30 m that takes account of vegetation, soil and topographic differences in the landscape (TOPLATS, Famiglietti and Wood, 1994). Data from two days were analysed for both statistical self-similarity and for the threshold scale at which a statistical representation can be considered an adequate replacement for the actual pattern of variability (the 'representative elementary area', REA, of Wood *et al.*, 1988, 1990; see also Famiglietti and Wood, 1995; Beven, 1995, Woods *et al.*, 1995). The REA is defined as the scale at which the variance in some characteristic response variable between areas of a given scale stabilises with increasing scale. In the Little Washita, it was found that the data did not follow a simple scaling law but rather showed multiscaling behaviour, while the REA threshold was at a scale of 5 - 10 km, rather than the 1 km found for previous modelling studies.

There is no *a priori* reason why a fixed REA scale should exist (see discussions in Blöschl *et al.*, 1995; Beven, 1995; Fan and Bras, 1995). Indeed, process considerations suggest that it may be expected to vary between storms with the correlation length scale of the inputs. The concept is convenient in modelling terms in that it suggests that an area can be represented by the statistical distribution of the responses within it rather than having to consider every point in space in detail. It does, however, assume that the aggregation of those sub-REA responses will be a linear integration, even though the responses themselves may be nonlinear. In this, it has much in common with some of the statistical-dynamical, patch or tiling schemes suggested for the aggregation of sub-grid land surface fluxes (see for example Entekhabi and Eagleson, 1989; Dolman, 1992; Koster and Suarez, 1992; Avissar, 1992; Quinn *et al.*, 1995) as an alternative to attempts to define grid scale effective parameters for a lumped model representation. The landscape representation is then a (linear) sum of the (nonlinear) patch responses, ignoring edge effects and boundary layer interactions (see discussion in Jarvis, 1995). Numerical modelling of the interactions of heterogeneous terrain with the atmosphere suggest that there is a maximum effective scale for real patches at which the statistical-dynamical approach will be acceptable (e.g. Blyth *et al.*, 1993).

## REMOTE SENSING AND SCALING OF EVAPOTRANSPIRATION

There are now a number of studies that have attempted to estimate evapotranspiration over landscape scales using a combination of remote sensing data and ground-based meteorological measurements. A methodology was outlined by Holwill and Stewart (1992) and similar approaches, requiring differing amounts of information, have since been used by Bastiaanssen *et al.*, (1994) and Wood and Laksmi (1993), and Wallace *et al.* (1991). In making these calculations it is necessary to make some fairly dubious assumptions about the extrapolation of the meteorological variables and resistance terms from the measurement points to other pixels in the landscape and the relationship of the remotely-sensed infra-red temperature to the effective surface temperature for sensible heat exchange (see for example Hall *et al.*, 1992 and Kalma and Jupp, 1990). The estimates are therefore perhaps better considered to be an index of actual evapotranspiration rates, but it has never before been possible to generate such fields of evapotranspiration estimates at the landscape scale. In every case, very strong differences have been revealed, which might in reality be reduced by processes in the boundary layer associated with advection and humidity feedbacks. The effects of man in irrigating the surface show up very strongly in the results of Bastiaanssen *et al.* (1994). This would clearly have an important effect in perturbing any natural scaling law.

Scaling of such evapotranspiration estimates (as well as the remote sensing derived normalised vegetation index, NDVI) has been analysed by Wood and Laksmi (1993) for the 15 km by 15 km FIFE area. It was found that for this data set, based on a single image, the NDVI and evapotranspiration estimates scale linearly from the finest 120 m resolution at least up to a scale of 3 km. Whether this result is general and how far it is dependent on the assumptions used in deriving the estimates of both quantities is unclear until further analyses are carried out for other areas.

## REMOTE SENSING AND SNOW

Snow is an important component of the water balance in many parts of the world. The imaging of snow cover has been one of the most successful applications of remote sensing in hydrology. We will refer here only to the recent papers of Rango (1995) and Kite (1995) as examples of the incorporation of remotely-sensed snow data into hydrological models. Scaling of snow cover assessments has been generally treated linearly, although clearly difficulties can arise where partial snow covers give rise to mixed pixels in the sensor images. Snow water equivalent can be measured using either passive microwave or gamma wave sensors; estimates are more uncertain but for large scale applications may be a useful replacement for snow course data (e.g. Kite, 1995)

## REMOTE SENSING AND HYDROLOGICAL VARIABILITY: THE CASE OF THE SLAPTON WOOD CATCHMENT, DEVON, UK

The small (0.94 km$^2$) Slapton Wood catchment lies on the south coast of Devon, UK, between Plymouth and Torbay. It is a tributary of the larger River Gara, which flows into Slapton Ley, the largest natural freshwater lake in the south-west of England. Slapton Ley is an important nature reserve and has been the subject of research on the impacts of the nutrient loads from the surrounding 46 km$^2$ of farmland since 1969 (see, for example, Burt and Butcher, 1983; Burt *et al.*, 1985). There has been a recent intensive period of measurements aimed at providing data for verifying the predictions of distributed hydrological models funded by UK Nirex Ltd. Average annual rainfall at the nearby Slapton Ley Field Studies Centre is 1035 mm. Land use is a mixture of permanent pasture (40%), arable, including ley pasture (49%) and deciduous woodland (11%). The woodland is maintained as a nature reserve. The soils of the catchment are primarily freely drained acid brown earths, overlying periglacial head deposits with the bedrock of fractured Dartmouth Devonian slates found at depths ranging from 0.5 to more than 4 metres. It is thought that the fractures impart some permeability to what would otherwise be an impermeable bedrock. The relief in the catchment has a range of 22 to 130 m above ordnance datum, with relatively flat interfluves and three marked hollows on the western side of the catchment (Figure 1).

The Slapton Wood catchment has been the subject of a number of remote sensing studies using different sensors with a view to classifying vegetation patterns and estimating spatial patterns of soil moisture for use in model calibration in a way that might allow extension to the larger scale catchments in the area. Airborne sensors included the 11-band Daedalus 1268 Airborne Thematic Mapper (ATM) scanner and Daedalus 1230 thermal infrared sensor deployed on the NERC aircraft and the JPL 3-band (P, L and C) airborne synthetic aperture radar (AIRSAR) system deployed as part of the NASA MAC-Europe 1991 campaign. Pixel sizes were of the

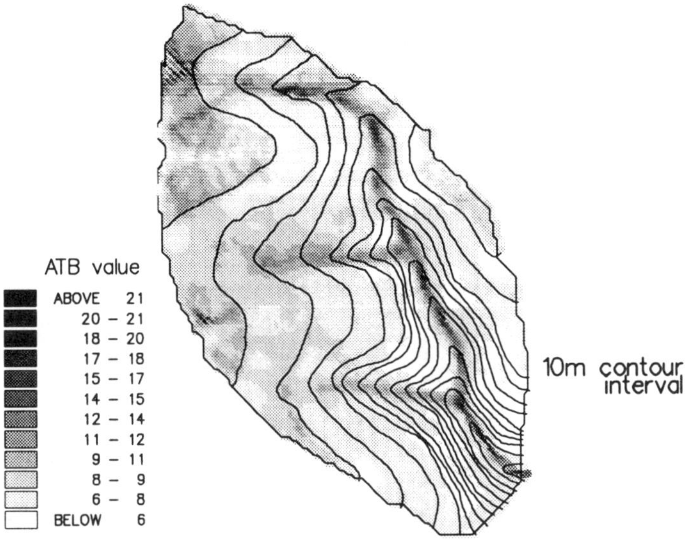


**Figure 1** Digital terrain model of the Slapton Wood catchment, Devon, UK (grid resolution 10 m)

order of 2 m for the ATM sensor and 3.3-6.1 m for the AIRSAR sensor. In addition, Roberts (1993) has reported the use of a hand-held IRIS spectroradiometer. Ground-truth data on surface soil moisture were collected using both gravimetric samples and the Institute of Hydrology near-surface capacitance probe (Dean *et al.*, 1987; Robinson and Dean, 1993).

A number of quite different techniques have been used in the analysis of these data in an attempt to correlate the remotely-sensed images with soil moisture data. Lin *et al.* (1994) make use of a backscattering model for the L-band AIRSAR signal, having determined that the most sensitive parameter in controlling the signal is the standard deviation of the roughness height, s. Averaging both remote sensing signals and ground truth data to the field scale, they calibrated the value of s for five fields in which measured surface soil moisture data were available. This revealed that the best-fit s values varied by a factor of two between different grass-covered fields, and by a factor of two between HH and VV polarisations of the L-band signals, thus demonstrating limitations of the backscattering model. Further regression analysis showed that the backscattering coefficients for the two polarisations were related to variables such as incident angle, canopy thickness, % sand and bulk density as well as moisture volume fraction. It is worth noting that conditions at the time of the overflight were such that the range of soil moistures sampled was limited to 27-58%. The result was essentially an empirical model that allowed soil moisture values to be derived at the field scale, given both the radar signal and values of the other parameters. This allowed field scale estimates of soil moisture to be extrapolated across the catchment, at least for the grass fields (approximately 60% of the catchment) for two dates in June and July 1991. The algorithms could not be used for the other land use types in the catchment. Surface saturation can also be a complicating factor in active microwave measurements (Merot *et al.*, 1994).

Roberts (1993) has reported on comparisons of 5 cm and 10 cm capacitance probe data with ATM sensor images in different wavelength bands for grassland sites at Slapton. Images of Normalised Difference Vegetation Index (NDVI; ATM (band7 - band5)/(band7 + band5)) and Moisture Stress Index (band9/band7) were also used in the comparison. Data from two days in April and July 1990 were examined in detail but the results were disappointing. The data indicated that there were differences between permanent and ley pasture but any correlations within these groups were poor. This was, in part, explained by the small ranges in near-surface soil moisture found in samples taken across the catchment on these days (0.11-0.19 in April; 0.11-0.17 in July). Remote sensing was useful in the classification of different vegetation types within the catchment.

A further data set from June 1990 was used by Fisher (1995). This included daytime and night-time flights of the Daedalus 1230 sensor in thermal imaging mode (8-14 m band). The aim was to use the diurnal temperature range as an index of surface soil moisture. Ground surface thermal inertia should be inversely related to soil moisture (see for example, Idso *et al.*, 1975; Heilman and Moore, 1980; van de Griend *et al.*, 1985). These data were supplemented by those obtained with a hand-held radiometer which gave point thermal temperature measurements. Gravimetric samples were obtained from a number of transects in the catchment to determine near-surface soil moisture. Land uses sampled included permanent pasture, ley pasture and arable land. Due to differences in altitude for the day and night flight paths, considerable work was required to geometrically correct and register the different images. Pixel resolution was reduced in each case to the scale of the 10 m resolution of the digital terrain data on which model predictions would be based. The daily maximum temperature estimates were then further normalised for the effect of topography on spatial distribution of incoming radiation using the methods of Varley *et al.* (1993). The pattern of the normalised diurnal temperature range is shown in Figure 2.

A comparison of the gravimetric soil moisture fractions (in this case with a range of 0.1 to 0.4) with the normalised diurnal temperature range was not particularly convincing (Figure 3a). One particular transect, on the steep slope to the east of the mainstream, showed a stronger relationship for which a simple linear regression between the variables explained 68% of the variance in the data (Figure 3b). It was found that the data had little useful information content when used in the calibration of a version of TOPMODEL applied to the Slapton catchment when the model parameters had already been conditioned on discharge data (see Fisher, 1995). This should not be unexpected: for the majority of the summer period surface soil moisture data should reflect evapotranspiration processes rather than runoff processes and even then may be very dependent on the nature of the surface cover and the pattern of extraction of moisture within the root zone.

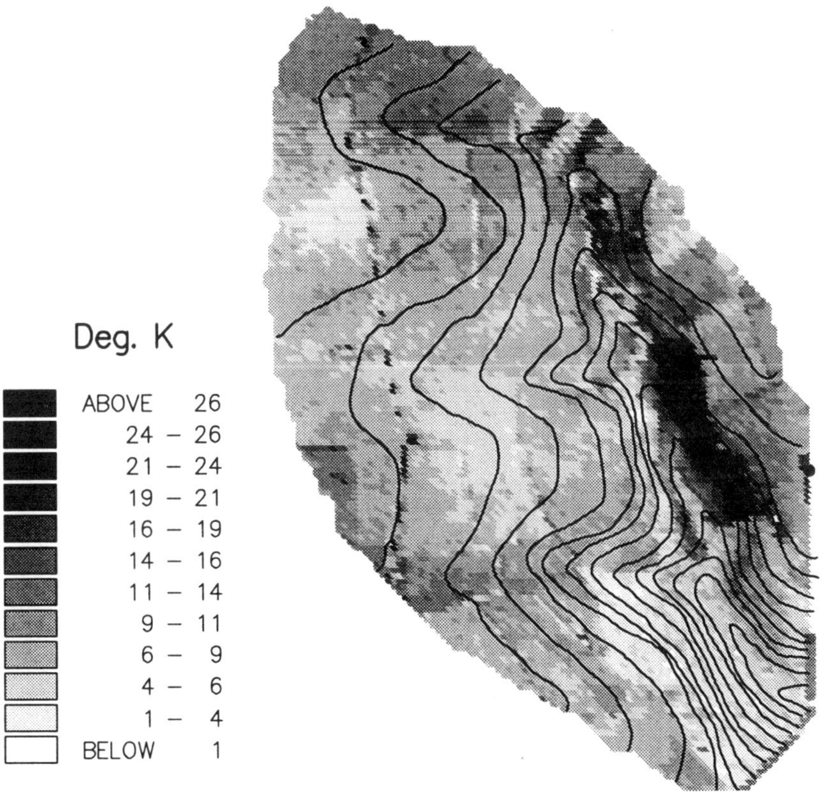


**Figure 2** Diurnal temperature range derived from day and night time thermal infra-red images, normalised for estimated net radiation receipt, Slapton Wood catchment, Devon.

To summarise the experiences from these studies, despite considerable effort remote sensing has not been useful in the extrapolation of hydrological variables from small scale detailed studies to larger scales at this site. The differences in vegetation type dominate the images so any pattern in surface soil moisture is a second-order effect. It is also perhaps questionable as to whether the pattern of surface soil moisture would indeed be useful in terms of either runoff or evapotranspiration prediction. Its utility will depend on further models that can use the surface soil moisture estimates either as a boundary condition in inferring the evolution of soil moisture profiles and therefore evapotranspiration estimates, or in the conditioning of distributed catchment scale models of runoff for which surface soil moisture may not strongly reflect the importance of subsurface flow processes.

A vegetated site in the south of England without strong spatial patterns in input rainfalls or evapotranspiration forcing may not be the best site in which to perform such tests. Remote sensing might prove to be a much more valuable tool in

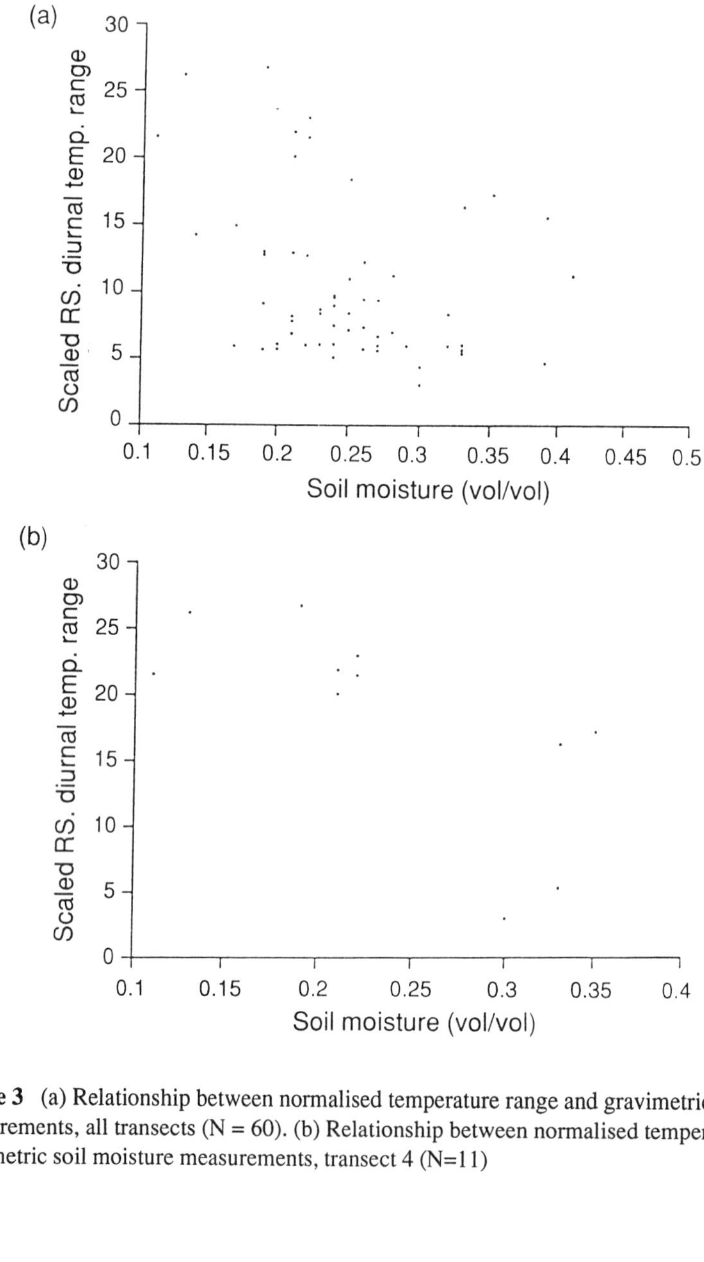


**Figure 3** (a) Relationship between normalised temperature range and gravimetric soil moisture measurements, all transects (N = 60). (b) Relationship between normalised temperature range and gravimetric soil moisture measurements, transect 4 (N=11)

extrapolating to larger scales in more semi-arid regions where patterns of rainfall, irrigation water or the maintenance of near-surface water content by subsurface flow to valley bottoms may have a much more significant effect, although even then more work is required to fully understand the factors affecting the sensor image (see for example, Wang *et al.*, 1992).

## REMOTE SENSING AND SOME FUNDAMENTAL HYDROLOGICAL SCIENCE

Remote sensing may well be useful in some environments. This raises the question of what are the most important sources of hydrological variability in a given environment and can they be detected by remote sensing. Within the UK, for example, remote sensing is now used routinely for land use classification at resolutions of 25 m (ITE, 1993). This can be combined within a GIS system with other spatial data sets (including topographic data) derived by other techniques, for most of the country at 50 m resolution, and a hydrological classification of soil type (Boorman *et al.*, 1994). The only hydrological variable for which remote sensing is used routinely is rainfall, with 15-minute images at a resolution of 2 km being produced routinely for most of the UK by the Meteorological Office, although calibration of these images in terms of volumes of rainfall at ground level is still the subject of active research (e.g. Lord *et al.*, 1995). Most water balance and flood forecasting calculations are still based on 'point' scale rainfall measurements.

How will such data help answer some of the fundamental hydrological questions of variability in the water balance and the partitioning of rainfall into discharge and evapotranspiration? The multispectral sensor data provide no direct estimates of hydrological variables. Other techniques, such as active microwave and SAR sensors, are also dependent on the parameterisation of surface roughness and emissivity before any estimation of hydrological variables can be achieved. This is true also of estimates of surface reflectance, estimates of leaf area index, and estimates of surface resistance and evapotranspiration. All these derived estimates of hydrologically relevant variables and parameters depend themselves on models, models that have parameters that must be calibrated. Recent work suggests that the calibration of even simple models may not be very robust; that the values determined for particular parameters may be dependent on the other parameters in the model as well on the model structure and input data set (see for example Beven, 1993; Freer *et al.*, 1996). There is not just one parameter set (or model structure) that is compatible with the data but many.

The implication of this work is that any physical interpretation of parameter values derived in this way must be made with extreme care, and that extrapolation of such values to other circumstances may be difficult. It will not therefore be easy to assign parameter values on the basis of vegetation type, for example, as overlap should be expected between the parameter sets consistent with the data for different vegetation types. Since vegetation type might be considered to be a primary variable in terms of derivation from remote sensing, secondary variables such as soil moisture and evapotranspiration estimates should, in general, be expected to be highly uncertain. This uncertainty does not bode well for any solution to the *scaling* problem which will ultimately depend on knowledge of patterns of hydrologically relevant variables for both theory development and verification.

## SCALE DEPENDENT HYDROLOGICAL MODELS

This leaves the *scale* problem which, as noted above, is concerned with the formulation of scale-dependent hydrological models. Beven (1995) has contrasted two approaches to the scale problem in hydrology: aggregation and disaggregation approaches. In aggregation approaches the type of GIS outlined above can be used to overlay multiple catchment or landscape characteristics including geology, soil type, elevation and elevation maps. Parameters are assigned to each joint occurrence of the various classifications. This approach is used, for example, in the Hydrological Response Units (HRUs) of Leavesley and Stannard (1984) and the Aggregated Simulation Areas (ASAs) of Kite (1995). The stochastic-dynamic macroscale model of Famiglietti and Wood (1994) based on the topographic index approach of TOPMODEL can also be interpreted in this way, if the parameters are estimated *a priori*. The application of distributed models, such as SHE, at large scales also falls into this category. In all these cases, it is again necessary to estimate **effective** parameter values at the fundamental element scale (pixel, grid, HRU or ASA) of the model. Uncertainty associated with hydrological variables and parameters at the point and pixel scales implies that an aggregation approach may be difficult to apply in practice. Kite *et al.* (1994), for example, use remote sensing to determine vegetation types for ASAs in the Mackenzie Basin, Canada (1,600,000 $km^2$) at different scales. They show how the use of the SLURP hydrological model with GCM-derived input data and a daily time-step provides a greatly improved simulation of mean monthly runoff in the basin. The SLURP model requires that 12 parameters be specified for each ASA unit, of which eight are calibrated against measured discharges. Potentially, at least, these parameters can be varied for each of five land use categories in five different simulation areas.

Objectives and principles of a disaggregation approach have been outlined in Beven (1995) in very general terms that could be met by many different model formulations. A disaggregation approach requires only a parameterisation of the smaller scale heterogeneity. This might be very simple (such as assuming that lumped large scale effective parameter values are adequate — as in current GCM hydrological formulations), or it may be more complex (as in the modified GCM hydrologies suggested by Henderson-Sellers and Pitman, 1992, or Avissar, 1992). Indeed, it can be argued that the various aggregation strategies described above might more properly be described as examples of **disaggregation** approaches since the effective parameters required at their element scale are not easily related to measurable quantities. Again, uncertainty in any estimates of the heterogeneity of hydrological responses may make it very difficult to differentiate between model formulations or parameter sets within a given formulation. However, consider the following possibility.

If different model formulations have similar functional responses then it may not be necessary to differentiate between them (unless there is good reason to choose one rather than another). A scale-dependent hydrological model might then consist of a weighted sum of an appropriate set of functional responses which covers the

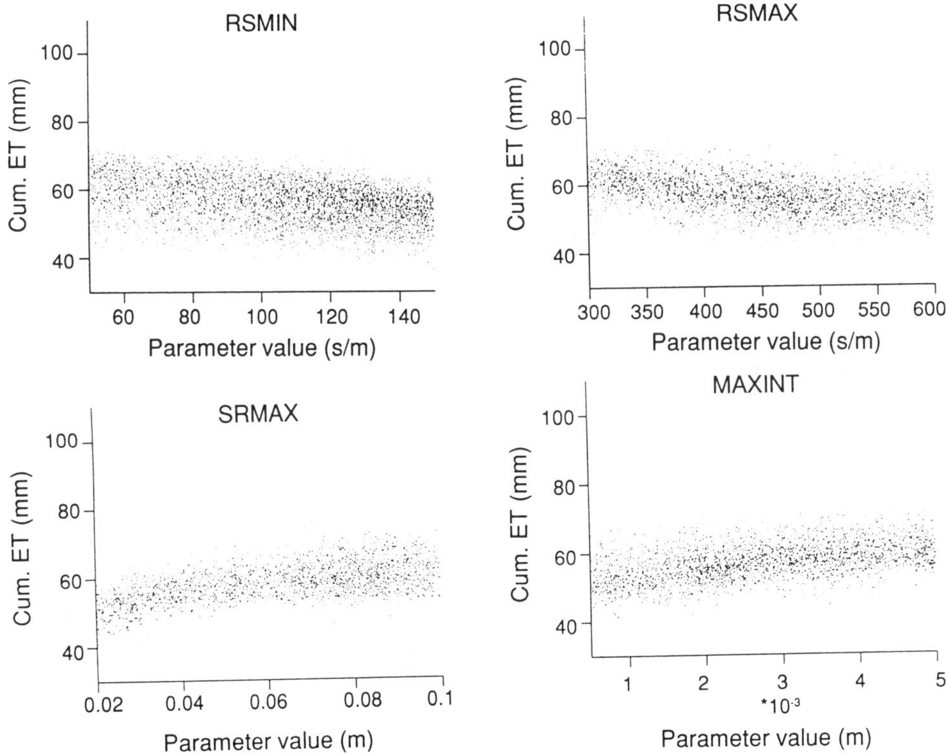


**Figure 4** Range of predicted cumulative evapotranspiration during intensive field campaign 3 (IFC3) of FIFE for random chosen parameter sets in the TOPUP model of Beven and Quinn (1994). Note the similar range in behaviour across the range of individual parameter values. Explanation of parameters: RSMIN (minimum dry canopy surface resistance), RSMAX (maximum surface resistance), SRMAX (available water capacity of root zone), MAXINT (interception and near surface water storage capacity).

possible range of hydrological behaviours to be expected at the scale of interest. Multiple computer runs using different *feasible* models and parameter sets within those models can be used to map the space of model responses and to determine which of those models might be equivalent in their response (see for example Figure 4). In some situations (e.g. humid, temperate, low altitude catchments) the range of possible responses may not be great: in other circumstances (e.g. semi-arid hillslopes with irregular rainfalls and humid valley bottoms supplied by groundwater or irrigation) the range may be much greater. The skill of the hydrologist is then required to match the range of model functional responses to the conditions in the landscape; i.e. to determine the *important* hydrological differences in the landscape and the range of functional responses that might be associated with those differences or, alternatively, to recognise the **similarity** in function of different areas.

It is worth emphasising that these "functional HRUs" need not be defined purely in terms of their characteristics as in more traditional approaches. It is quite possible that different functional responses exist within a single GIS-defined HRU or ASA or areas with similar topographic index due to inter-element and sub-element variability in certain parameters, variability that may not be easily understood (e.g. Hjelmfelt and Burwell, 1984). It is also possible that areas that are differentiated within a GIS classification are simulated as having very similar hydrological function. At its simplest, the scale-dependent model is then a set of (possibly fuzzy) weights for a chosen set of significantly different **model** responses (Beven, 1995). The determination of such weights is not, however, a well-posed problem: data are required to impose constraints on the solution.

One such source of data will be measured stream discharges. Any disaggregation parameterisation should be consistent with measured discharges (most land surface sub-grid parameterisations used in current GCMs are not, for a variety of reasons). The calibration of any aggregation strategy can prove difficult in this respect, since it will not be clear which parameters, of all those that have been specified at the element scale, should be changed to improve the reproduction of discharges. Indeed, accuracy in reproducing discharge data will not alone be a sufficient constraint on the type of functional disaggregation suggested, in that there will be multiple (in fact, an infinite number of) weighting functions that could be consistent with the point discharge measurements. However, taking account of the multiple possibilities, for example by Monte Carlo simulation, would give a range of predictions: a disaggregation model with uncertainty.

The difficulty in such an approach would appear to be accounting for the hydrological effects of landscape changes; it would not be known *a priori* how the weights would vary. It might, however, be possible within the fuzzy weighting approach to compare the possible contribution of the characteristics to the functional response weights from the partial area subject to change, with the possible contribution to be expected post-change. This would give some basis for changing the weighting functions while still retaining the effects of the discharge conditioning. The approach is primarily based on model responses; it therefore depends on using a model that is reasonably realistic in its predictions. Any such predictions can, however, be checked for compatibility with data that might be available at different scales.

## IN SUMMARY

Remote sensing data provide a source of spatial information that, for the first time, allows scaling issues in hydrology to be addressed in a way that is not purely hypothetical or model based. The studies to date have been small in number and limited in scope. This is, in part, a result of the logistics of working with remotely-sensed images. It is also a result of remote sensing data having only limited direct relevance to hydrological variables. Even soil moisture estimates require an intermediate model for the interpretation of remotely-sensed fields which then introduces significant uncertainty. The same is true for evapotranspiration estimates. Also, surface soil moisture fields have only limited hydrological significance in many environments and, in semi-arid areas where there is the probably the best chance of obtaining accurate surface soil moisture estimates from any of the available techniques, accurate measurement of rainfall inputs appears to dominate any information gained from knowledge of the variability of soil moisture (e.g. Goodrich *et al.*, 1994). The limitations of remotely-sensed data in a humid temperate catchment were illustrated by the studies of the Slapton Wood catchment in Devon, UK.

There is still much to learn about what are the dominant or important sources of hydrological variability in different environments. A suggested strategy to develop scale-dependent hydrological models from a disaggregation viewpoint has been outlined. Attention is focused on the identification, either by measurement or modelling, of functionally similar or different areas within an area of interest. It has not shown that the proposed approach will work — this is the subject of ongoing research — but it would appear to have more chance of providing a simplified macroscale parameterisation than any other approach. A particular outstanding question is an appropriate definition of functional similarity.

We are pessimistic about the prospects of a working scaling theory for hydrology that will allow information gained at small scales to be used at larger scales, or vice-versa. In certain restricted areas of study scaling rules may yet be discovered, for example, in the structure of rainfalls (see Schertzer and Lovejoy, 1995) but it may be very difficult to extend the scaling parameters to other time periods or areas. What is clear is that the development of scale-dependent models and scaling rules depends crucially on data, far more so than on theory. In this respect remote sensing will remain essential in providing information on spatial patterns, even if the information can only be considered to be of index type in hydrological terms.

## ACKNOWLEDGEMENTS

The evolution of the ideas contained in this paper has taken place over a long period of time. Discussions with Eric Wood, Paul Quinn, and various members of the SVAT consortium of the NERC TIGER Programme which has supported recent work on this topic must be acknowledged. Figure 4 was produced by Stewart Franks.

## REFERENCES

Avissar, R, 1992, Conceptual aspects of a statistical-dynamical approach to represent landscape subgrid-scale heterogeneities in atmospheric models, *J. Geophys. Res.*, **97**(D3), 2729-2742.

Bastiaanssen, W G M, Hoekman, D H and Roebeling, 1994, A methodology for the assessment of surface resistance and soil water storage variability at mesoscale based on remote sensing measurements. IAHS Special Publ. No. 2, Wallingford, UK.

Beven, K.J. 1989. Changing ideas in hydrology: the case of physically-based models. *J Hydrol.*, **105**, 157-172.

Beven, K.J. 1993. Prophecy, reality and uncertainty in distributed hydrological modelling. *Adv. in Wat. Resour.*, **16**, 41-51.

Beven, K.J. 1995. Linking parameters across scales: subgrid parameterisations and scale dependent hydrological models. *Hydrol. Process.*, **9**, 507-525; also in J.D. Kalma and M. Sivapalan (eds.), *Scale Issues in Hydrological Modelling*, Wiley, Chichester.

Beven, K.J. and Binley, A.M. 1992. The future of distributed models: model calibration and uncertainty prediction. *Hydrol. Process.*, **6**, 279-298.

Beven, K.J. and Quinn, P.F. 1994. Similarity and scale effects in the water balance of heterogeneous areas. In T. Keane and E. Daly (eds.), *The Balance of Water - Present and Future*. Proc. AGMET conference, Trinity College, Dublin, 69-86.

Binley, A.M., Beven, K.J. and Elgy, J. 1989. A physically-based model of heterogeneous hillslopes. 2. Effective hydraulic conductivities. *Wat. Resour. Res.*, **25**, 1227-1233.

Blöschl, G., Grayson, R.B. and Sivapalan, M. 1995. On the representative elementary area (REA) concept and its utility for distributed rainfall-runoff modelling, *Hydrol. Process.* 9, 313-330; also in J.D. Kalma and M. Sivapalan (eds.), *Scale Issues in Hydrological Modelling*, Wiley, Chichester.

Blyth, E.M., Dolman, A.J. and Wood, N. 1993. Effective resistance of sensible and latent heat flux in heterogeneous terrain. *Q. J. Roy. Met. Soc.* **119**, 423-442.

Boorman, D.B., Gannon, B., Gustard, A., Hollis, J.M. and Lilley, A. 1994. *Hydrological aspects of the HOST classification of soils*. Institute of Hydrology Report No. 126, Wallingford.

Burt, T.P. and Butcher, D.P. 1983. The natural history of Slapton Ley Nature Reserve. XV. Hydrological processes in the Slapton Wood catchment. *Field Studies*, **5**, 731-752.

Burt, T.P., Butcher, D.P., Coles, N. and Thomas, A.D. 1985. On the generation of delayed peaks in stream discharge. *J. Hydrol.*, **78**, 361-378.

Charpentier, M.A. and Goffman, P.M., 1992. Soil moisture variability within remote sensing pixels. *J. Geophys. Res.*, **97**(D17), 18987-18995.

Dean, T.J., Bell, J.P. and Baty, A.J.B. 1987. Soil moisture measurement by an improved capacitance technique. Part 1. Sensor design and performance. *J. Hydrol.*, **93**, 67-78

Dagan, G. 1986. Statistical theory of groundwater flow and transport: pore to laboratory, laboratory to formation and formation to regional scale. *Water Resour. Res.*, **22**(9), 120s-134s.

Dolman, A.J. 1992. A note on areally averaged evaporation and the values of effective surface conductance. *J. Hydrol.*, **138**, 583-589.

Dooge, J.C.I. 1994. Climate change and a new water balance. In: T. Keane and E. Daly (eds.), *The Balance of Water - Present and Future*. Proc. AGMET conference, Trinity College, Dublin, 215-224.

Drayton, R.S., Wilde, B.M. and Harris, J.H.K. 1992. Geographical Information System approach to distributed modelling,. In: K.J. Beven and I.D. Moore (eds.) *Terrain Analysis and Distributed Modelling in Hydrology*. Wiley, Chichester, 193-202.

Duan, Q., Sorooshian, S. and Gupta, V. 1992. Effective and efficient global optimisation for conceptual rainfall-runoff models. *Wat. Resour. Res.*, **28**, 1015-1031.

Engman, E.T. and Gurney, R.J. 1991. *Remote Sensing in Hydrology*. Chapman and Hall, London, 1991.

Engman, E.T., Angus, G. and Kustas, W.P. 1989. Relationships between the hydrologic balance of a small watershed and remotely-sensed soil moisture. In: Remote Sensing and Large Scale Global Processes. *IAHS Publ. No. 186*, 75-84.

Entekhabi, D. and Eagleson, P.S., 1989. Land surface hydrology parameterization for atmospheric general circulation models including subgrid scale spatial variability. *J. Climate*, **2**, 816-831.

Famiglietti, J.S., and Wood, E.F. 1994. Multiscale modeling of spatially variable water and energy balance processes. *Water Resour. Res.*, **30**, 3061-3078.

Fan, Y. and Bras, R.L. 1995. On the concept of a representative elementary area (REA) in catchment runoff. *Hydrol. Process.* **9**, 821-832.

Fisher, J. 1995. The use of remote sensing and other system state estimates in the calibration of a distributed hydrological model. Unpublished PhD thesis, University of Lancaster.

Freer, J. Beven, K.J. and Ambroise, B. 1996. Bayesian estimation of uncertainty in runoff prediction and the value of data: an application of the GLUE approach. *Wat. Resour. Res.* In press.

Goodrich, D.C., Schmugge, T.J., Jackson, T.J., Unkrich, C.L., Keefer, T.O., Parry, R., Bach, L.B. and Amer, S.A. 1994. Runoff simulation sensitivity to remotely-sensed initial soil moisture content. *Water Resour. Res.*, **30**, 1393-1405.

Hall, F.G., Huemmrich, K.F, Goetz, S.J., Sellers, P.J. and Nickeson, J.E. 1992. Satellite remote sensing of surface energy balance: success, failures and unresolved issues in FIFE. *J. Geophys. Res.,97 (D17), 19,061-19,089.*

Heilman, J.L. and Moore, D.G. 1980. Thermography for estimating near surface soil moisture under developing crop canopies. *J. Appl. Meteorol.*, **19**, 324-328.

Henderson-Sellers, A. and Pitman, A.J. 1992. Land-surface schemes for future climate models: specification, aggregation and heterogeneity. *J. Geophys. Res.*, **97**(D3), 2687-2696.

Hjelmfelt, A.T. and Burwell, R.E. 1984. Spatial variability of runoff. *J. Irrig. Drain. Eng., ASCE*, **110**, 46-54.

Holwill, C.J. and Stewart, J.B. 1992. Spatial variability of evapotranspiration derived from aircraft and groundbased data. *J. Geophys. Res.* **97**(D17), 18673-18680.

Idso, S.B., Schmugge, T., Jackson, R.D. and Reginato, R.J. 1975. The utility of surface temperature measurements for the remote sensing of surface soil moisture status. *J. Geophys. Res.*, **80**, 3044-3049.

ITE. 1993. *ITE Land Cover Map of Great Britain. Land Cover Classes: A description.* Institute of Terrestrial Ecology, Monks Wood, Cambs.

Jarvis, P.G. 1995. Scaling processes and problems. *Plant, Cell and Environment*, **18**, 1079-1089.

Jolley, T. and Wheater, H.S. 1993. Macromodelling of the River Severn, in W B Wilkinson (ed.), Macroscale Modelling of the Hydrosphere. *IAHS Publ. No.* 214, 91-100

Kalma, J.D. and Jupp, D.L.B. 1990. Estimating evaporation from pasture using infra-red thermometry: evaluation of a one-layer resistance model. *Agric. Forest. Meteorol.*, **51**, 23-246.

Kite, G.W. 1995. Scaling of input data for hydrologic modeling. *Water Resour. Res.*, **31**, 2769-2781.

Kite, G.W., Dalton, A. and Dion, K. 1994. Simulation of streamflow in a macroscale watershed using general circulation model data. *Wat. Resour. Research*, **30**, 1546-1560.

Koster, R.D. and Suarez, M.J. 1992. Modelling the land surface boundary in climate models as a composite of independent vegetation stands. *J. Geophys. Res.*, **97**(D3), 2697-2715.

Leavesley, G.H. and Stannard, L.G. 1984. Application of remotely-sensed data in a distributed parameter watershed model. In: G.W. Kite and A. Wankiewicz (eds.), *Proc. of Workshop on Applications of Remote Sensing in Hydrology*, NHRI Symposium 5, 47-68, Natl. Hydrol. Res. Inst. Saskatoon, Sask.

Lhomme, J-P., Chehbouni, A. and Monteny, B. 1994. Effective parameters of surface energy balance in heterogeneous landscape. *Bound. Layer Meteorol.*, **71**, 297-309.

Lin, D.S., Wood, E.F., Beven, K.J. and Saatchi, S. 1994. Soil moisture estimation over grass-covered areas using AIRSAR. *Int. J. Remote Sensing*, **15**, 2323-2343.

Lord, M.E., Young, P.C. and Goodhew, R.C. 1995. Adaptive radar calibration using rain gauges. In K.A. Tilford (ed.), *Hydrological Uses of Weather Radar*, Brit. Hydrol. Soc. Occasional. Paper No. 5, 10-25.

Merot, P., Crave, A., Gascuel-Odoux, C. and Louhala, S. 1994. Effect of saturated areas on backscattering coefficient of the ERS-1 synthetic aperture radar: first results. *Wat. Resour. Res.*, **30**, 175-179.

Nemani, R., Pierce, L.L., Running, S.W. and Band, L.E. 1993. Ecological processes at the watershed scale: sensitivity to remotely-sensed leaf area index estimates. *J. Appl. Meteorol.*, **28**, 276-294.

Quinn, P.F., Beven, K.J. and Culf, A. 1995. The introduction of macroscale hydrological complexity into land surface-atmospheric transfer function models and the effect on planetary boundary layer development. *J. Hydrol.*, **166**, 421-445.

Rango, A. 1995. The Snowmelt Runoff Model (SRM). In V.P. Singh (ed.), *Computer Models of Watershed Hydrology*. Water Resource Publications, Colorado, 477-520.

Raupach, M.R. and Finnigan, J.J. 1995. Scale issues in boundary-layer meteorology: surface energy balances in heterogeneous terrain. *Hydrol. Process.*, **9**, 589-612; also in J.D. Kalma and M. Sivapalan (eds.), *Scale Issues in Hydrological Modelling*. Wiley, Chichester.

Refsgaard, J.C., Seth, S.M., Bathurst, J.C., Erlich, M., Storm, B., Jørgensen, G.H. and Chandra, S. 1992. Application of SHE to catchments in India. Part 1. General results, *J. Hydrol.*, **140**, 1-23.

Roberts, G. 1993. The Slapton Ley NERC Airborne campaign - The results of the image analysis and their relevance to the hydrology of the catchment. *Unpubl. rept.*, Institute of Hydrology Report, Wallingford.

Robinson, M. and Dean, T.J. 1993. Measurement of near surface soil water content using a capacitance probe. *Hydrol. Process.*, **7**, 77-86.

Schertzer, D. and Lovejoy, S. 1995. From scalar cascades to lie cascades: joint multifractal analyis of rain and cloud processes. In: R.A. Feddes (Ed.)*Space and time scale variability and interdependencies in hydrological procsses*. Cambridge University Press. 153-174.

Sloan, P.G. and Moore, I.D. 1984. Modeling subsurface stormflow on steeply sloping forested watersheds. *Water Resour. Res.,* **20**, 1815-1822.

Todini, E.M. 1988. Rainfall-runoff modelling - past, present and future. *J. Hydrol.*, **100**, 341-352.

Varley, M.J., Beven, K.J. and Oliver, H.R. 1993. A method for predicting spatial distribution of evaporation using simple meteorological data. In H.J. Bolle, R.A. Feddes and J.D. Kalma (eds.), *Exchange Processes at the Land Surface for a Range of Space and Time Scales,* IAHS Publ. No. 212, 619-626.

Van de Griend, A.A. and Engman, E.T. 1985. Partial area hydrology and remote sensing, *J. Hydrol.*, **81**, 211-251.

Van de Griend, A.A., Camillo, P.J. and Gurney, R.J. 1985. Discrimination of soil physical parameters, thermal inertia and soil moisture from diurnal surface temperature fluctuations. *Wat. Resour. Res.*, **21**, 997-1009.

Wallace, J.S., Wright, I.R., Stewart, J.B. and Holwill, C.J. 1991. The Sahelian Energy Balance Experiment (SEBEX): ground based measurements and their potential for spatial extrapolation using satellite data. *Adv. in Space Res.*, **11**, 3131-3141.

Wang, J.R., Gogineni, S.P. and Ampe, 1992. Active and passive microwave measurement of soil moisture in FIFE. *J. Geophys. Res.* **97**(D17), 18979-18985.

Wood, E.F. 1995. Scaling behaviour of hydrological fluxes and variables: empirical studies using a hydrological model and remote sensing. *Hydrol. Process.*, **9**, 331-347; also in J.D. Kalma and M. Sivapalan (eds.), *Scale Issues in Hydrological Modelling*, Wiley, Chichester.

Wood, E.F., Sivapalan, M., Beven, K.J, and Band, L. 1988. Effects of spatial variability and scale with implications to hydrological modelling. *J. Hydrol.*, **102**, 29-47.

Wood, E.F., Sivapalan, M. and Beven, K.J. 1990. Similarity and scale in catchment storm response, *Rev. Geophys.,* **28**, 1-18.

Wood, E.F. and Laksmi, V. 1993. Scaling water and energy fluxes in climate systems: three land-atmospheric modeling experiments. *J. Climate*, **6**, 839-857.

Woods, R.A., Sivapalan, M. and Duncan, M. 1995. Investigating the representative elementary area concept: an approach based on field data. *Hydrological Processes*, **9**, 291-312; also in J.D. Kalma and M. Sivapalan (eds.), *Scale Issues in Hydrological Modelling*, Wiley, Chichester.

# 2 Effects of surface heterogeneity on thermal remote sensing of land parameters

E.G. NJOKU[1], S.J. HOOK[1] and A. CHEHBOUNI[2]
[1]*Jet Propulsion Laboratory, California, USA*
[2]*ORSTOM Laboratoire d'Hydrologie, Montpellier, France*

## INTRODUCTION

Remote sensing observations, used in conjunction with *in situ* measurements, can provide information of value in modelling and monitoring the biophysical processes governing balances of energy and water at the land surface. Observations from space are uniquely suited to studying the spatial and temporal variabilities of these processes over a wide range of scales due to their wide-swath mapping and orbital sampling capabilities. Knowledge of the fluxes of energy and water at the surface, the manner in which these fluxes are affected by changing atmospheric and surface characteristics, and the way they vary in different climatic regimes is important, since the fluxes represent boundary conditions for dynamical models of atmospheric circulation and water movement in the soil (Bougeault, 1991; Famiglietti *et al.* 1992). The fluxes are also important controlling influences in dynamical models of ecosystem functioning and crop growth (Running, 1990; Lo Seen *et al.*, 1996).

A number of important issues must be addressed in the use of remote sensing observations as inputs for the development and operation of these models. Parameterisations of surface characteristics and fluxes must be matched to parameterisations of radiative transfer. This requires formulation of relationships between the parameters of the biophysical and observational models and methods to interpolate their different space-time sampling. The sensitivities of the models to their descriptive parameters must be assessed, so that assimilation methods may be developed that are resistant to the effects of uncertainties and noise inherent in the observational data, and in order to select optimal channel configurations and develop retrieval methods for maximum information transfer. Remote sensing observations must be considered as an adjunct to, not as a replacement for, *in situ* data in such studies.

In addressing linkages between surface fluxes and remote sensing observations, most studies either assume spatial homogeneity (single component) within the remote sensor footprint (pixel), or they assume that the observations are related to

*Scaling up in hydrology using remote sensing.* Edited by J.B. Stewart, E.T. Engman, R.A. Feddes and Y. Kerr

area-weighted averages of the surface parameters ("composite" parameters), in which the weights are the fractional coverages of the different surface types. The implications of these assumptions have not been adequately investigated. Thus, studies using thermal infrared sensing to retrieve surface temperatures (Wan and Dozier, 1989; Becker and Li, 1990; Sobrino *et al.*, 1991; Kerr *et al.*, 1992; Prata, 1993; and Kealy and Hook, 1993) and to obtain surface fluxes (e.g. Kustas *et al.*, 1989; Holwill and Stewart, 1992) have commonly assumed that surface temperature and emissivity are constant within the observation pixel (~1-km dimension from space). Similarly, in studies of vegetation and soil moisture using passive microwave sensing, homogeneous or area-weighted surface parameters are usually assumed within the footprint (Choudhury *et al.*, 1987; Owe *et al.*, 1992). However, land surfaces exhibit significant heterogeneity at all spatial scales, and the coupling of radiation from different surface components into the sensor receiving aperture, as expressed by the radiative transfer and sensor reception equations, is nonlinear. Hence, unexpected biases may occur in retrievals of surface parameters and fluxes using remotely-sensed data if these heterogeneities are not investigated and understood. In many cases these biases may be negligible; in some they may be unacceptably large. Recent investigations have begun to address these aspects and to quantify the extent of naturally-occurring heterogeneity (Dozier, 1981; Caselles and Sobrino, 1989; Labed and Stoll, 1991; Becker and Li, 1994; Humes *et al.*, 1994).

We use here a simplified approach to investigate the effects of surface heterogeneity on thermal infrared and microwave remote sensing observations, and on the parameterizations by which they are linked to land surface energy fluxes. Most rough soil and vegetated surfaces impose a three-dimensional heterogeneity at the land-atmosphere interface. However, for our purposes we shall consider only the two-dimensional heterogeneity of the radiation field just above the soil or vegetation, thus avoiding consideration of radiation interactions within the volume or cavities of the surface. The discussion is somewhat idealized, but nevertheless can be used to illustrate the magnitudes of the effects of interest. To illustrate the effects, a variety of two-component surfaces are simulated with parameters ranging over extents typical of some natural surfaces. The wavelength regions considered are 9-12 μm in the infrared, and 3-24 cm (10-1.25 GHz) in the microwave, since these are the regions most commonly used for thermal sensing of land parameters.

## MODEL PARAMETERISATIONS

The two basic challenges to be addressed in assimilating remote sensing data into hydrological and atmospheric circulation models are: (1) defining appropriate parameterisations for linking the remote sensing and surface flux models so that the remotely-sensed data can provide information directly to the models of biophysical processes in the soil and atmosphere; and (2) accounting for the differences in spatial and temporal resolutions at which the remote sensing measurements are made and at which the models of processes in the soil and atmosphere operate. This is particularly important in regions of strong surface heterogeneity, where high spatial and temporal variability of surface fluxes occur, such as are often encountered in semi-arid and some agricultural environments. Such environments are important for study, since they cover a large fraction of the globe, and a significant proportion of the human population depends on them for survival. Adverse climate change or surface degradation in these regions can have critical impacts on food and water supplies. Errors in remote sensing and flux modelling caused by ignoring the effects of surface heterogeneity are largest in these regions.

The focus of this chapter is to illustrate the effects of heterogeneity primarily from a remote sensing viewpoint, thus the issue of determining appropriate parameterizations for aggregating surface fluxes over heterogeneous terrain is not addressed here (see for example Chehbouni *et al.*, 1995). However, the basic expressions for surface energy fluxes are provided below as a starting point to illustrate the manner in which remote sensing parameterisations are related to parameterizations of surface fluxes.

## SURFACE FLUX PARAMETERIZATIONS

A useful summary of surface flux parameterisation schemes applicable to climate and hydrological modelling is provided by Rowntree (1991). The energy flux boundary conditions are expressed by the surface energy balance equation:

$$R_n = H + LE + G \tag{1}$$

where $R_n$ is the net downward radiation flux at the surface, H and LE are the upward sensible and latent heat surface fluxes, respectively, (the latent heat flux LE is the product of the evaporation flux E and the latent heat of vaporisation L), and G is the downward heat flux into the soil. Units are in W m$^{-2}$.

Assuming a homogeneous, single interface, the following expressions may be used to describe the energy fluxes (Brutsaert (1984), Rowntree (1991)):

$$R_n = (1 - \alpha) R_s + \varepsilon (R_l - \sigma T_s^4) \tag{2}$$

$$H = \rho c_p \frac{(T_s - T_a)}{(r_a)} \tag{3}$$

$$LE = \frac{\rho c_p}{\gamma} \left[ \frac{e^*(T_s) - e_a}{r_a + r_s} \right] \tag{4}$$

$$G = \left( \lambda_g \frac{\partial T(z)}{\partial z} \right)_{z=0} \tag{5}$$

In these expressions, $R_s$ and $R_l$ are, respectively, the downward shortwave (solar) and longwave (thermal atmospheric) radiation at the surface, $\alpha$ and $\varepsilon$ are the (broadband) shortwave albedo and longwave emissivity, respectively, (corresponding to $R_s$ and $R_l$), and $\sigma$ is the Stefan-Boltzmann constant. $\rho$ and $c_p$ are the air density and specific heat at constant pressure, respectively, $T_s$ is the surface temperature, $T_a$ and $e_a$ are the temperature and vapour pressure of the air, respectively, at a reference height z above the surface, and $r_a$ is the atmospheric (or "aerodynamic") resistance to transfer from the surface (soil or vegetation) to the atmosphere, defined with respect to height z. $\gamma$ is the psychrometric constant, $e^*(T_s)$ is the saturated vapour pressure at temperature $T_s$, $r_s$ is the surface resistance to transfer from the soil (or the stomatal resistance in the case of vegetation), $\lambda_g$ is the soil conductivity, and T(z) is the soil temperature profile (z positive upwards).

In Equations 2-5 the surface temperature $T_s$ appears explicitly as a parameter. However, for many vegetated and heterogeneous surfaces it is difficult to define this temperature precisely. If the remotely-sensed infrared (or "radiative") skin surface temperature is used for $T_s$ then the definition of $r_a$ must be adjusted accordingly by including a supplementary resistance to heat transfer (Prevot *et al.*, 1994). Since the microwave-derived surface temperature represents a deeper layer in the surface (Njoku, 1995), a different modification to the resistance definition may be necessary in this case.

Soil moisture enters into the flux equations less explicitly, either in simplified expressions for the latent heat flux, or in more detailed treatments through expressions for the surface resistance $r_s$. According to the surface cover type and environmental conditions, the surface resistance includes dependences on atmospheric water vapour deficit, surface soil moisture, and water availability in the root zone. Various parameterisations of the surface resistance, involving soil moisture in the shallow surface layer and in the deeper root zone layer, have appeared in the literature (Rowntree, 1991; Bougeault, 1991). Microwave remote sensing can provide information directly on soil moisture in the surface layer (~1 cm). For bare soils, moisture in the deeper layers may be obtained indirectly from the surface soil moisture by combining surface measurements with a soil heat and moisture flux model (Entekhabi *et al.*, 1994). For vegetated soils, there does not appear to be a straightforward relationship between vegetation water content, as measured by a microwave radiometer, and soil moisture in the root zone. However, the vegetation water content may be a useful parameter in helping to specify the fractional vegetation cover, and for monitoring vegetation growth, in climate and ecosystem models.

## RADIATIVE TRANSFER PARAMETERISATIONS

The thermal spectral radiance from the Earth $I_{\lambda o}$, as observed by a spaceborne sensor at infrared or microwave wavelengths, can be expressed as:

$$I_{\lambda o} = \exp(-\tau_\lambda)\,\{\varepsilon_\lambda B_\lambda(T_s) + [1 - \varepsilon_\lambda]\; I_{\lambda d} + I_{\lambda u} \tag{6}$$

where the subscript $\lambda$ is the wavelength and refers to spectral quantities. $I_{\lambda u}$ is the upwelling atmospheric radiance, $\tau_\lambda$ is the atmospheric opacity, and $\varepsilon_\lambda$ is the surface emissivity, (all directional quantities in the viewing direction of the sensor). $I_{\lambda d}$ is the downwelling atmospheric radiance (assumed isotropic for simplicity), and $B_\lambda(T_s)$ is the blackbody radiance at surface temperature $T_s$. (The term $\exp(-\tau_\lambda)$ is often written as the "transmittance", particularly in infrared sensing. However it is written here in terms of "opacity" for consistency with usage elsewhere in microwave modelling). Equation 6 is valid for a non-scattering atmosphere (aerosol- and cloud-free), and assumes azimuthal isotropy. (Dependence on $\theta$, the zenith viewing angle, is understood). Ideally the expression should be integrated over the detector bandwidth centred at wavelength $\lambda$, but this effect is small and will be ignored in this analysis. The Planck function $B_\lambda(T)$ is given by the expression:

$$B_\lambda(T) = \frac{C_1}{\pi\lambda^5\left[\exp(C_2/\lambda T) - 1\right]} \tag{7}$$

where, $C_1 = 2\pi hc^2$ ( $= 3.74183 \times 10^{-16}$ W m$^{-2}$) and $C_2 = hc/k$ ( $= 1.43879 \times 10^{-2}$ m K) are radiation constants, (h and k are Planck's and Boltzmann's constants, respectively, and c is the speed of light), $\lambda$ and T are in units of m and K, respectively, and $B_\lambda(T)$ is in units of W m$^{-2}$ m$^{-1}$ sr$^{-1}$ (Goody and Yung (1989)).

The spectral, directional emissivity $\varepsilon_\lambda$ is not the same as the emissivity $\varepsilon$ appearing in Equation 2, which is equivalent to $\varepsilon_\lambda$ integrated over the upper hemispherical solid angle, and over the broad thermal infrared spectral region. However, these quantities are often assumed to be the same, and for some surfaces the difference is indeed small. However, insufficient data exist to characterise this fully for many naturally occurring surfaces. Measurements of infrared emissivity have been reported in the literature both using laboratory samples and in natural environments (Salisbury and D'Aria, 1992; Hipps, 1989; van de Griend *et al.*, 1991; and Labed and Stoll, 1991). (The laboratory measurements are usually done as reflectance and have to be converted to emissivity, and the field measurements must be corrected for stray radiation artifacts. Thus, both procedures contain some measurement uncertainties.) Typical values range from about 0.9-0.96 for bare soils to about 0.985 for vegetation in the 9-12 μm waveband.

In the microwave region, the emissivity of bare soil $\varepsilon_{\lambda s}$ decreases approximately linearly as a function of the soil water content m in the top 1-2 cm of soil, i.e.:

$$\varepsilon_{\lambda s} = \varepsilon_{\lambda o} + a_\lambda (m - m_o); \qquad (m \geq m_o) \tag{8}$$

where $\varepsilon_{\lambda o}$ and $m_o$ are the emissivity and moisture content of dry soil, respectively, and $a_\lambda$ is the slope of the emissivity-soil moisture relationship ($a_\lambda$ takes negative values). $\varepsilon_{\lambda o}$ and $a_\lambda$ are functions of wavelength, known by measurement and through their dependence on the dielectric constant of water, and depend also on the polarisation of the radiation (vertical or horizontal). A more complete description of the relationships between soil moisture, dielectric constant, and emissivity, for smooth and rough surfaces, and soils of different textures, is given in Njoku and Entekhabi (1995). Equation 8, though approximate, suffices for our purpose here. Representative values for $m_o$, $\varepsilon_{\lambda o}$, and $a_\lambda$ in the wavelength range 3-30 cm are available from measurements reported in the literature. In this study we shall use the following values, typical of a smooth, sandy loam soil, at horizontal polarization, and 20° zenith viewing angle (Wang *et al.* (1983)): $m_o = 0.05$ g cm$^{-3}$, $\varepsilon_{\lambda o} = 0.87$, and $a_\lambda = -1.5$ g$^{-1}$ cm$^3$. The emissivity can be seen to range, under these assumptions, from a high of approximately 0.87 for a dry soil, to a low of less than 0.5 for a wet soil ($m = 0.3$ g cm$^{-3}$).

When vegetation is present, it may be modelled at long microwave wavelengths as a uniform, non-scattering layer above the soil. Assuming that the soil is at the same temperature as the overlying canopy, the emissivity of the vegetation-plus-soil medium $\varepsilon_{\lambda c}$ may be expressed as:

$$\varepsilon_{\lambda c} = 1 - \exp(-\tau_{\lambda c}) [ 1 - \varepsilon_{\lambda s} ] \qquad (9)$$

where $\tau_{\lambda c}$ is the canopy opacity, which depends linearly on the vegetation water content $W_c$ (kg m$^{-2}$):

$$\tau_{\lambda c} = bW_c/\cos\theta \qquad (10)$$

The dependence of the coefficient b on wavelength and vegetation type is subject to some uncertainty. However, measurements indicate that b varies approximately inversely with wavelength, and is not highly dependent on vegetation type. A value of $b = 0.1$ kg$^{-1}$ m$^2$ at $\lambda = 20$ cm is typical (Jackson and Schmugge, 1991). Equations 9 and 10 show that the presence of vegetation above the soil increases the surface emissivity from the value for bare soil to a value of ~1 when the vegetation is very dense. (The single scattering albedo of the vegetation reduces this slightly but is ignored here (Kerr and Wigneron, 1995)). Values of $W_c$ range typically from approximately 0.5 kg m$^{-2}$ for grass to 5.0 kg m$^{-2}$ for dense crops and >10 kg m$^{-2}$ for forests.

## REMOTELY-SENSED EFFECTIVE PARAMETERS

Effective parameters have been proposed as a means for applying equations developed for homogeneous or point-scale processes to processes occurring at larger spatial scales over heterogeneous surfaces. Thus, for example, Shuttleworth (1991), Chehbouni *et al.* (1995), and others have studied the use of effective parameters in representing energy flux processes over heterogeneous surfaces. One method for deriving expressions for the effective parameters is to express the energy balance equation separately in terms of effective and component surface parameters, and then to match the expressions term by term using some initial assumption about which fluxes will be 'conserved' in going from the small to the large scale. The assumptions about which fluxes or parameters are conserved, and uncertainties as to whether the forms of the bulk parameterisation equations remain valid at large scales over heterogeneous terrain, lead to some ambiguity in interpreting such effective parameters. However, much insight can be obtained by this method into the effects of aggregating nonlinear processes to larger scales.

A similar aggregation procedure can be used in applying the radiative transfer equation over large scales. In this case the situation is more clear-cut since radiometer receiver theory defines how the radiation from a heterogeneous surface is aggregated as it is measured by the radiometer. This has recently been discussed by Becker and Li (1994). Following this approach, the observed radiance from a heterogeneous surface is expressed as the solid angle weighted average $\bar{I}_{\lambda o}$ of the radiance from each of the N components of the surface within the observed pixel or footprint:

$$\bar{I}_{\lambda o} = \sum_{i=1}^{N} \xi_i I_{\lambda o_i} \tag{11}$$

where $\xi_i = (d\Omega_i/\Omega)$ is the fractional cover of component i, $d\Omega_i$ is the solid angle subtended by component i at the sensor, $\Omega$ is the solid angle subtended by the entire pixel at the sensor, and the fractional covers satisfy the constraint:

$$\sum_{i=1}^{N} \xi_i = 1$$

are the radiance components at the horizontal two-dimensional plane just above the surface (thus avoiding consideration of radiation interactions within the cavities of the rough or vegetated surface). Strictly speaking, the expression for the weights $\xi_i$ should include the effects of the sensor angular reception characteristics (i.e. the antenna pattern for a microwave radiometer). However this additional effect is ignored here, and it is assumed that the weights can be computed directly from the horizontal spatial areas of the surface components and pixels in the two-dimensional horizontal plane, and the zenith viewing angle. Applying Equation 11 to Equation 6 we obtain:

$$\bar{I}_{\lambda o} = \exp(-\tau_\lambda) \{ \sum_{i=1}^{N} \xi_i \varepsilon_{\lambda i} B_\lambda(T_{si}) + [1 - \sum_{i=1}^{N} \xi_i \varepsilon_{\lambda i}] I_{\lambda d} \} + I_{\lambda a} \tag{12}$$

where it is assumed that the atmosphere is homogeneous over the pixel.

The radiance observations $\bar{I}_{\lambda o}$ are normally used in remote sensing retrieval algorithms to estimate the sensor-averaged or 'effective' emissivity and temperature over the pixel $\bar{\varepsilon}_\lambda$ and $\bar{T}_s$, using an expression equivalent to Equation 6, i.e:

$$\bar{I}_{\lambda o} = \exp(-\tau_\lambda) \{ \bar{\varepsilon}_\lambda B_\lambda(\bar{T}_s) + [1 - \bar{\varepsilon}_\lambda] I_{\lambda d} \} + I_{\lambda a} \tag{13}$$

Thus, from a remote sensing retrieval point of view, the following relationships between component and effective parameters are defined by matching terms in Equations. 12 and 13:

$$\overline{\varepsilon}_\lambda = \sum_{i=1}^{N} \xi_i \, \varepsilon_{\lambda i} \tag{14}$$

and,

$$\overline{\varepsilon}_\lambda \, B_\lambda(\overline{T}_s) = \sum_{i=1}^{N} \xi_i \, \varepsilon_{\lambda i} \, B_\lambda(T_{si}) \tag{15}$$

or, equivalently, using Equation 7:

$$\overline{T}_s = (C_2/\lambda) \left\{ \ln \left[ 1 + \frac{\overline{\varepsilon}_\lambda \, (C_1/\pi\lambda^5)}{\sum_{i=1}^{N} \xi_i \, \varepsilon_{\lambda i} \, B_\lambda(T_{si})} \right] \right\}^{-1} \tag{16}$$

In the microwave region of the spectrum, and for temperatures characteristic of Earth radiation, the Rayleigh-Jeans approximation ( $hc \ll \lambda kT$ ) is valid, so that for a given wavelength the blackbody radiance is linear in temperature. In this case, Equation 16 takes the simpler form:

$$\overline{T}_s = \frac{1}{\overline{\varepsilon}_\lambda} \sum_{i=1}^{N} \xi_i \, \varepsilon_{\lambda i} \, T_{si} \tag{17}$$

In the microwave region, we can take the analysis of emissivity a step further by combining Equations 8-10 to get:

$$\varepsilon_\lambda = 1 - \exp[-bW_c/\cos\theta] \, [1 - \varepsilon_{\lambda o} - a_\lambda(m - m_o)] \tag{18}$$

and thus, using Equation 14:

$$\overline{\varepsilon}_\lambda = 1 - [1 - \varepsilon_{\lambda o} + a_\lambda m_o] \sum_{i=1}^{N} \xi_i \, \exp[-bW_{ci}/\cos\theta] + a_\lambda \sum_{i=1}^{N} \xi_i m_i \, \exp[-bW_{ci}/\cos\theta] \tag{19}$$

The modelled sources of heterogeneity in this case are the soil moisture and vegetation water content. Other sources of heterogeneity such as soil texture, surface roughness, and vegetation type also play a role but their effects are considered secondary here. From Equations 18 (treating the parameters as effective parameters) and 19, and matching terms as before, we find:

$$\overline{W}_c = -\frac{\cos\theta}{b} \ln \left\{ \sum_{i=1}^{N} \xi_i \, \exp[-bW_{ci}/\cos\theta] \right\} \tag{20}$$

and,

$$\overline{m} = \frac{\sum_{i=1}^{N} \xi_i m_i \exp[-bW_{ci}/\cos\theta]}{\exp[-b\overline{W}_c/\cos\theta]} \qquad (21)$$

Equations 14, 16, 17, 20, and 21 are the derived expressions for remotely-sensed effective parameters. It can be seen that only in the case of emissivity are the effective parameters simple averages of the component parameters weighted by their fractional coverage areas as is often assumed. Except for emissivity, the weights are also functions of wavelength. This means that for multichannel retrieval algorithms, the weights for the retrieved effective parameters will be a combination of those at each wavelength. Furthermore, since microwave sensors often have different footprint sizes at different wavelengths, the type and magnitude of surface heterogeneity observed at each wavelength may be different. Also, in a more detailed analysis, the effects of heterogeneity should be evaluated in the context of atmospheric and sensor noise, so that the manner in which these influence the effects of heterogeneity in the retrievals may be assessed.

In the simulations of the next section, we restrict the discussion to two-component surfaces (i.e. N = 2), and allow the characteristics of the component surfaces to vary over realistic ranges. No constraints are provided on the independent variability of the surface parameters or on the juxtaposition of surface types as might be desired in a more realistic simulation. (However, the combinations of surface types used are quite representative of natural conditions.) The intent is to investigate the magnitudes of the nonlinear effects of heterogeneity, and to place bounds on these effects. It is of most interest to examine the differences between the effective parameters p, (where p may represent $\overline{\varepsilon}_l$, $\overline{T}_s$, $\overline{W}_c$, or $\overline{m}$), and those which would be obtained if an area-weighted average were applied — referred to as 'composite' parameters. We therefore define composite parameters p*, by the expression:

$$\overline{p}^* = \sum_{i=1}^{N} \xi_i p_i \qquad (22)$$

and examine, via simulation, the difference quantities:

$$\Delta\overline{p} = \overline{p} - \overline{p}^* \qquad (23)$$

# SIMULATIONS

## TEMPERATURE AND EMISSIVITY

The simulated surface pixel is considered to be made up of two components, of fractional surface areas $\xi_1$ and $\xi_2$ (= 1-$\xi_1$), at temperatures $T_{s1}$ and $T_{s2}$, and with emissivities $\varepsilon_{\lambda 1}$ and $\varepsilon_{\lambda 2}$, respectively. The fractional cover $\xi_1$ is varied from 0 to 1 and the emissivities are either considered uniform or are assigned specific values. In the infrared, the assigned values are $\varepsilon_{\lambda 1} = 0.985$ (typical of vegetation) and $\varepsilon_{\lambda 2} = 0.93$ (typical of bare soil) (Hipps, 1989; van de Griend *et al.*, 1991). In the microwave, the emissivities are computed for specific values of soil moisture and vegetation water content. The temperature $T_{s1}$ is fixed at 300 K, while $T_{s2}$ is varied from 280-320 K such that the difference $\Delta T_{s1,2} = (T_{s2} \pm T_{s1})$ varies 20 °C. The positive range (up to +20 °C) is typical of peak daytime temperature contrasts between bare and vegetated soil in semi-arid environments (Humes *et al.*, 1994), while the negative temperature extreme (up to –20 °C) is somewhat high for peak night-time conditions but is included for the sake of illustration. The two components may be considered as separate and spatially distinct, with a single boundary between them, or as mixed, such as with patchy grasses or clumps of shrubs or trees.

The aggregation formula for emissivities (Equation 14) is a simple area-weighting, hence the effective emissivity is the same as the composite emissivity, i.e. the difference $\Delta\bar{\varepsilon}_\lambda = \bar{\varepsilon}_\lambda - \bar{\varepsilon}_\lambda{}^*$ is zero. The interpretation of effective emissivity is therefore straightforward. Below, we focus mainly on interpreting the formulations for effective temperature.

*Infrared*

Equation 16 is the aggregation formula for remotely-sensed infrared effective surface temperature. Due to the nonlinearity of the Planck function, the infrared effective temperature differs from the composite temperature, i.e. $\Delta\bar{T}_s = \bar{T}_s - \bar{T}_s{}^* =$ 0. For the uniform emissivity case (i.e. $\varepsilon_{\lambda 1} = \varepsilon_{\lambda 2}$), Figures 1(a) and 1(b) show the dependence of on the temperature contrast. The curves are shown for two different wavelengths, $\Delta\bar{T}_s$ 12 μm and 9 μm, with fractional cover as a parameter. It is seen that $\Delta\bar{T}_s$ is always positive, i.e. the effective temperature is always greater than the composite temperature, and increases with magnitude of the component temperature contrast (the nonlinearity of the Planck function always biases the effective temperature towards the higher temperature component). $\Delta\bar{T}_s$ is largest for fractional cover $\xi_1$ near 0.5, and decreases to zero as expected for $\xi_1 = 0$ or 1, representing a homogeneous surface. This can be seen more clearly by plotting the curves versus fractional cover as in Figure 1(c). The effects are not symmetrical with $\Delta\bar{T}_{s1,2}$ or with $\xi_1$. The maximum value of $\Delta\bar{T}_s$ for the ranges shown is about 0.4 °C at 12 μm and 0.6 °C at 9μm. The maximum value increases towards shorter wavelengths, and has a value of 1.7 °C in the 4 μm atmospheric window (not shown).

Figure 1(d) shows the situation when the two components have different emissivities. The lower emissivity of the simulated bare soil (component 2) has a compensatory effect for the bias towards its higher temperature — hence $\Delta\bar{T}_s$ is lower for positive $\Delta\bar{T}_{s1,2}$ than in Figure 1(a). The converse is true for negative $\Delta\bar{T}_{s1,2}$. In addition, $\Delta\bar{T}_s$ takes negative values over a portion of the range.

An implication of these results is that differences can be expected between remotely-sensed (effective) surface temperatures and composite surface temperatures (such as used for validation purposes or as an assumption for remotely-sensed surface temperatures in flux models). In the 10-12 μm region, and for naturally-occurring combinations of temperature and emissivity, these differences may range from a few to several tenths °C — the specific values will depend also on the details of the multichannel algorithm used to retrieve the effective temperatures. One should consider carefully, therefore, the combinations of surface and observational

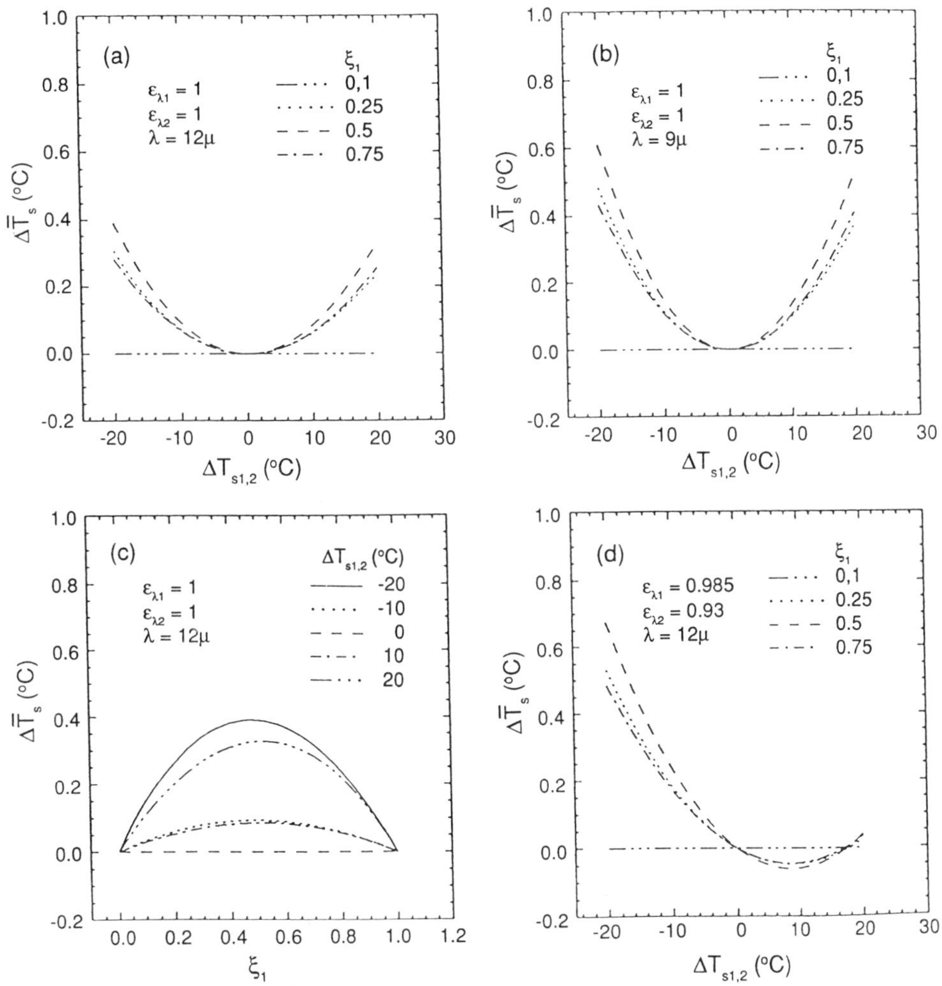


**Figure 1** Difference $\Delta\overline{T}_s$ (°C), at infrared wavelengths, between effective (remotely-sensed) and composite surface temperatures for a two-component surface: (a) as a function of temperature contrast $\Delta\overline{T}_{s1,2}$ between the two components, for uniform emissivity and a wavelength of 12 μm — curves are shown for different values of fractional cover $\xi_1$ of component 1; (b) same as (a), but for a wavelength of 9 μm; (c) same as (a), but shown as a function of fractional cover $\xi_1$, with temperature contrast $\Delta\overline{T}_{s1,2}$ as a parameter; (d) same as (a), but for components with different emissivities $\xi_{\lambda 1}$ = 0.985 (typical of vegetation) and $\xi_{\lambda 2}$ = 0.93 (typical of bare soil).

conditions described here that give rise to the larger biases, especially at shorter wavelengths.

*Microwave*

For uniform emissivity, in the microwave case, Equation 17 reduces to simple area-weighting. Thus, the microwave effective and composite surface temperatures are

**Table 1** Simulated two-component surfaces for analysis of effects of microwave-derived effective parameters. ($W_c$ is vegetation water content in kg m$^{-2}$; m is surface soil moisture in g cm$^{-3}$; $\varepsilon_\lambda$ is modelled emissivity, where $\lambda$ is wavelength in cm)

| Case | Component | Description | $W_c$ | m | $\varepsilon_{24}$ | $\varepsilon_{12}$ | $\varepsilon_6$ | $\varepsilon_3$ |
|---|---|---|---|---|---|---|---|---|
| 1 | 1 | Moist soil | 0 | .1 | .795 | .795 | .795 | .795 |
| | 2 | Dry soil | 0 | .05 | .870 | .870 | .870 | .870 |
| 2 | 1 | Grass | .5 | .1 | .804 | .812 | .828 | .856 |
| | 2 | Dry soil | 0 | .05 | .870 | .870 | .870 | .870 |
| 3 | 1 | Crops | 2 | .1 | .828 | .856 | .899 | .950 |
| | 2 | Dry soil | 0 | .05 | .870 | .870 | .870 | .870 |
| 4 | 1 | Crops | 2 | .2 | .703 | .751 | .825 | .914 |
| | 2 | Wet soil | 0 | .15 | .720 | .720 | .720 | .720 |

the same, and $\Delta\overline{T}_s = \overline{T}_s - \overline{T}_s^* = 0$.

In order to describe, somewhat realistically, the effects of non-uniform microwave emissivity, four cases of two-component surface types are considered as listed in Table 1. For each case, surface soil moisture and vegetation water content values are specified for each component, and corresponding emissivities are computed using Equations 8-10, assuming horizontal polarization and 20° zenith viewing angle, for wavelengths of 24, 12, 6, and 3 cm. These wavelengths correspond to frequencies 1.25, 2.5, 5, and 10 GHz, which fall within the designated L, S, C, and X microwave wave-bands. The m and $W_c$ parameter contrasts between components 1 and 2 have been chosen to be reasonably realistic, and extreme cases have not been considered.

Figure 2 shows the dependence of $\Delta\overline{T}_s$ on temperature contrast $\Delta T_{s1,2}$ for each of the four cases. The curves in Figure 2(a) (Case 1) are plotted for different values of fractional cover $\xi_1$, (there is no modelled wavelength dependence in this case), while in Figures 2(b)-(d) (Cases 2-4) they are plotted for different wavelengths using a fixed value of $\xi_1 = 0.5$. As before, the differences $\Delta\overline{T}_s$ are largest for fractional cover $\xi_1 \cong 0.5$. The differences can take positive or negative values. The differences are positive when $\Delta T_{s1,2}$ is positive and the emissivity in component 2 is greater than that in component 1 (as expected from the form of Equation 17). For a given temperature contrast $\Delta T_{s1,2}$, the difference $\Delta\overline{T}_s$ is largest when the emissivity contrast is greatest. This can occur at different wavelengths depending on the surface types. Heterogeneous surfaces containing wet, bare soil and dense vegetation exhibit the largest emissivity contrasts and hence differences, $\Delta\overline{T}_s$. (The portions of the temperature contrast range $\Delta T_{s1.2} = \pm 20$ °C that are realistic to consider depend on the component surface types. No attempt is made here to limit the $\Delta T_{s1,2}$ values to realistic ranges for each case.)

Figure 3 shows the difference between effective and composite vegetation water

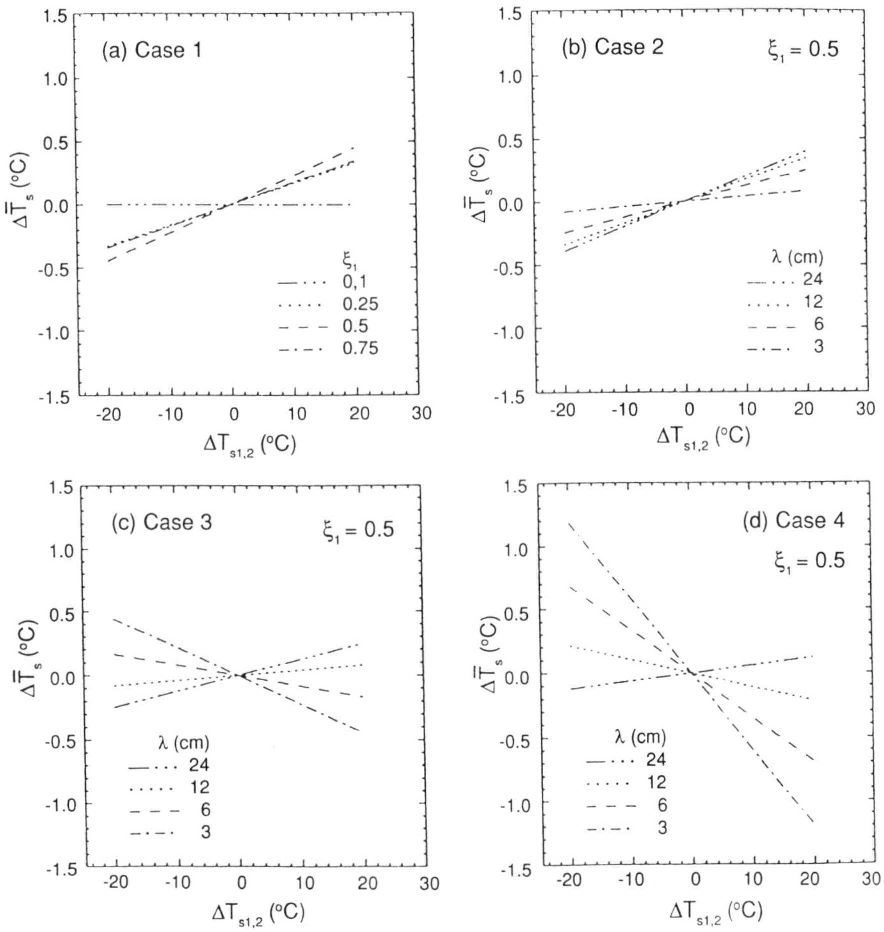


**Figure 2** Difference $\Delta T_s$ (°C), at microwave wavelengths, between effective (remotely-sensed) and composite surface temperatures for a two-component surface, displayed as a function of temperature contrast $\Delta T_{s1,2}$ between components. The emissivity contrasts between components for each case are given in Table 1. Values shown are for horizontal polarization at a zenith viewing angle of 20°. (a) Case 1: shown for different values of fractional cover $\xi_1$ of component 1. (b) Case 2: shown for different wavelengths and for fractional cover $\xi_1 = 0.5$. (c) Case 3: shown as in (b). (d) Case 4: shown as in (b).

contents, $\Delta\overline{W}_c$, based on Equations 20 and 22. The difference is always negative, indicating that the mean (area-averaged) vegetation water content over a heterogeneous footprint is underestimated in the microwave retrieval. The underestimation is greater for greater vegetation water contrasts, and for shorter wavelengths, at which the non-linear effects of vegetation on microwave brightness temperature are greater. The effect is greatest at intermediate values of fractional cover, but there is increasing asymmetry as a function of fractional cover at shorter wavelengths, biasing the occurrence of the maximum difference towards smaller values of fractional cover of the lower-vegetation component.

Figure 4 shows the difference between effective and composite surface soil moisture contents, $\Delta\overline{m}$, based on Equations. 21 and 22. In Figures 4(a) and 4(b) the difference is shown as a function of fractional cover $\xi_1$ for values of moisture $m_1 = 0.1\ \mathrm{g\,cm^{-3}}$ and $m_2 = 0.2\ \mathrm{g\,cm^{-3}}$ in components 1 and 2 respectively, i.e. a value of $\Delta m_{1,2}$

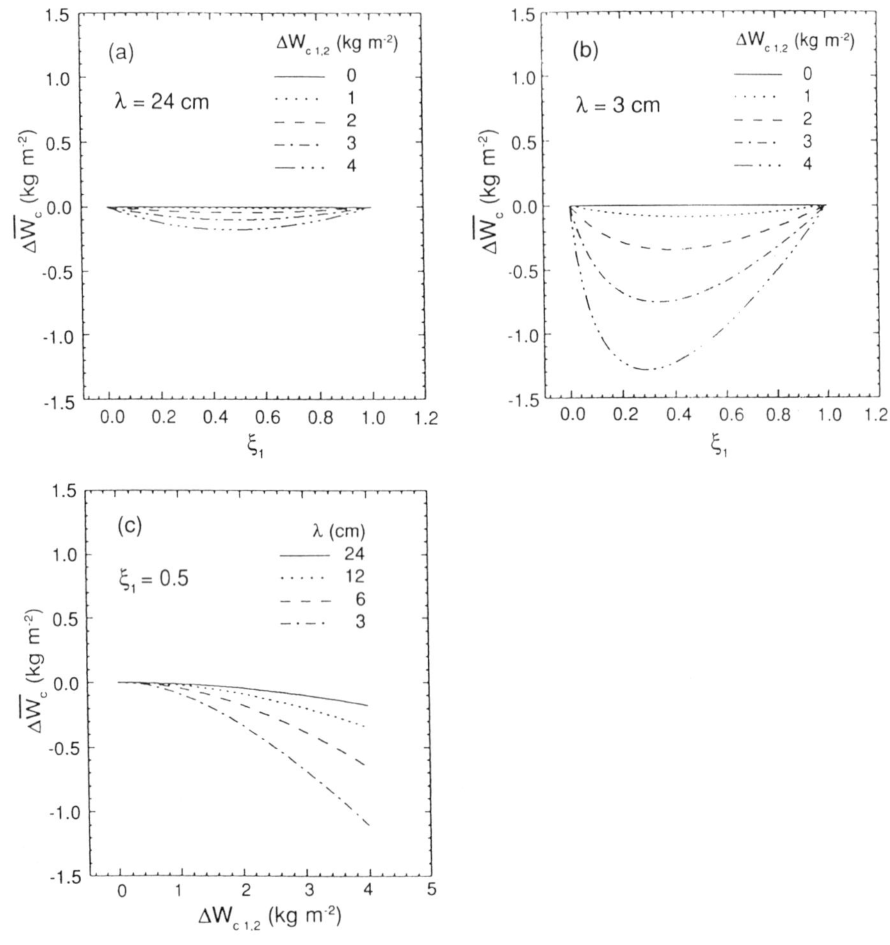


**Figure 3** Difference $\Delta\overline{W}_c$ (kg m$^{-2}$), at microwave wavelengths, between effective (remotely-sensed) and composite vegetation water contents for a two-component surface: (a) as a function of fractional cover $\xi_1$, with vegetation water content contrast $\Delta W_{c1,2}$ as a parameter, at a wavelength of 24 cm; (b) same as (a) but for a wavelength of 3 cm; (c) as a function of vegetation water content contrast $\Delta W_{c1,2}$, with wavelength as a parameter, and fractional cover $\xi_1 = 0.5$.

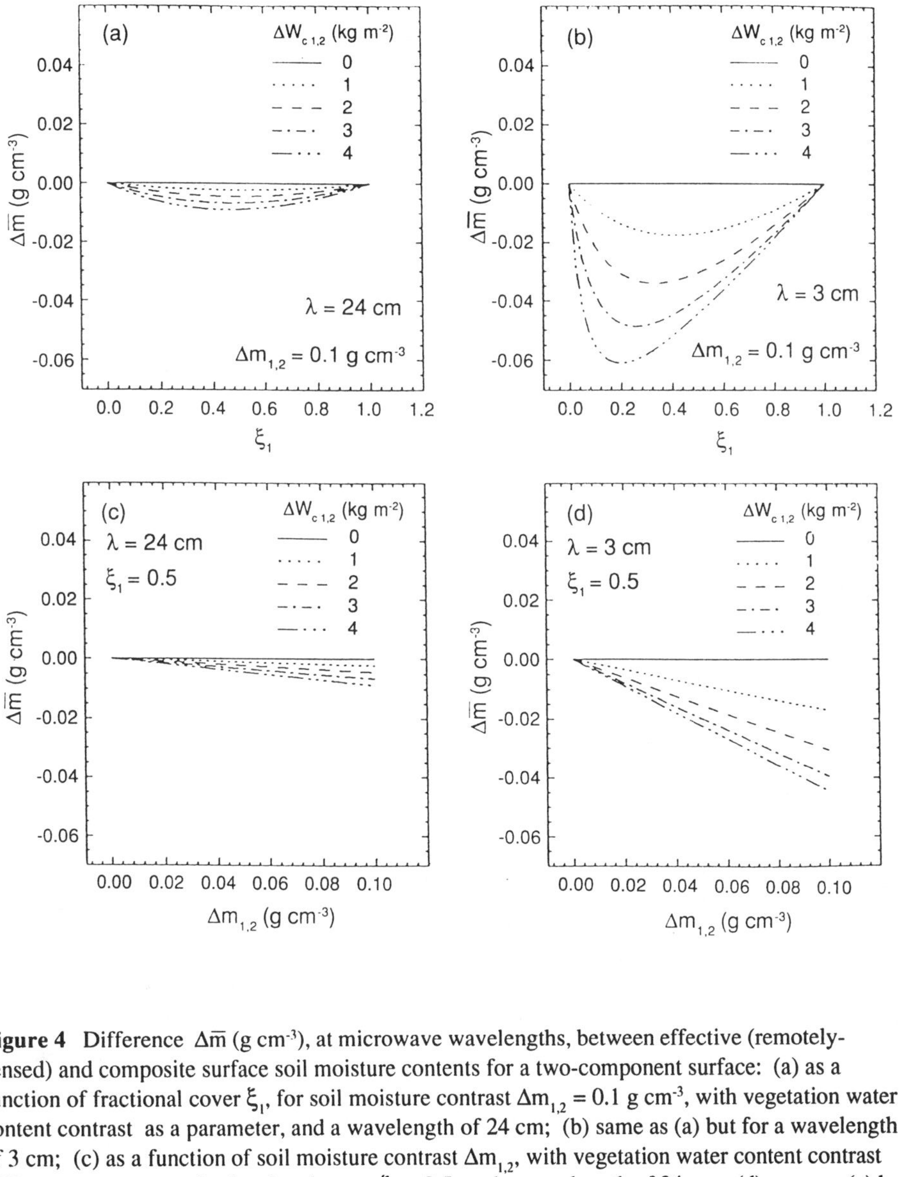


**Figure 4** Difference $\Delta\overline{m}$ (g cm$^{-3}$), at microwave wavelengths, between effective (remotely-sensed) and composite surface soil moisture contents for a two-component surface: (a) as a function of fractional cover $\xi_1$, for soil moisture contrast $\Delta m_{1,2} = 0.1$ g cm$^{-3}$, with vegetation water content contrast as a parameter, and a wavelength of 24 cm; (b) same as (a) but for a wavelength of 3 cm; (c) as a function of soil moisture contrast $\Delta m_{1,2}$, with vegetation water content contrast $\Delta W_{c1,2}$ as a parameter, for fractional cover $\xi_1 = 0.5$, and a wavelength of 24 cm; (d) same as (c) but for a wavelength of 3 cm.

= 0.1 g cm$^{-3}$. Component 1 is bare soil, and the different curves correspond to increasing amounts of vegetation in component 2. The difference is quite small at 24 cm wavelength but increases to large values at shorter wavelengths. The difference is shown in Figures 4(c) and (d) as a function of moisture contrast $\Delta m_{1,2}$ for fractional cover $\xi_1 = 0.5$.

## SURFACE FLUX EFFECTIVE PARAMETERS

We may contrast the expressions obtained in the earlier section on remotely-sensed effective parameters with those derived for application to flux modelling. One set of these, as derived by Chehbouni *et al.* (1995), is:

$$\bar{\varepsilon} = \sum_{i=1}^{N} \xi_i \varepsilon_i \tag{24}$$

$$\bar{T}_s = \frac{1}{\bar{w}} \sum_{i=1}^{N} \frac{\xi_i T_{si}}{w_i} \tag{25}$$

$$\frac{1}{\bar{w}} = \sum_{i=1}^{N} \frac{\xi_i}{w_i} \tag{26}$$

$$\frac{1}{\bar{r}_a + \bar{r}_s} = \sum_{i=1}^{N} \frac{\xi_i}{r_{ai} + r_{si}} \tag{27}$$

where, $\varepsilon$, $r_a$, and $r_s$, were defined earlier, and w is a function of $r_a$ and $r_s$ and other surface parameters. We may note that while the expressions for effective emissivity (Equations 14 and 24) are the same, the expressions for effective surface temperature (Equations 16 or 17 and 25) are not. Chehbouni *et al.* show that the difference between effective temperature defined by Equation 25 and composite temperature can range from approximately –4 to 1 °C for typical two-component surface contrasts. Future work should explore further the relationships between effective parameters defined by remote sensing and those defined by flux aggregation, under realistic conditions, in order to improve the applicability of remote sensing data to surface flux studies.

## CONCLUSIONS

Simplified radiative transfer models and two-component simulations have been used to show the effects of surface heterogeneity within the observation footprint on remotely-sensed parameters in the thermal infrared and microwave spectral regions. The simulations have shown that remote sensing observations of heterogeneous terrain yield estimates of sensor-averaged or 'effective' parameters that may be significantly different from the simple area-averaged or 'composite' parameters that are often assumed to be estimated by the remote sensors. These differences arise from different sources in the infrared and microwave domains but are a result in both cases of the nonlinear radiative transfer relationships between surface parameters and radiance received by the sensor. In the infrared case, for surface temperature sensing, the nonlinearity is caused by the Planck function. In the microwave case the nonlinearity is caused primarily by the vegetation opacity, such that the effective surface temperature, surface soil moisture, and vegetation water content always differ from their composite counterparts when vegetation is present. The magnitude of these differences depends on wavelength, the nonlinearities increasing in both infrared and microwave regions as the wavelengths decrease. The differences also depend on the relative fractional covers of the components and on their parameter contrasts, typically being largest at intermediate values of fractional cover. In many situations, the differences between effective and composite surface parameters may be small and can be safely neglected. However, in some cases, particularly in semi-arid environments or agricultural areas where large parameter contrasts exist between bare and vegetated surfaces, unexpectedly large differences may occur that need to be addressed.

The results obtained here use a simplified approach, and do not take into account the details of specific multichannel parameter retrieval algorithms or other practical issues such as variable instrument bandwidths, antenna beamwidths, atmospheric and instrument noise, multicomponent (more than two) footprints, and other factors. These should be the subject of further study. The results are indicative, however, of the magnitudes of non-linear effects to be anticipated in estimating land surface parameters over heterogeneous terrain. The results may also lead to further insights into the use of visible and near-infrared (VNIR) remote sensing data, in combination with thermal infrared and microwave data, for estimating and monitoring surface fluxes. The VNIR data may provide information on vegetation fractional cover within the thermal infrared and microwave footprints, thus providing a means to correct for nonlinearities in the remotely-sensed effective temperature, and to relate the remotely-sensed effective temperature more closely to the effective temperature appropriate for large-area flux models.

## ACKNOWLEDGMENTS

This research was conducted at the Jet Propulsion Laboratory, California Institute of Technology, under contract with the National Aeronautics and Space Administration.

## REFERENCES

Becker, F. and Li, Z.-L. 1990. Towards a local split window method over land surfaces. *Int. J. Rem. Sens.*, **11**, 369-393.

Becker, F. and Li, Z.-L. 1994. Surface temperature and emissivity at various scales: definition, measurement and related problems. In: *Proceedings of the Workshop on Thermal Remote Sensing of the Energy and Water Balance over Vegetation in Conjunction with Other Sensors, La Londe*

*Les Maures, France, 20-23 September 1993*, CEMAGREF Editions, Montpellier, France, 35-60.

Bougeault, P. 1991. Parameterization schemes of land-surface processes for mesoscale atmospheric models. In: Schmugge, T.J. and Andre, J.-C. (eds). *Land surface evapouration.* Springer-Verlag, New York, 93-120.

Brutsaert, W. 1984. *Evapouration into the atmosphere.* Kluwer Academic Publishers, Dordrecht, The Netherlands.

Caselles, V. and Sobrino, J.A. 1989. Determination of frosts in orange groves from NOAA-9 data. *Rem. Sens. Environ.*, **29**, 135-146.

Chehbouni, A., Njoku, E.G., Lhomme, J.P. and Kerr, Y.H. 1995. Approaches for averaging surface parameters and fluxes over heterogeneous terrain. *J. Climate*, **8**, 1386-1393.

Choudhury, B.J., Tucker, C.J., Golus, R.E. and Newcomb, W.W. 1987. Monitoring vegetation using Nimbus-7 scanning multichannel microwave radiometer data. *Int. J. Rem. Sens.*, **8**, 533-538.

Dozier, J. 1981. A method for satellite identification of surface temperature fields of subpixel resolution. *Rem. Sens. Environ.*, **11**,221-229.

Entekhabi, D., Nakamura, H. and Njoku, E.G. 1994. Solving the inverse problem for soil moisture and temperature profiles by the sequential assimilation of multifrequency remotely-sensed observations. *IEEE Trans. Geosci. Rem. Sens.*, **32**, 438-448.

Famiglietti, J.S., Wood, E.F., Sivapalan, M. and Thongs, D.J. 1992. A catchment scale water balance model for FIFE. *J. Geophys. Res.*, **97**(D17), 18997-19007.

Goody, R.M. and Yung, Y.L. 1989. *Atmospheric radiation. Theoretical basis.* Oxford University Press, Oxford, UK.

Hipps, L.E. 1989. The infrared emissivities of soil and *Artemisia tridentata* and subsequent temperature corrections in a shrub-steppe ecosystem. *Rem. Sens. Environ.*, **27**, 337-342.

Holwill, C.J. and Stewart, J.B. 1992. Spatial variability of evaporation derived from aircraft and ground-based data. *J. Geophys. Res.*, **97**(D17), 18,673-18,680.

Humes, K.S., Kustas, W.P., Moran, M.S., Nichols, W.D. and Weltz, M.A. 1994. Variability of emissivity and surface temperature over sparsely vegetated surfaces. *Wat. Resour. Res.*, **30**, 1299-1310.

Jackson, T.J. and Schmugge, T.J. 1991. Vegetation effects on the microwave emission of soils. *Rem. Sens. Environ.*, **36,** 203-212.

Kealy, P.S. and Hook, S.J. 1993. Separating temperature and emissivity in thermal infrared multispectral scanner data: Implications for recovering land surface temperatures. *IEEE Trans. Geosci. Rem. Sens.*, **31**, 1155-1164.

Kerr, Y.H., Lagouarde, J.P. and Imbernon, J. 1992. Accurate land surface retrieval from AVHRR data with use of an improved split window algorithm. *Rem. Sens. Environ.*, **41**, 197-209.

Kerr, Y.H. and Wigneron, J.P. 1995. Vegetation models and observations - a review. In: Choudhury, B.J., Kerr, Y.H., Njoku, E.G., and Pampaloni, P. (eds). *Passive microwave remote sensing of land-atmosphere interactions.* VSP Utrecht, The Netherlands, 317-344.

Kustas, W.P., Choudhury, B.J., Moran, M.S., Reginato, R.J., Jackson, R.D., Gay, L.W. and Weaver, H.L. 1989. Determination of sensible heat flux over sparse canopy using thermal infrared data. *Agric. Forest Meteorol.*, **44,** 197-216.

Labed, J. and Stoll, M.P. 1991. Spatial variability of land surface emissivity in the thermal infrared band: Spectral signature and effective surface temperature. *Rem. Sens. Environ.*, **38**, 1-17.

Lo Seen, D., Chehbouni, A., Njoku, E., Saatchi, S., Mougin, E. and Monteny, B. 1995. An approach to couple vegetation functioning and SVAT models for semi-arid grasslands during the HAPEX-Sahel experiment. *Agric. Forest Meteorol.* In press.

Njoku, E.G. 1995. Surface temperature estimation over land using satellite microwave radiometry. In: Choudhury, B.J., Kerr, Y.H., Njoku, E.G. and Pampaloni, P. (eds). *Passive microwave remote sensing of land-atmosphere interactions.* VSP Utrecht, The Netherlands, 509-530.

Njoku, E.G. and Entekhabi, D. 1995. Passive microwave remote sensing of soil moisture. *J. Hydrol.* (In press).

Owe, M., van de Griend, A.A. and Chang, A.T.C. 1992. Surface moisture and satellite microwave observations in semi-arid southern Africa. *Wat. Resour. Res.*, **28**, 829-839.

Prata, A.J. 1993. Land surface temperatures derived from the Advanced Very High Resolution Radiometer and the Along-Track Scanning Radiometer 1. Theory. *J. Geophys. Res.*, **98**(D9), 16,689-16,702.

Prevot, L., Brunet, Y., Paw U, K.T. and Seguin, B. 1994. Canopy modelling for estimating sensible heat flux from thermal infrared measurements. In: *Proc. Workshop on Thermal Remote Sensing of the Energy and Water Balance over Vegetation in Conjunction with Other Sensors, La Londe Les Maures, France, 20-23 September 1993*, CEMAGREF Editions, Montpellier, France, 17-22.

Rowntree, P.R. 1991. Atmospheric parameterization schemes for evaporation over land: basic concepts and climate modelling aspects. In: Schmugge, T.J. and Andre, J.-C. (eds). *Land surface evapouration.* Springer-Verlag, New York, 5-29.

Running, S.W. 1990. Estimating terrestrial primary productivity by combining remote sensing and ecosystem simulation. In: Hobbs, R. and Mooney, H. (eds). *Remote sensing of biosphere functioning*. Springer-Verlag, New York, 65-86.

Salisbury, J.W. and D'Aria, D. 1992. Emissivity of terrestrial materials in the 8-14 μm atmospheric window. *Rem. Sens. Environ.*, **42**, 83-106.

Shuttleworth, W.J. 1991. The modellion concept. *Rev. Geophys.*, **29**, 585-606.

Sobrino, J.A., Coll, C. and Caselles, V. 1991. Atmospheric correction for land surface temperature using NOAA-11 AVHRR channels 4 and 5. *Rem. Sens. Environ.*, **38**, 19-34.

van de Griend, A.A., Owe, M., Groen, M. and Stoll, M.P. 1991. Measurement and spatial variation of thermal infrared spatial emissivity in a savanna environment. *Wat. Resour. Res.*, **27**, 371-379.

Wan, Z. and Dozier, J. 1989. Land surface temperature measurement from space: Physical principles and inverse modelling. *IEEE Trans. Geosci. Rem. Sens.*, **27**, 268-278.

Wang, J.R., O'Neill, P.E., Jackson, T.J. and Engman, E.T. 1983. Multifrequency measurements of the effects of soil moisture, soil texture, and surface roughness. *IEEE Trans. Geosci. Rem. Sens.*, GE-**21**, 44-51.

# 3 Estimation of aerodynamic roughness at various spatial scales

**M. MENENTI[1], J.C. RITCHIE[2], K.S. HUMES[3], R. PARRY[2], Y. PACHEPSKY[4], D. GIMENEZ[2] and S. LEGUIZAMON[5]**
[1]*DLO-Winand Staring Centre, Wageningen, The Netherlands*
[2]*USDA ARS Hydrology Laboratory, Beltsville, USA*
[3]*University of Oklahoma, Norman, USA*
[4]*USDA ARS Remote Sensing and Modeling Laboratory, Beltsville, USA*
[5]*University of Mendoza, Argentina*

## INTRODUCTION

Addressing land surface processes at a variety of length scales without establishing a measure of these length scales is difficult. In principle, remote sensing measurements may provide an experimental basis from which to apply such a definition in practice. The estimation of an effective roughness length ($z_{0m}$) for momentum is used as a paradigm to illustrate how length scales can be defined and measured and how $z_{0m}$ can be estimated at increasing length scales for a heterogeneous landscape.

Topographic differences and vegetation patterns are integral parts of the landscape. These are properties which have to be measured and evaluated if we are to understand hydrological and climatic effects on natural and agricultural resources on a large scale. Airborne laser altimeter data have been used to quantify landscape topography, gully and stream cross-sections, and vegetation canopy properties. Measurements of the micro and macro topography can help quantify water retention, infiltration, evaporation, and movement from landscape surfaces into channels and across the flood plains. Development of channel and gully development and degradation can be measured and evaluated to estimate soil loss, potential degradation of water quality, and flow patterns. Measurements of canopy properties and their distribution across the landscape and their effect on water movement and aerodynamic roughness can contribute to a better understanding of the evaporative loss, infiltration and surface water movement. Airborne laser altimeters offer the potential to measure landscape properties over large areas quickly and easily. Such measurements will improve our understanding of the effects of these factors on hydrology and climate systems of heterogenous landscapes.

The basic measurement made with an airborne laser to study small landscape elements, such as vegetation stands, is the frequency distribution of objects hit by the laser beam versus their height above the ground. Profiles of ground elevation are

*Scaling up in hydrology using remote sensing*. Edited by J.B. Stewart, E.T. Engman, R.A. Feddes and Y. Kerr

also obtained with the laser measurements along the track as the lower envelope of the actual detailed observations.

*Partition of shear stress*

Means of simple logarithmic wind profiles can describe airflow (mean horizontal wind speed) over composite land surfaces. Different surface roughness parameters must be applied, however, at different heights. Except for the lowest few metres over patches having homogeneous cover and morphology, surface stress over horizontal length scales affects the wind profile by one or two orders of magnitude larger than the height under consideration. Flow over composite land surfaces should be parameterised so that the total force applied by airflow to the land surface is conserved. The general expression for surface drag (F) illustrates this:

$$F = \rho u^2 A C_m \quad (\mathrm{kg\ m^{-1}\ s^{-2}}) \tag{1}$$

where $\rho$ is air density, u is mean horizontal wind speed, A is surface area per unit horizontal area and $C_m$ is the momentum transfer coefficient. As shown in Equation 1, surface drag is proportional to the area of roughness elements per unit horizontal area. Traditionally, a distinction is made between 'skin friction' (= local drag over flat homogeneous surfaces) and 'form drag' (= drag resulting from the effects of relief on airflow). Arya (1975) gave the equation

$$S_d + F_d = \tau_t \tag{2}$$

where $S_d$ is skin friction, $F_d$ is form drag and $\tau_t$ is the total stress. Menenti & Ritchie (1994) gave a review of the theory and experimental results about the observation and modelling of aerodynamic roughness length of composite landscapes. Equation 2 implies the existence of roughness elements having distinctly different length scales in which obstructions (form drag) and some intervening surfaces (skin friction) are present. Topographic effects must be considered since several different length scales are likely to be present in the relief of a composite landscape. Menenti & Ritchie (1994) devised a procedure to calculate an effective aerodynamic roughness of a composite landscape using laser measurements. In their case study (MONSOON'90 on Walnut Gulch Watershed, Arizona) three landscape elements having different length scales were considered: grasses (the intervening surfaces); shrubs and trees (obstructions); and topography (having just one characteristic length scale). We present here the results of a closer examination of the determination and of the separability of length scales.

*Landscape and roughness elements at different length scales*

Most of the literature on aerodynamic roughness length relates to complete canopies placed on flat topography. A widely used practical rule is that $z_{0m} = 0.1$ h, where h is vegetation height. For partial canopies the ratio $z_{0m}/h$ is not constant and depends on the leaf area index (LAI) which can be considered as a measure of the concentration of roughness elements. In composite landscapes the complete and partial canopy schematisation may be applicable up to length scales established for each individual case.

At larger length scales, the effect on airflow over taller plants and other obstructions with large intervening spaces has to be combined with the effect of canopies occupying the spaces. Later, an equation to estimate an effective roughness will be given. Here it suffices to say that the parameterisation of the aerodynamic effect of obstructions requires more detailed measurements of surface geometry relative to complete and partial canopies. Again, the schematization implies the assumption that the mixture of canopies and obstructions is found on flat topography. In a composite landscape characterized by different length scales, the canopy with obstacle schematization applies at length scales much smaller than those characterizing topography.

In simple cases (e.g. the periodic topography of Taylor *et al.*, 1989), a suitable measure of the length scale topography (hillocks and ridges) is the wavelength of an elevation profile. A more detailed review of this case can be found in Menenti & Ritchie (1994).

*Experimental evidence for the existence of a height dependent vertical length scale*

The previous discussion has two implications:

(i) measurements of shear stress at different heights above a given point in a composite landscape would give different values of $z_{0m}$, and

(ii) when horizontal length scales of the three landscape elements (canopy, obstructions, and topography) are very different and precisely identified, then constant shear stress is expected for layers at different heights.

Some laboratory (Wooding *et al.*, 1973) and field experiments (Thompson, 1978; Kustas & Brutsaert, 1986; Beljaars & Holtslag, 1991) support these concepts, especially the existence of a constant momentum flux layer at higher elevations in the planetary boundary layer (PBL). In each constant momentum flux layer, the wind profile is scaled by different vertical scales ($z_{0m}$) which are related to surface geometry at horizontal length scales related to the elevation of the constant flux layer. Kustas & Brutsaert (1986) found reasonable agreement of $z_{0m}$ measurements obtained with radiosonde wind profiles with calculations based on the shape and slope of ridges in an alpine watershed.

*Constant momentum flux layers and blending heights*

The existence of distinct constant layers implies that a transition zone should be observable, at least in principle, between these layers. These transition zones have been termed *blending height(s)* by Wieringa (1986). The preceding arguments suggest that the existence of two or more pairs can characterize the land surface-atmosphere interaction for composite landscapes of varying horizontal and vertical scales.

*Horizontal and vertical length scales*

Finnigan *et al.* (1990) addressed the interdependence of horizontal and vertical length scales in relation to PBL fluxes by relating horizontal (X) to vertical length scales as:

$$X=\frac{u}{u_*}h_i \tag{3}$$

where $u_*$ is the friction velocity and $h_i$ is the depth of the inner layer under neutral conditions.

A preliminary conclusion may be drawn at this stage. To parameterise momentum flux over composite landscapes, it is first necessary to determine the relevant horizontal scales. This will lead to the identification of significant reference PBL elevations and provide information about how to aggregate land surface properties and fluxes from large to small horizontal and vertical length scales.

*Measurements of horizontal length scales*

The shear stress partition theory (Equation 2) indicates a precise relationship between land surface geometry of a complex landscape at different length scales and vertical transport of momentum. It appears that measuring these length scales are necessary because:

(i) the shear stress partition theory applies if landscape elements have very different scales. The intervening land cover is homogeneous at length scales much smaller than the typical size obstructions and these are much smaller than the length scales of the topography, and

(ii) given the relationship between vertical and horizontal length scales, the interpretation of the momentum flux measurements at a given height requires knowledge of the horizontal length scales.

Both (i) and (ii) have obvious implications for modelling and remote sensing studies of land surface-atmosphere interactions such as the required parameterisation of land surface momentum and heat transfer processes and for the horizontal resolution of remote sensing observations used in heat balance studies. When attention is focused on momentum flux, this leads to the necessity of precise measurement of land surface geometry for a range of length scales.

Laser measurements provide an accurate and very detailed description of the geometry of heterogenous land surfaces. Several statistical techniques can be easily applied to evaluate characteristics of land surface structure, as results from the following techniques show:

- *Autocorrelation function* is used to assess at which length scales the surface structure is homogeneous, i.e. when a simple statistic such as the standard deviation of observed elevations for a segment of laser measurements does not depend on the length of the segment.
- *Fourier analysis* is used to establish the dominant 'wavelength' in the structure of the topography.

- *Fractal dimensions* are used to establish which resolutions are necessary to describe the land surface structure. The objective is to evaluate whether the observed topography is statistically homogeneous and at (up to) which length scales this applies. If different segments of laser measurements have different fractal dimensions, the segments are heterogeneous. This applies to segments of different lengths, so that estimating relevant length scales from the statistics is possible, e.g. plotting the variance of a segment versus its length.
- *Wavelets transform* is used to establish which wavelengths are relevant for different segments along an elevation profile.

## MATERIALS

The results presented here have been obtained with data sets collected with the same airborne laser altimeter in the framework of different experiments carried out between 1992 and 1995. These experiments were MONSOON'90 in the Walnut Gulch watershed in 1990 and 1991, Arizona, HAPEX-Sahel 1992 in Niger, and JORNEX'95 in 1995 in the Jornada Long Term Ecological Reserve (LTER), New Mexico.

A laser altimeter mounted on a low flying aircraft was used to measure the distance between the aircraft and the landscape surface. The landscape surface was defined by any object (i.e. soil, rock, vegetation, man-made structure) reflecting the laser pulse (Ritchie & Jackson, 1989). Landscape profiles were measured from altitudes between 150 and 200 m and at nominal speeds of 60 metres per second. The laser was a pulsed gallium arsenide diode laser, transmitting and receiving laser signals at a wavelength of 0.904 μm with a field-of-view of 0.6 milliradians. The laser operated at a rate of 4000 pulses per second and recorded with a vertical resolution of 5 cm for a single pulse. Digital data for the laser were recorded directly with a portable personal computer. Data from a gyroscope and accelerometer allowed the laser measurements to be corrected for aircraft motion. A video camera borehole sighted with the laser beam, recorded an image of the flight line. Each video frame (60 per second) was electronically numbered and this number was recorded with the laser data to allow the comparison of the laser data and the video data.

Backscatter can occur when the laser beam hits a target smaller that the footprint. Usually this backscattered signal will be sufficient to trigger the timer in the instrument and to measure the aircraft-to-object range for objects smaller than the footprint. This has been observed in the studies for agricultural crops and natural vegetation.

Low resolution radar altimetry measurements from space platforms for studies at large length scales were also investigated. Measurements made with the SEASAT radar altimeter in 1979 were used. A number of segments covering Niger and Mali were retrieved from the SEASAT archive maintained at NASA/GSFC in Greenbelt, Maryland.

## METHOD

*Effective roughness length of a composite landscape*

Menenti & Ritchie (1994) applied the following equations in sequence to estimate the effective aerodynamic roughness length of the MONSOON'90 data set. Local aerodynamic roughness length for momentum ($z_{0m}$) due to the intervening complete and partial canopies are expressed by

$$z_{0m} = \frac{1}{N}\sum_{1}^{n} \left[\frac{(\sigma_{h,i} - \sigma_0)}{h_i}\right]\bar{h} \quad (m) \tag{4}$$

where n is the number of samples in a segment of laser measurements, $\sigma_{h,i}$ is the standard deviation of the vegetation height in the i-th sample of the h-th segment, $\sigma_0$ is instrument noise, $h_i$ is mean vegetation height in the i-th sample and h is the h-th segment.

Effective aerodynamic roughness length ($Z_{0m}$) of the intervening canopy and of obstructions is defined by (Arya, 1975):

$$\ln\frac{Z_{0m}}{z_{0m}} = \frac{4}{5}\left[\ln\frac{h}{z_{0m}} + \ln\left(\frac{1}{\lambda} - \frac{b}{h} - \frac{B}{h}\right)\right] \cdot \left\{1 - \left[lm\lambda - \frac{C_{dh}\lambda}{2k^2}\left(\ln\frac{h}{z_{0m}}\right)^2\right]^{-\frac{1}{2}}\right\} \tag{5}$$

where k is von Karman's constant, m=(b/h)+(B/h)+(L/2h), b is the base width of shrubs, h is their height, B is the base width of the region with separated airflow behind obstructions, $\lambda$=h/s with s the spacing of shrubs and L is a restoration length of the wind profile behind obstructions. The parameters (b, h, and s) were obtained with the laser measurements, while B and L were estimated from the literature.

Effective aerodynamic roughness ($Z'_{0m}$) of interspersed cover, of obstructions and of topography is defined by (Taylor *et al.*, 1989)

$$\ln\left(\frac{Z'_{0m}}{Z_{0m}}\right) = 3.5\left(a\frac{2\pi}{\lambda}\right)^2 \ln\left(\frac{\lambda}{Z_{0m}}\right) \tag{6}$$

where a is the amplitude of the (assumed periodic) relief and $\lambda$ its wavelength.

*Measurement of length scales of a landscape*

A first measure of the length scales in a landscape is obtained by calculating the spatial *autocorrelation* function of laser measurements using

$$H(\gamma) = \frac{\int H(x+\gamma)H(x)dx}{\int H(x)\cdot H(x)dx} \tag{7}$$

In practice, Equation 7 estimates the autocorrelation function as a summation in which $\gamma$ is a spatial lag. Breakpoints in H($\gamma$) indicate changes in the spatial structure of the topography and may be used as an estimate of length scales.

When dealing with signals, which are some combinations of an autocorrelated noise with a periodic function, the autocorrelation function decreases for a small $\gamma$ but does not become negligibly small at large $\gamma$ due to periodic components. In these cases the autocorrelation function cannot be used to characterize the observations, although it does provide a length scale for the breakpoint where the noise becomes smaller than the periodic component. The periodic portion of a spatial function $H(\gamma)$ does indeed provide information on length scales but this information can be extracted by calculating the Fourier transform $H(\kappa)$ of $H(x)$

$$H(\kappa) = \int H(\kappa) e^{i\kappa x} dx \tag{8}$$

where $\kappa = 2\pi/\lambda$ is the wavevector and $\lambda$ the wavelength. Note the $\lambda$ in Equation 8 is a continuous variable and $H(\kappa)$ a continuous function. In practice, $H(\kappa)$ is calculated for a discrete set of $\lambda$-s and Equation 8 written as a summation of a finite number of periodic functions having different wavelengths (Fourier series). If the observed topography can be approximated by one or more terms in this summation, the wavelengths of these terms can be considered as length scales of the observed landscape. The power spectrum of $H(x)$ is the squared magnitude of $H(\kappa)$.

Sometimes neither the autocorrelated function nor the Fourier transform provide useful information about length scales. One may look at how simple statistics describe the length of the observed elevation profile at different spatial sampling intervals (resolutions). Description of terrain and estimation of roughness parameters using *fractal techniques* has been used to define landscape parameters.

Fractal scaling was found applicable to many data sets related to geomorphology (Malinverno, 1989 and 1990; Klinkenberg & Goodchild, 1992; Ouchi & Matsushita, 1992). Parameters of the fractal model were used for surface roughness assessment (Roy & Robert, 1989; Rees & Muller, 1990), for improvement of digital elevation models (Rees & Muller, 1990; Polidori *et al.* 1991), and for surface reflectance estimations (Jaggi *et al.* 1992; de Jong & Burrough, 1995). Remotely-sensed data of different types were analysed to reveal scaling properties. Fractal scaling was an appropriate tool to model sonar data on sea floor topography (Malinverno, 1989 and 1990), airborne multispectral scanner data (Jaggi *et al.*, 1993), radar remote sensing data (Shepard *et al.*, 1995), and LANDSAT Thematic Mapper images (de Jong & Burrough, 1995). Comparison of fractal dimensions may show changes in surface geometry induced by factors under study (Hallet, 1990; de Jong & Burrough, 1995).

Airborne laser altimetry data have not been analysed to date using fractal scaling. The objective of this study was to establish whether fractal scaling is applicable to airborne laser altimetry data as implemented by Menenti & Ritchie (1994). Several techniques can be used to estimate fractal parameters of a line (Russ, 1994; Shepard *et al.*, 1995). Not all of them are applicable to geomorphological data because these data are self-affine, i.e. have different scaling in vertical and in horizontal directions (Russ, 1994). We selected three techniques appropriate for self-affine lines, namely Korcak's technique, the root-mean-square technique, and the power spectrum techniques, and applied these techniques to data collected on fifteen airborne laser altimetry profiles.

Korcak's technique uses a cumulative plot of the number of horizontal intercepts as a function of length for any arbitrary selected elevation value. A log-log plot of the number of distances that exceeds a value K as a function of K has a slope that is exactly one less than the fractal dimension of the profile.

Root-mean-square (RMS) techniques presume a calculation of the standard deviation of elevation values within a window of width W. Within a window the points are 'de-trended' by subtracting a best fit straight line. The remaining RMS deviation or roughness parameter will depend on window width (W). The slope of the log of RMS versus log of W plot gives the fractal dimension.

Power spectrum techniques use the Fourier transform of the elevation data. Squares of magnitudes are related to corresponding frequencies and a plot of the log of the squared magnitude versus log of the frequency has a slope that is equal to 2D-4, where D notes the fractal dimension. We used the software provided on the diskette with the monograph of Russ (1994) to calculate the power spectrum.

Data were collected (MONSOON'90; HAPEX-Sahel 1992) during flights with the spacing between measurements of approximately 1.5 cm. We analysed the raw data after block averaging had removed the noise. Ground elevation trends were removed by subtracting linear trends between minimum elevations interpreted to be soil surface elevations.

A different approach was applied to the more recent data collected in the Jornada LTER during the first Intensive Observation Period (JORNEX'95, May 1995). The data sets correspond to two N-S and two E-W lines flown over two sites:

(A) a grass-dominated site with flat terrain, and
(B) a shrub-dominated site with sand dunes.

The data were analysed using a method proposed by Malinverno (1990), who suggested that the root mean square deviation (d) of observed elevations along a horizontal distance (x) is related to x and to the fractal dimension D as

$$d(x) = Ax^{2-D} \tag{9}$$

where A is a constant. Observed elevations are de-trended and then d(x) is calculated as

$$d(x) = \frac{1}{n_x}\sum_{1}^{n_x}\sqrt{\frac{1}{m_i-2}\sum (z_j-\bar{z})^2} \tag{10}$$

where $n_x$ is the total number of windows of length x, $m_i$ is the number of points, $z_j$ is elevation and z is mean elevation in the I-th window.

The observed d(x) values were fitted as

$$d(x) = b_2\left(\frac{x_i}{b_1}\right) + b_3\left(\frac{x_i}{b_1}\right)^{b_4} \qquad x_i < b_1 \tag{11}$$

The *wavelets transform* combines the capabilities of the Fourier transform with multi-resolution analysis of data. This is a different approach to measuring length scales in a complex landscape by modelling observed elevation profiles using wavelets transform (David, 1992). Wavelets can be used to define a multi-resolution orthonormal basis (Mallat, 1989). A set of independent functions, a linear combination of which can be used to "reproduce" any other function is defined in the same space. A wavelets transform can be seen as an extension of the Fourier transform. In quantitative terms one may say that wavelets are periodic over finite intervals of the spatial coordinates only and as such may provide a better description of quasi-periodic functions. For a correct comparison of wavelets and Fourier transform, David (1992) gave the following: "vague interpretation". For each j, $P_j$ is the orthogonal projection operator of $V_j$; Pf—>0 when j—>–infinity and Pf—>f when j—> +infinity. Each j corresponds roughly to the size $2^{-j}$ and the projection Pf will give details about f up to the size equivalent to $2^{-j}$, so that Pf should give ever more precise details about f.

The application presented here is very preliminary and simply illustrative. The Daubechies (1988) wavelets were applied to laser measurements to illustrate how precisely they can be described by wavelets at a given resolution.

## DISCUSSION OF RESULTS

The frequency distribution of heights of objects hit by the laser beam is the basic measurement. A validation data set was collected on the ground in the field during the HAPEX-Sahel experiment. The results (Figure 1) indicate that the laser observations are accurate, even for low and sparse vegetation (Ritchie *et al.*, 1996).

Estimates of the local $z_{0m}$ were obtained (Menenti & Ritchie, 1994) using Equation 4 with full resolution laser measurements for each of eight MONSOON'90 intensive study sites (Kustas & Goodrich, 1994) in which four segments (each 30 m) of laser measurements were analyzed. These estimates compared well (Table 1) with ground measurements and estimates of $z_{0m}$ obtained by wind profile observation under near neutral conditions (18 cases at site 1 and 62 cases at site 5). At sites 2, 3, 4, 6, 7 and 8 no reference observations of $z_{0m}$ were available and the values given in Table 1 are estimates based on qualitative similarities in vegetation cover. The site 5 value was assigned to site 4 and site 1 values to sites 2 and 3. At sites 6, 7, and 8 a different method was used based on the determination of the variance of vertical wind velocity, $\sigma_w$, described by Kustas *et al.* (1994).

Values of the effective roughness length $Z_{0m}$ (Equation 5) and $Z'_{0m}$ (Equation 6) were calculated, although no field measurements were available for validation. To apply Equation 6, the observed elevation profile was approximated as a periodic

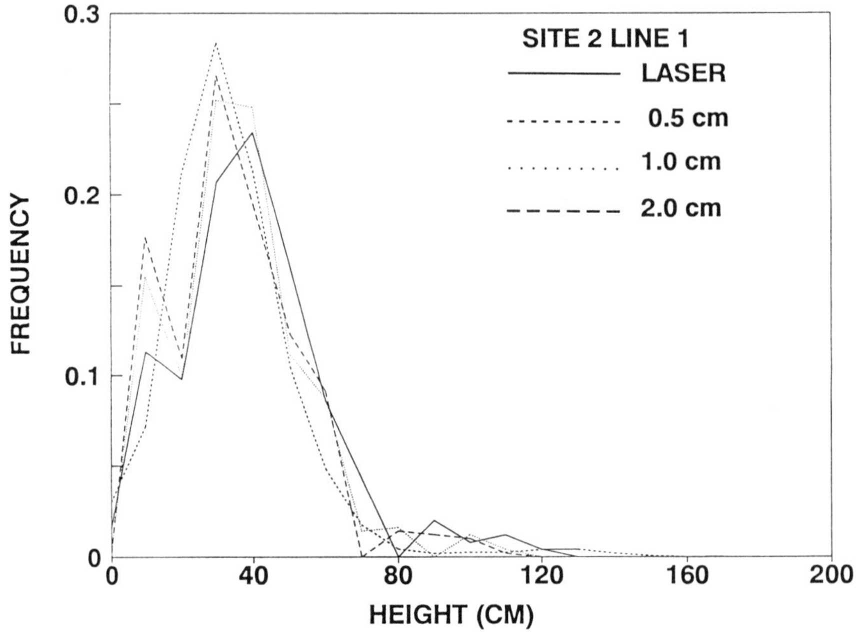


**Figure 1** Frequency distribution of vegetation elements as measured by the airborne laser altimeter and obtained with ground measurements using three different resolutions, 5, 10 and 20 mm, HAPEX Sahel, Niger 1992.

function having just one wavelength. This gave $\alpha = 39$ m and $\lambda = 2410$ m and $Z_{0m} =$ 0.22 m and $Z'_{0m} = 0.84$ m. It should be noted that this procedure relies on the assumption that $z_{0m}$, $Z_{0m}$ and $Z'_{0m}$ apply to distinct and increasing length scales.

The same approach was applied to low resolution elevation measurements obtained with the radar altimeter onboard the SEASAT for a 1978 data set. These data (located in Mali and Niger) were analyzed and sensor-related artifacts were

**Table 1** Values of aerodymanic roughness length for momentum; A) measured values for near neutral wind profiles (Kustas *et al.*, 1994); B) estimates used in model calculations (Kustas *et al.*, 1994); C) estimates from Equation 4 from laser measurements MONSOON'90 Walnut Gulch watershed, 1991.

| Site | $z_{0m}$ A (m) | $z_{0m}$ B (m) | $z_{0m}$ C (m) |
|---|---|---|---|
| 1 | 0.06 | 0.04 | 0.07 |
| 2 | - | 0.04 | 0.05 |
| 3 | - | 0.04 | 0.06 |
| 4 | - | 0.01 | 0.05 |
| 5 | 0.013 | 0.01 | 0.02 |
| 6 | - | 0.04 | 0.05 |
| 7 | - | 0.03 | 0.045 |
| 8 | - | 0.05 | 0.08 |

noted in a number of cases: therefore no detailed account of results is given here. Note however that the values of the ratio $\sigma_x/h$ calculated with data having a spatial sampling interval of 6500 m were comparable to values obtained using the laser measurements at much higher resolution and applied to partial canopies.

In this context it should be mentioned that reference measurements are not important for validation. *Local roughness, form drag* and *topography* are relative rather than absolute notions and may be applied to elevation measurements done at low resolution such as with radar altimeters. At very large horizontal length scales, however, the parameterisation of momentum flux based on the wind profile may not always be applicable and experiments are necessary to establish whether this schematization on land surface-atmosphere interaction is meaningful.

The results indicate that the spatial properties of the elevation profiles measured with the airborne and satellite altimeters need to be analyzed in more detail. The laser measurements (Figure 2) provide excellent material for these studies, since a variety of length scales is evident by simple visual inspection. Although the profiles considered are relatively short, the superposition of the vegetation fine structure with topography at shorter and longer length scales can be clearly appreciated. The issue is how to measure these length scales prior to the determination of the parameters required by Equations 4 - 6.

A first measure of length scale can be provided by the *autocorrelation function* of the elevation profiles. The correlation length $\gamma$, defined by $H(\gamma)=e_j H(\gamma=0)$, is a measure of the segment length required for statistical homogeneity. For profiles containing a periodic component, one might take the $\gamma$ value at which the periodic

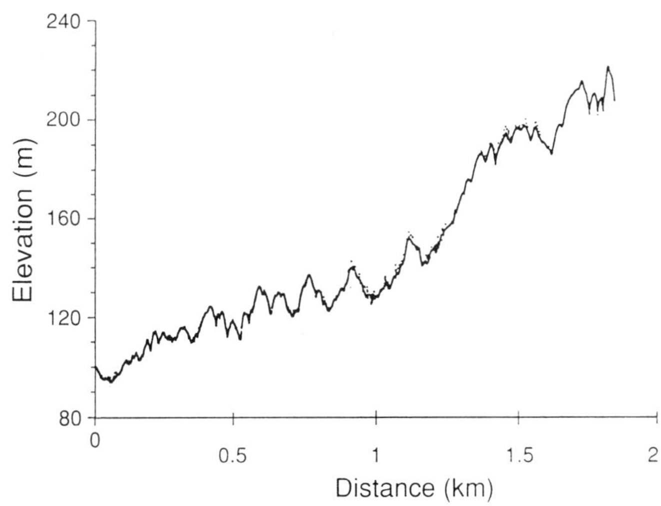


**Figure 2** Elevation profile measured with an airborne laser altimeter in west to east direction. Elevation is in meters above an arbitrary reference. Original measurements have been smoothed by block averaging over 16 data points, MONSOON'90, Walnut Gulch, AZ, 1991.

component becomes dominant as a length scale. The wavelength(s) of the periodic component(s) is another relevant length scale(s), but in this case is better dealt with by means of the Fourier and wavelets transforms.

The autocorrelation of elevation profiles such as the example in Figure 2 may provide trivial information, since the large scale topography dominates the signal (the trend from lower to upper portions of the watershed). So the data were de-trended by subtracting a baseline obtained as the lower envelop of the observations. The autocorrelation of the residual (Figure 3) at the larger $\gamma$ values indicates the presence of a periodic component. The correlation length of the observations is 98 m, while the periodic component becomes dominant at $\gamma$=120 m. It should be noted that such length scales cannot be defined very precisely and these results indicate that the length scale of this elevation profile lie somewhere between 100 m and 120 m.

To estimate the wavelength(s) of the periodic component of elevation profiles such as the one shown above, the *Fourier transform* of the laser measurements collected during the MONSOON'90 was calculated. This was applied to four E-W 8 km flight lines after block averaging (16 data points) the full resolution measurements.

The power spectra (Figure 4) decay linearly as expected from the autocorrelation function. The linear decay region starts at a wavelength of about 500 data points or 125 m that agrees with the correlation length obtained from the autocorrelation functions. The dominant components have wavelengths between 2000 (400 m) and 5000 (1250 m) data points. A closer examination of the amplitudes (Figure 5) indicates some distinct features in the four E-W elevation profiles. The four components in line 1 have rather similar and relatively small amplitudes. Elevation along line 5 is dominated by a periodic component at a frequency of 1184 m.

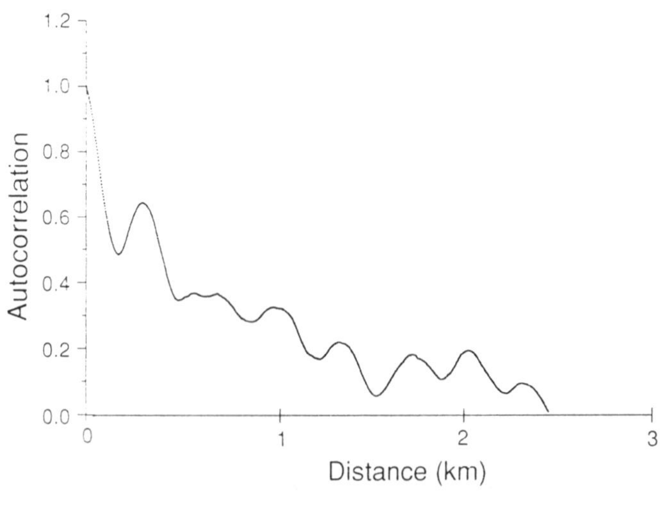


**Figure 3** Autocorrelation function of elevation profile in Figure 2 after de-trending.

Thus far attention has been focused on larger length scales and periodic components of the signal, while the evaluation of the laser measurements at smaller length scales has been limited to estimation of $z_{0m}$, $Z_{0m}$, and $Z'_{0m}$. The finer structure of the elevation profiles was studied by means of the *fractal techniques*. The objective of this analysis was two-fold:

(i) determine whether a fractal model can describe the elevation data such as;

$$l(\delta x) = B(\delta x)^{-D} \quad (12)$$

where $l(\delta x)$ is the length of the elevation profile as a function of the spatial sampling interval (yardstick, resolution, $\delta x$), $1<D<2$ is the fractal dimension of l and B is a constant, and

(ii) estimate length scales as breakpoints in a plot of $l(\delta x)$ versus $\delta x$.

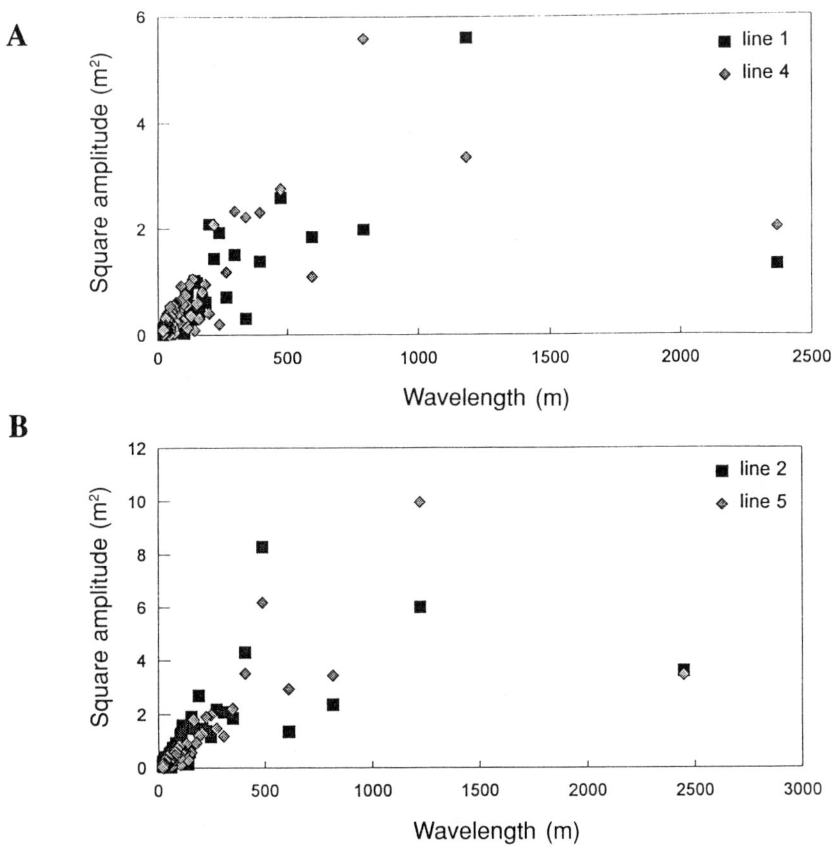


**Figure 4** Power spectrum of elevation profiles measured with an airborne laser. Original measurements have been smoothed by block averaging over 16 dat points and de-trending, MONSOON'90, Walnut Gulch, AZ 1991. 1000 data points = 245 m, (A) flight lines 1 and 4; (B) flight lines 2 and 5.

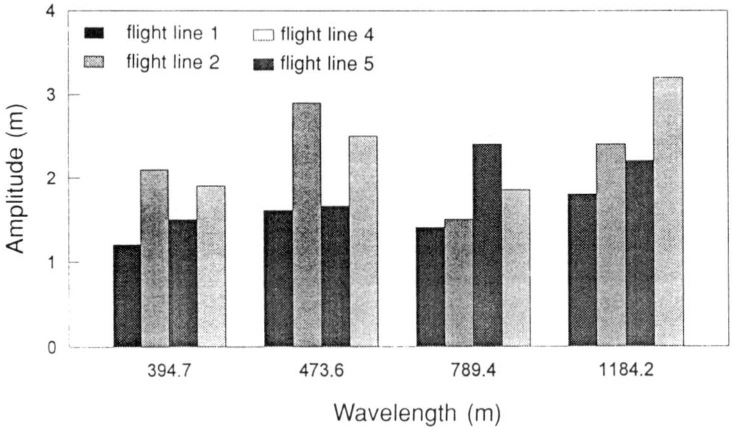


**Figure 5** Dominant components in the power spectra from Figure 4

If Equation 12 applies then $l(\delta x)$ is linearly related to $\delta x$ in a log-log plot (Figure 6), with the slope being –D. Breakpoints in this plot indicate typical length scales. The length of the elevation profile does not change with sampling intervals if $\delta x < \delta x_1$ or $\delta x > \delta x_2$. Elevation profiles having the same D-values are self-similar. These breakpoints indicate relevant length scales for momentum transfer. Since they are estimates of the typical smaller and larger size objects placed on the land surface, the presence of breakpoints indicates that the length of the profile (measured on the observed surface) does not change if the spatial sampling intervals are smaller than $\delta x_1$ or larger than $\delta x_2$.

In general, Equation 12 is applicable to data from the airborne laser altimeter (Table 2). Profiles where noise has been removed have fractal dimensions well within the theoretical range from 1 to 2. Data sets where noise has not been removed have calculated fractal dimensions close to 2. Different techniques result in different estimates of the fractal dimension. This is a common phenomenon (Milne, 1988) resulting from (i) data sets that are not ideally fractal, (ii) series of measurements that are inevitably short, and (iii) there exists a low resolution limit. Power spectrum and RMS techniques gave results that are in broad general agreement. Korcak's technique yields fractal dimensions that vary over a wider range, probably because of uncertainty in selection of the region of fitting as described by Milne (1988). The removal of the soil surface elevation did not change fractal dimensions.

This analysis provides estimates (Table 3) of the smaller and larger length scales for the same laser measurements used for the estimation of $z_{0m}$ (Table 1). Although the aerodynamic roughness length is a vertical length scale of the wind profile, there is a fundamental relationship between $z_{0m}$ and the placement density and size of roughness elements on the land surface. The detailed elevation profiles provided by airborne laser altimeter describe this relationship and is indicated by the fractal dimension of these profiles.

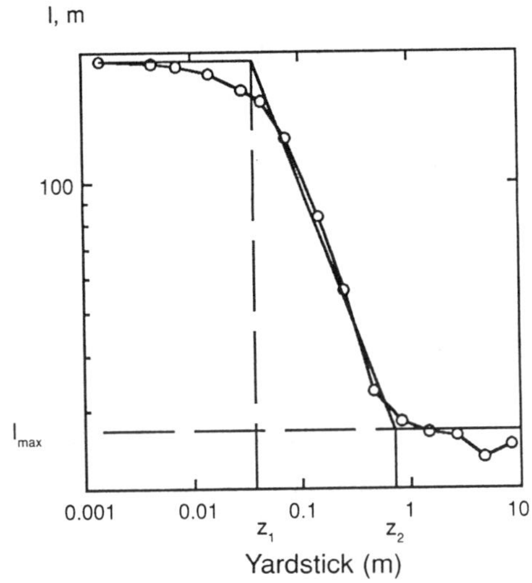


**Figure 6** Elevation profiles measured with an airborne laser altimeter. Determination of the fractal dimension and of smaller ($z_1$) and larger cutoff ($z_2$) resolution; l is the length of the elevation profile as a function of the spatial sampling interval (yardstick or resolution).

Values of $\delta x_1$ are similar to the $z_{0m}$ values in Table 1. Relative difference between site 1 and 5 agree with the results given there. For example, land cover at site 1 is rougher than site 5 although at site 5, $\delta x$ is larger than both $z_{0m}$-A and $z_{0m}$-C. The $\delta x_2$ values are similar to the typical size shrubs and to the estimates of the height and

**Table 2** Determination of the fractal dimension of elevation profiles from MONSOON'90 and HAPEX-Sahel airborne laser measurements. Noise/trend: none indicates that neither noise or trend have been removed.

| Date set | Korcak | RMS | Power spectrum | Noise/trend |
|---|---|---|---|---|
| Niger1 | 2.0 | 2.0 | 1.842 | none |
| Niger2 | 2.0 | 2.0 | 1.863 | none |
| Niger3 | 1.457 | 1.643 | 1.562 | block average |
| Niger4 | 2.0 | 2.0 | 1.911 | none |
| Niger5 | 1.518 | 1.643 | 1.594 | none |
| WG6s1 | 1.472 | 1.5 | 1.503 | block average |
| WG6s2 | 1.881 | 1.594 | 1.502 | block average |
| WG6s3 | 1.617 | 1.534 | 1.496 | block average |
| WG6s4 | 1.569 | 1.53 | 1.516 | block average |
| WG7s1 | 1.665 | 1.77 | 1.590 | block average |
| WG7s2 | 1.771 | 1.725 | 1.657 | block average |
| WG7s3 | 1.634 | 1.546 | 1.503 | block average |
| WG7s4 | 1.879 | 1.588 | 1.523 | block average |

width of shrubs obtained from laser measurements and field observations. There are cases where Equation 12 does not apply and no estimates of $z_{0m}$ can be obtained. This is a case of a short segment of laser measurements covering a large tree that is not included in the analysis leading to Table 1 and 3.

A different fractal model can also describe the laser measurements (Equation 10) which emphasises the relationship between vertical features and horizontal lengths. A more recent data set (JORNEX'95) was analysed using Equations 10 and 11. The results (Figure 7) indicate a distinct breakpoint at $W=b_1$ for both the grass-dominated and the shrub-dominated sites. Further it appears that Eqs. 10 and 11 satisfactorily describe the grass data (Figure 7) while more refinement is needed to describe the shrub observations.

The values of the breakpoint window length are given in Table 4. These lengths are estimates of the window length at which a distinct surface geometry appears while the linear shape of the log d(x) versus log x indicates that the fractal dimensions of d(x) are constant up to the breakpoint respectively. In particular, these plots indicate that roughness is quasi-constant and relatively small for $x<b_1$, while it increases beyond it due to trends in data and the inclusion of larger objects in the windows, so that $b_1$ is a significant length scale typical of this landscape. Note that the procedure applied to obtain $b_1$ gives an estimate of the error for the data given in Table 4 that is less than 1%. It appears that the length scale of the grass site after averaging is consistent and equal to 43 m. For the shrubs site some variability is observed due to the combination of sparser but larger roughness elements represented by shrubs and dunes.

**Table 3** Estimates of length scales based on fractal analysis of laser altimeter measurements. For explanation of symbols see Figure 6, MONSOON'90, Walnut Gulch 1991. Each data set is a segment measured at one of the eight sites.

| Site | $\delta x_1$ (m) | $\delta x_2$ (m) | $l_{max}(\delta x)$ (m) | Site | $\delta x_1$ (m) | $\delta x_2$ (m) | $l_{max}(\delta x)$ (m) |
|---|---|---|---|---|---|---|---|
| 1A | 0.037 | 0.717 | 27.083 | 5A | 0.035 | 0..712 | 26.815 |
| 1B | 0.055 | 0.932 | 26.506 | 5B | 0.038 | 0.638 | 27.083 |
| 1C | 0.057 | 1.004 | 26.506 | 5C | 0.036 | 0.706 | 26.815 |
| 1D | 0.061 | 1.173 | 26.506 | 5D | 0.037 | 0.756 | 26.815 |
| 2A | 0.036 | 1.075 | 26.506 | 6A | 0.048 | 0.665 | 27/346 |
| 2B | 0.048 | 0.728 | 26.815 | 6B | 0.037 | 0.731 | 26.815 |
| 2C | 0.052 | 0.703 | 27.083 | 6C | 0.037 | 0.727 | 26.815 |
| 2D | 0.051 | 0.724 | 27.346 | 6D | 0.052 | 0.740 | 27.083 |
| 3A | 0.047 | 0.764 | 27.083 | 7A | 0.048 | 0.623 | 27.083 |
| 3B | 0.050 | 0.702 | 26.815 | 7B | 0.049 | 1.261 | 35.386 |
| 3C | 0.048 | 0.742 | 26.815 | 7C | - | - | - |
| 3D | 0.050 | 0.767 | 26.815 | 7D | 0.066 | 2.182 | 28.913 |
| 4A | 0.035 | 0.862 | 26.506 | 8A | 0.053 | 0.897 | 26.506 |
| 4B | 0.050 | 0.615 | 26.815 | 8B | 0.055 | 0.999 | 26.506 |
| 4C | 0.048 | 0.731 | 26.815 | 8C | 0.056 | 0.936 | 26.506 |
| 4D | 0.037 | 0.681 | 26.815 | 8D | 0.054 | 0.951 | 26.506 |

**Table 4** Values of the breakpoints obtained (Fig. 7 and Equation 11) by plotting log d(x) versus log x for JORNEX'95 (Jornada Long Term Ecological Reserve, New Mexico) May 1995.

| Data Set | length (m) | Grass (m) | Shrub (m) |
|---|---|---|---|
| Grass.01 | 722 | 37 | 18 |
| Shrub.02 | 1716 | | |
| Grass.03 | 714 | 44 | 89 |
| Shrub.04 | 1717 | | |
| Grass.13 | 1152 | 50 | 58 |
| Shrub.09 | 512 | | |
| Grass.14 | 717 | 41 | 53 |
| Shrub.10 | 1285 | | |

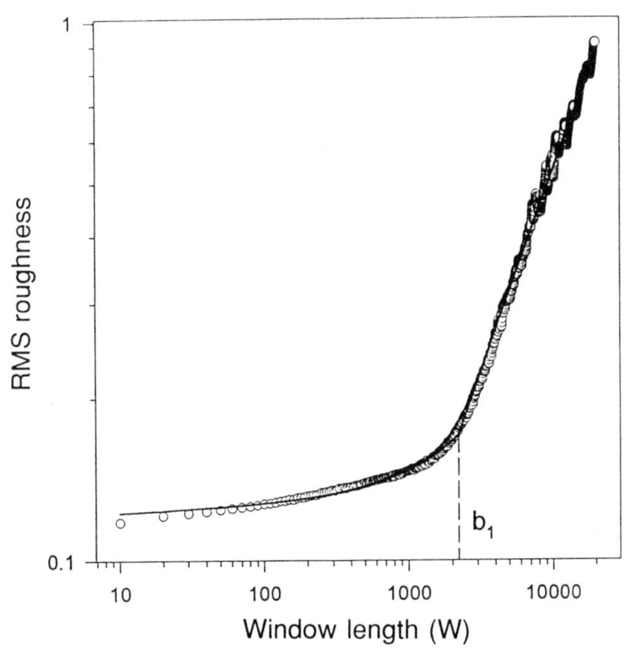


**Figure 7** RMS deviation of elevation with respect to mean elevation in a window of length W (number of data points) versus length of window for the grass dominated site, May 1995, Jornada Long term Ecological Reserve, New Mexico (JORNEX'95)

The *wavelets transform* can be applied to estimate the wavelength of the periodic component of the topography. The Daubechies (1988) wavelets were applied to the MONSOON'90 line 1 data (Figure 2). The data were reproduced by using two dominant components having wavelengths of about 1500 data points (425 m) and 2100 data points (525 m) and were in general agreement with the results given in Figure 5. While the analysis shows promise in estimating periodic components of the topography, more analyses are needed on the applicability of wavelets transform for defining such periodic components.

## CONCLUSION

In the issue of scaling up surface roughness, the properties of heterogeneous land surfaces are important for understanding and modelling land surface-atmosphere interactions from two different points of view:

(i) The estimation of aerodynamic roughness at increasing length scales using concepts, theory and Equations from the literature and measurements conducted by an airborne laser altimeter, and

(ii) the estimation of length scale of composite landscapes using the same altimeter measurements and a variety of statistical techniques.

The results of the analyses done in relation with (ii) indicated that calculations done with (i) should be based on a preliminary determination of length scales, since the sequence local, obstructions, and topography may not consist of neatly different length scales, as required by theory applied to calculated $Z'_{0m}$ from $Z_{0m}$ and $z_{0m}$. This applies particularly to the parameterisation of topography, which required multiple wavelengths for all the cases analyzed. On the other hand, the method applied here appears reliable and provides consistent estimates of length scales ranging from fine structure of observed elevation over short distances to the characterization of multispectral relief over distances of a few kilometres.

## ACKNOWLEDGMENTS

The manuscript did benefit significantly from the comments of an anonymous reviewer and, particularly, of Dr. John Prueger whose detailed remarks are gratefully acknowledged.

## REFERENCES

Arya, S.P.S. 1975. A drag partition theory for determining the large scale roughness parameter and wind stress in the Arctic pack ice. *J. Geophys. Res.*, **80**, 3447-3454.

Beljaars, A.C.M. and Holtslag, A.A. 1991. Flux parameterization over land surfaces for atmospheric models. *J. Appl. Meterol.*, **30**, 327-341.

David G. 1992. Wavelets and singular integrals on curve and surfaces. *Lecture Notes in Mathematics*

1465, Springer-Verlag, Heidelberg: 109 pp.

Daubechies, I. 1988. Orthonormal bases of compactly supported wavelets. *Comm. Pure Appl. Math.* 41: 909-996

Finnigan, J.J., Raupach, M.R., Bradley, E.F. and Aldis, G.K. 1990. A wind tunnel study of turbulent flow over a two-dimensional ridge. *Bound.-Layer Meterol.*, **50**, 277-317.

Hallet, B. 1990. Spatial self-organization in geomorphology: from periodic bedforms and patterned grounds to scale invariant topography. *Earth Sci. Rev.*, **29**,57-75.

Jaggi, S., Quatrochi, D.A. and Siu-Ngan Lam, N. 1992. Multi-resolution processing for fractal analysis of airborne remotely sensed data. In: JPL, Summaries of the *3rd Annual JPL, Airborne Geoscience Workshop* **2**, 25-27.

Jaggi, S., Quatrochi, D.A. and Siu-Ngan Lam, N. 1993. Implementation and operation of three fractal measurement algorithms for analysis of remote sensing data. *Computers and Geosci.*, **19**, 745-767.

Jong de, S.M. and Burrough, P. 1995. A fractal approach to the classification of Mediterranean vegetation types in remotely sensed images. *Photogram. Engr. and Remote Sens.*, **61**, 1041-1053.

Klinkenberg, B. and Goodchild, M.F. 1992. The fractal properties of topography: A comparison of methods. *Earth Surface Process and Landform*, **17**, 217-234.

Kustas, W. P. And Brutsaert, W. 1986. Wind profile constants in a neutral atmospheric boundary layer over complex terrain. *Bound.-Layer Meterol.*, **34**, 35-54.

Kustas, W.P. and Goodrich, D.C. 1994. Preface. MONSOON'90 Special issue. *Water Resour. Res.* **30**, 1211-1225.

Kustas, W.P., Blanford, J.H., Stannard, D.I., Daughtry, C.S.T., Nichols, W.D. and Weltz, M.A. 1994. Local energy flux estimates for unstable conditions using variance data in semi-arid rangelands. *Water Resour. Res.*, **30**,1241-1259.

Malinverno, A. 1989. Testing linear models of sea-floor topography. *Pure and Appl. Geophys.*, **131**, 139-155.

Malinverno, A. 1990. A simple method to estimate the fractal dimension of a self-affine series. *Geophys. Res. Letters*, **17**, 1953-1956.

Mallat, S.G. 1989. Multi resolution approximations and wavelet orthonormal bases of $L^2(R)$. *Trans. Am. Math. Soc.*, **315**, 69-87.

Menenti, M. and Ritchie, J.C. 1994. Estimations of effective aerodynamic roughness of Walnut Gulch watershed with laser altimeter measurements. *Wat. Resour. Res.*, **30**, 1329-1337.

Milne, B.T. 1988. Measuring the fractal geometry of landscapes. *Appl. Math. and Computers*, **27**, 67-79.

Ouchi, S. and Matsushita, M. 1992. Measurement of self affinity on surfaces as a trial application of fractal geometry to landform analysis. *Geomorphology*, **5**, 115-130.

Polidori, L., Chorowizs, J. and Guillande, R. 1991. Description of terrain as a fractal surface and application to digital elevation model quality assessment. *Photogram. Engr. and Remote Sens.*, **57**, 1329-1332.

Rees, D. and Muller, J.P. 1990. Surface roughness estimation using fractal variogram analysis. In: *Proc. IGARSS '90* **3**, 1951-1954.

Ritchie, J.C., Menenti, M. and Weltz, M.A. 1996. Measurements of land surface features using an airborne laser altimeter: the HAPEX-Sahel experiment. Submitted to *Int. J. Rem. Sens.*

Ritchie, J.C., and Jackson, T.J. 1989. Airborne laser measurements of the surface topography of simulated concentrated flow gullies. *Trans. Amer. Soc. Agr. Eng.*, **32**, 645-648.

Roy, A.G. and Robert, A. 1989. Fractal techniques and the surface roughness of talus slopes: a comment. *Earth Surface Processes and Landforms* **15**, 283-285.

Russ, J.C. 1994. *Fractal Surfaces.* Plenum Press, New York.

Shepard, M.K., Brackett, R.A. and Arvidson, R.E. 1995. Self-affine topography: Surface parameterization and radar scattering. *J. Geophys. Res.*, **100**,11,709-11,718.

Taylor, P.A., Sykes, R.I. and Mason, P.J. 1989. On the parameterization of drag over small scale topography in neutrally stratified boundary layer flow. *Bound-Layer Meteorol.*, **48**, 409-422.

Thompson, R.S. 1978. Note on the aerodynamic roughness length for complex terrain. *J. Appl.*

*Meterol.*, **17**, 1402-1403.

Wieringa, J. 1986. Roughness dependent geographical interpolation of surface wind speed averages. *Q. J. R. Meterol. Soc.*, **112**, 867-889.

Wooding, R.A., Bradley, E.F. and Marshall, J.K. 1973. Drag due to regular arrays of roughness elements on varying geometry. *Bound.-Layer Meterol.*, **5**, 285-308.

# 4 Areal average surface fluxes from mesoscale meteorological models: the application of remote sensing

**R.J. HARDING, C.M. TAYLOR and J.W. FINCH**
*Institute of Hydrology, Wallingford, UK*

## INTRODUCTION

Mesoscale Meteorological Models (MMMs) are used extensively for research and weather forecasting on the regional scale. The models run typically with a grid cell 10 to 20 km square with a domain size of approximately 1000 km. The models are run for approximately 24 hours forced at the boundaries with output from a global forecast model or analysis. There is increasing recognition that, even at this small spatial scale and at integration times of a few hours, the description of the land surface is important. Blyth *et al.* (1994) showed in a case study in SW France that the predicted total rainfall from a frontal system could change by a third depending on the characteristics of the land surface. At a slightly larger scale Beljaars *et al.* (1996) found that the anomalous rainfall which occurred in the USA in 1993 could only be correctly simulated with a fully interactive soil moisture scheme within the forecast model.

Mesoscale meteorological models are powerful tools to investigate the effects of surface variability at a scale smaller than that of a grid cell of a GCM. The variability which can be included is due to landscape variability (vegetation, soil type, topography) and surface state variability (soil moisture and surface wetness). The MMMs will also include the feedback effects between the surface and the boundary layer and additional transport due to circulations within the boundary layer (see e.g. Pielke *et al.* 1991).

Existing MMMs, especially those used for weather forecasting, frequently use mean soil moisture fields as a lower boundary (such as that described by Willmott *et al.*, 1985). Such maps have a number of limitations: first, they cannot represent the soil moisture pattern in a particular season, this is important in the Sahel where the intra-annual variation of rainfall is high. Second, a single figure for one month can cause large errors. It is possible to construct a realistic soil moisture field by carrying forward the soil moisture fields from the successive runs of the forecast model over a season. This approach has a number of problems due to errors in forecast rainfall, which propagate into the soil moisture and to inadequacies in the

*Scaling up in hydrology using remote sensing*. Edited by J.B. Stewart, E.T. Engman, R.A Feddes and Y. Kerr soil moisture models used (for example simple bucket type models tend to dry out too quickly). One solution to the inadequate rainfall estimation is to incorporate rainfall measurements in a four dimensional assimilation system (see e.g. Mitchell 1994). However this latter approach can only be undertaken when areal average rainfall is measured routinely to a high degree of accuracy and in many areas of the world this is not the case.

Remote sensing can help in the specification of the required vegetation and rainfall fields. In theory they could also help with surface soil moisture, parameter fields (such as roughness and albedo) and vegetation state (using leaf area index and FPAR [= Fraction of absorbed Photosynthetically Active Radiation] ) although this potential has yet to be realised.

We present here a case study of the implementation of a MMM of a region in west Africa, centred on the HAPEX-Sahel 1° by 1° square (Goutorbe *et al.*, 1994). The model used is the United Kingdom Meteorological Office (UKMO) forecast model. The purpose of this implementation is not weather forecasting but the study of causes of variability of surface fluxes and strategies for the calculation of average fluxes. The location of this analysis in a semi-arid region was deliberately chosen because of the considerable spatial variability of the rainfall and vegetation. It is also a region where the partition of the surface fluxes is generally controlled by the soil moisture. The HAPEX-Sahel measurements provide an opportunity to calibrate the land surface model used and to validate the model output. However, outside the HAPEX-Sahel 1° by 1° square and the two-month intensive observation period, the surface data are extremely limited. In particular, rainfall shows a substantial variability which cannot be resolved by the manual raingauge network. To provide the model with realistic estimates of the surface parameters, both mean and spatial patterns, the use of remote sensing is essential.

The surface parameter fields for the MMM were constructed using a combination of satellite data, surface measurements and land surface models. This process is described in detail below, together with an example of the use of the output of a mesoscale simulation for a single day to assess the effects of heterogeneity.

## SURFACE FIELDS FOR THE MESOSCALE MODEL

### INTRODUCTION

The mesoscale model requires fields of soil and vegetation parameters and soil moisture as its lower boundary. These parameters need to be specified for each of the 92 by 92 ten km grid cells; preferably as a realistic average over each cell. The approach adopted below has been to construct a 1 km resolution land cover map for the domain from satellite imagery and relate the individual parameter values to land cover classes based on surface observations, taken in HAPEX-Sahel and elsewhere. Unlike the majority of the soil and vegetation parameters, the soil moisture varies from day to day and thus a good map of soil moisture is obviously a requirement if a realistic meteorological simulation of the Sahelian region is to be made. The United Kingdom Meteorological Office forecast model uses a climatological average monthly soil moisture map (Willmott *et al.* 1985). However, for the Sahelian region these patterns of monthly soil moisture appear to be unrealistic probably because of the problems associated with the use of mean climatological soil moisture fields as discussed above.

## SURFACE COVER MAP

A simple land cover map of the region was created by carrying out a supervised classification (Settle and Briggs, 1987) of the first two bands of an AVHRR image acquired on 17 June, 1991. The training areas for the classification were defined by using a classified Landsat TM image of part of the HAPEX-Sahel square (Prince, *pers. comm.*). The two data sets had been geo-referenced so it was a simple matter to import the classified Landsat TM areas as the training areas. The land cover classes used were: open natural forest (known locally as tiger bush), sparse tiger bush, savannah, dense millet, sparse millet and bare soil. For the purposes of the soil moisture map for October, the dense and sparse millet classes were combined because at this time both classes would in reality be bare stubble with weeds.

## SOIL MOISTURE FIELDS

A simple, pragmatic soil moisture model has been developed and used here to provide the necessary inputs to the mesoscale model. The soil moisture fields were constructed by modelling the soil moisture over the months preceding the day of the simulation. For each grid cell of the MMM a time series of soil moisture is constructed for each vegetation class using a simple soil moisture accounting model. The final soil moisture is taken as the average of the estimates for the individual cover types, weighted according to the observed fractional coverage of each class in each cell. The soil model is driven by distributed rainfall patterns constructed from satellite and surface observations.

The required domain is 1000 by 1000 km centred on Niamey, Niger with a 10 kilometre grid size. Any method used in this region will be limited by the data available and therefore a complex evaporation or soil model is neither practical nor desirable. Wherever possible the soil model has been based on the existing land surface scheme of the Unified Model (Warrilow *et al.*, 1986), to provide compatibility with the eventual usage of the output.

The model is a daily soil moisture accounting model with input of daily rainfall and potential evaporation and output of daily soil moisture storage and evaporation:

$$S_{i+1} = S_i + R_i - E_i - D_i \tag{1}$$

where $S_i$ is the total soil moisture storage in the rooting zone (in mm) on day i, $R_i$ is the rainfall, $E_i$ is the evaporation and $D_i$ is the drainage.

The soil moisture storage is split into two sections (or layers): a surface store and rooting zone store. These differ in the way evaporation is calculated (see below) and their drainage characteristics. Rain first enters the surface store and drains into the root store when this is full. Evaporation comes preferentially from the surface store and only depletes the root store when the surface store is empty.

*Evaporation*
For the surface store:

$$E = E_p * F_s \tag{2}$$

where $E_p$ is the potential evaporation (see below) and $F_s$ is the surface evaporation factor, which in this implementation is calibrated against observations. The surface store is assumed to evaporate at a constant proportion of the potential evaporation until the store is empty; the evaporation from this store is then zero.

The root store has a two stage evaporation model determined by the mean volumetric water contents within the store ($\theta$):

$$\theta = S/Z_r \tag{3}$$

where $Z_r$ is the rooting depth (mm). If $\theta$ is above a critical value, $\theta_c$, the evaporation is given by:

$$E = E_p * F_r \tag{4}$$

where $F_r$ is a root zone evaporation factor, which is likely to be related to the fractional coverage of the vegetation but in this case is calibrated against observations. If $\theta$ is equal to the wilting point, $\theta_w$, the evaporation is zero. Between the critical value of the soil moisture $\theta_c$ and the wilting point $\theta_w$, the evaporation is linearly interpolated thus:

$$E = E_p * F_r \qquad \text{if } \theta_s > \theta > \theta_c \tag{5a}$$

$$E = E_p * F_r * \beta \qquad \text{if } \theta_c > \theta > \theta_w \tag{5b}$$

$$E = 0 \qquad \text{if } \theta = \theta_w. \tag{5c}$$

where $\beta = (\theta - \theta_w)/(\theta_c - \theta_w)$ and $\theta_s$ is the volumetric water content at saturation. This root zone component is similar to the UKMO single store soil model.

If during the time step (one day) the calculated evaporation from the surface store is greater than the surface store, the surface store is put to zero and the residual potential evaporation is used to calculate additional evaporation from the root store.

*Drainage*

The drainage model is similar to that used in the UKMO model:

$$D = K\left(\frac{\theta - \theta_w}{\theta_s - \theta_w}\right)^C \tag{6}$$

where K is the saturated hydraulic conductivity (kg $m^{-2}$ $s^{-1}$) and C a constant.

*Potential evaporation*

The potential evaporation is calculated using the Priestley-Taylor formula. The input used is the mean climatological monthly values for cloud coverage and temperature described by Leemans and Cramer (1991). These data are on a 0.5° by 0.5° grid. Mean net radiation is calculated from the cloud amount and air temperature and input into the Priestley-Taylor formula. It should be noted that these values are mean values on a fairly coarse grid; however, Calder *et al.* (1983) and Anderson and Harding (1992), using soil moisture data from the UK and Sweden, concluded that the use of mean climatological potential evaporation in simple soil moisture models can give good estimates of day-to-day changes in soil moisture storage.

In conditions when evaporation is limited by soil moisture, the potential evaporation is underestimated by the Priestley-Taylor evaporation, $E_{PT}$. A comparison with the Penman-Monteith estimate of potential evaporation (using a surface resistance appropriate to a wet soil and meteorological observations from the region) suggested the formula:

$$E_P = E_{PT}*(2.7 - 1.7*\beta) \tag{7}$$

This modification increases the potential evaporation above $E_{PT}$ when $\beta$ drops below one.

*Rainfall*

Unlike potential evaporation, it is absolutely vital to input accurate rainfall figures into soil moisture models. This is a particular problem in the Sahelian region where the rainfall variability is high (both year-to-year and spatially) and the raingauge network is sparse. Three sources of daily rainfall data were used:

(1) The TAMSAT (Tropical Applications in Meteorology Satellite and other data) rainfall estimates. These are based on estimates of cloud top temperatures from METEOSAT (Milford and Dugdale, 1992). The calibrated product is generally available as 10-day means but for modelling purposes was disaggregated using the raw daily cold cloud duration estimates. The TAMSAT data give a good spatial pattern of rainfall on a 10 by 10 km grid.

(2) The EPSAT (Estimation des Pluies par Satellite) grid. This is a short term network of gauges operated 1990-1993. It consists of over 100 raingauges

bounded by 1°40'E and 3°E Lat. and 13°N and 14°N Long. and therefore containing within it the HAPEX-Sahel square (Lebel *et al.*, 1996).

3. Monthly totals from West Africa. This is a data set collected by Reading University for calibration of the TAMSAT data.

Within the EPSAT domain a gridded version of the EPSAT raingauge data was used. Outside this domain the TAMSAT data were used, gridded onto the mesoscale grid. The TAMSAT estimates were recalibrated using all the available raingauge measurements from 1992. Separate calibration factors were used for each month and for three latitude bands.

*Soil and vegetation parameters*
The soil parameters ($\theta_s$, $\theta_c$, $\theta_w$ and C) are taken from the UKMO database; this uses three soils (fine, medium and coarse) and is based on Wilson and Henderson-Sellers (1985) and Warrilow *et al.* (1986). However, in the absence of a detailed digital soil map of this region, all soils are assumed to be coarse i.e. sand (Table 1).

**Table 1** Soil parameters (wilting, $\theta_w$, critical, $\theta_c$, and saturated, $\theta_s$ , volumetric water contents) used in the UKMO model

| Texture | $\theta_w$ | $\theta_c$ | $\theta_s$ | C |
|---|---|---|---|---|
| Coarse | 0.055 | 0.087 | 0.420 | 5 |

The rooting depths in the UKMO data base are too small for this region. Values given for the HAPEX-Sahel region are typically 0.8 m, whereas the soil moisture and evaporation measurements suggest water extraction by roots from at least the top 1.5 m and almost certainly from deeper depths. For this study more realistic values were estimated using the evaporation measurements from HAPEX-Sahel (see below).

*Calibration*
The evaporation factors, $F_s$ and $F_r$, have been calibrated against evaporation measurements from the three vegetation covers at the HAPEX-Sahel Southern Super Site, SSS, (savannah, tiger bush and millet) and from two covers at the East Central Site, ECS, (savannah and millet). Daily totals of evaporation measured by eddy correlation are available for the HAPEX-Sahel Intensive Observation Period (IOP), 18 August to 8 October, Gash *et al.* 1996; only days with more than 20 hours of data in the day were used in the analysis. The data from the Automatic Weather Station (AWS) on each site were used to calculate the potential evaporation using the Priestley-Taylor formula; however, the model was run from 1 May, using the

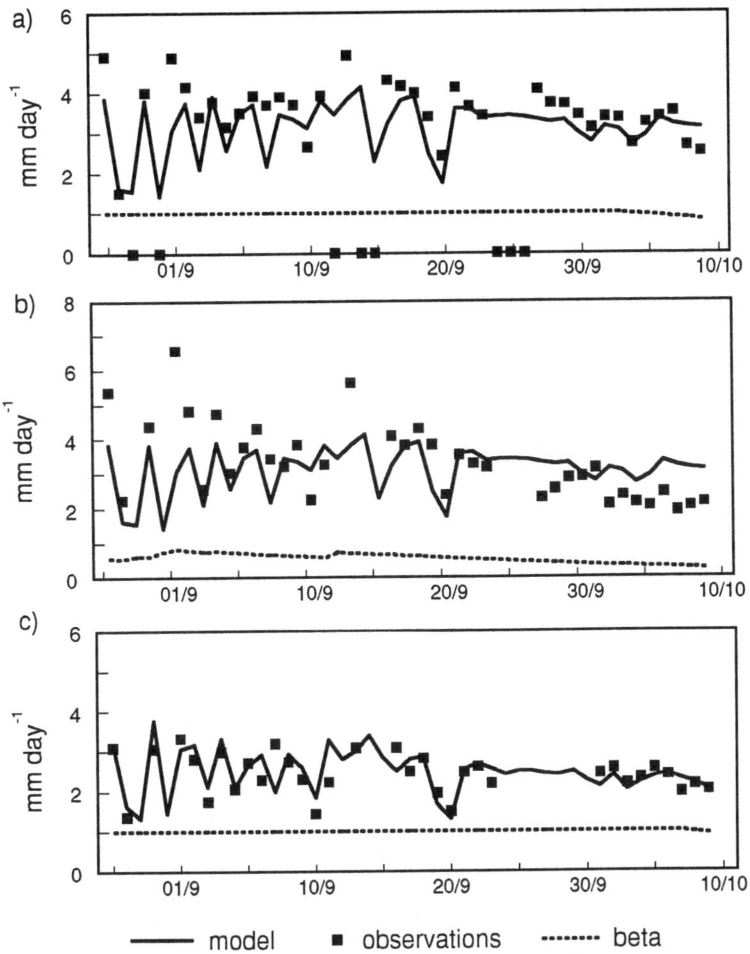


**Figure 1** Daily evaporation totals, measured and modelled with simple soil moisture model for the three land cover types in the Southern Super Site (SSS): (a). savannah, (b)tiger bush and (c) millet, for the HAPEX-Sahel IOP. Also shown is the soil water stress factor, β.

mean climatological estimates of the Priestley-Taylor potential evaporation for the period prior to the IOP. Local rainfall data for each site was used. The values were optimised by minimising the sums of squares of the difference between the modelled and measured evaporation after 17 September (day number 260).

Although the optimisations were made using only data from the dry down period, the fit between measured and fitted values of daily evaporation are good for the whole period of measurement (Figures 1 and 2). Optimised values for $F_r$ are 0.7, 0.85 and 0.7 for the savannah, tiger bush and millet respectively at the SSS and 0.65 and 0.5 for the savannah and millet at the ECS. These values are consistent with the sparser vegetation found at the ECS. Values for $F_s$ are generally close to $F_r$.

A simple optimisation procedure was used to identify effective values for the

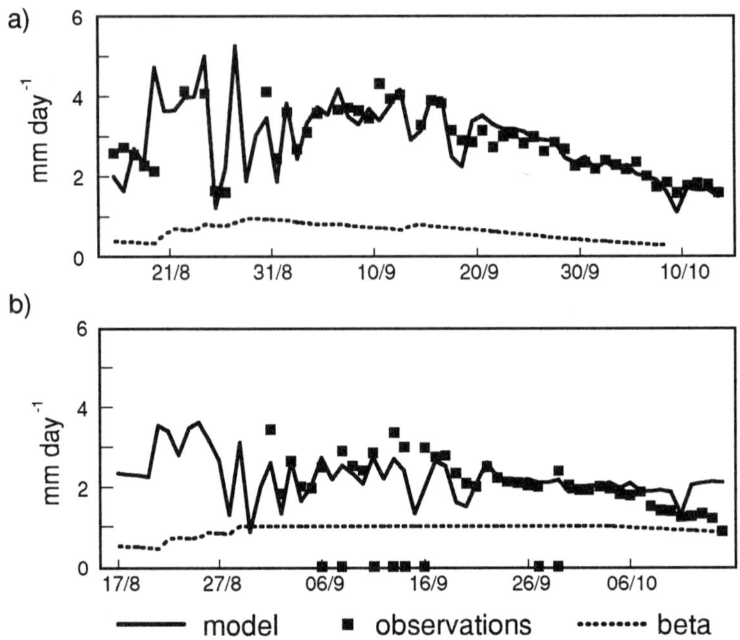


**Figure 2** As figure 1 but for the East Central Site (ECS): (a) savannah and (b) millet

hydraulic conductivity (K) and rooting depths. The value of K routinely used is not necessarily correct for the daily time step of this model or the soils considered here. Inspection of a series of model runs suggested that a value of K of 14900 kg m$^{-2}$ day$^{-1}$ fitted the evaporation measurements and was therefore used for all the sites in the subsequent optimisations. Values of the rooting depth between two and five metres were tried. Values which gave the best fits with the evaporation measurements were 5 m for tiger bush, 4 m for savannah and 2 m for millet. These values are consistent with deep soil moisture measurements taken in years subsequent to 1992 (S. Gaze, *pers. comm.1996*).

The soil model was run four times over each mesoscale model grid point using the gridded rainfall series. Each run assumes a different uniform vegetation cover (tiger bush, savannah, millet and dense vegetation), from 1 May to 8 October. The dense vegetation class occurs only in the south of the domain and takes a value of $F_r$ of 1 (i.e. the potential evaporation equals the Priestley-Taylor estimate) and a rooting depth of 4 m. The potential evaporation is also modified for each cover type using the difference in observed albedo. Initial values of soil moisture on 1 May are taken from the climatology of Willmott *et al.* (1985).

## SOIL AND VEGETATION PARAMETERS

The mesoscale model requires a single set of surface parameters to describe the surface over every 10 by 10 km cell. Table 2 shows the complete set of surface parameters required and values used for the individual cover types. As described above, the rooting depths were estimated from the daily evaporation time series. The albedo and roughness lengths were taken from the literature (Allen *et al.* 1994; Dolman *et al., 1993* and Lloyd *et al.* 1992). A dependence of surface resistance ($r_s$) on soil moisture, solar radiation, β and humidity deficit was developed using the hourly evaporation measurements from the five HAPEX-Sahel flux sites described above and the functional forms presented by Huntingford *et al.* (1995) and thus the value of $r_{smin}$ quoted in Table 2.

For sparse canopies, typical of the Sahel, the roughness length of heat transfer is generally smaller than that of momentum. This is a consequence of using a single source description for what is essentially a two source system (Blyth and Dolman 1995). The ratio of the roughness lengths will depend on the length scales and energy balances of the soil and vegetation (Blyth and Harding 1995); in the absence of a suitable model to specify the ratio, in this study $z_{0T}$ is made 1000 times smaller that $z_{0M}$. This is in broad agreement with published values for sparse canopies (Mahli 1996, Beljaars & Holtslag 1991).

As with the soil moisture model, the soil parameters ($\theta_s$, $\theta_c$, $\theta_w$ and C) are taken from the UKMO data base, assuming coarse sand throughout (Table 1). The soil thermal properties are again based on Warrilow *et al.* (1986) but with the introduction of soil moisture dependence.

**Table 2** Vegetation parameters used in the simulation

| Variable | Bare soil | Sparse millet | Dense millet | Fallow savannah | Sparse tiger bush | Dense tiger bush | Dense vegetation |
|---|---|---|---|---|---|---|---|
| Rooting depth (m) | - | 2.0 | 2.0 | 4.0 | 5.0 | 5.0 | 4.0 |
| Albedo | 0.3 | 0.3 | 0.3 | 0.18 | 0.25 | 0.19 | 0.27 |
| Roughness length (m) | 0.01 | 0.05 | 0.1 | 0.17 | 0.45 | 0.45 | 0.4 |
| $r_{s\,min}$ (s m$^{-1}$) | 210 | 210 | 210 | 105 | 120 | 80 | 80 |

To calculate a single set of parameters for each grid cell, in the case of β, albedo and the soil thermal properties an average weighted on the proportions of the different land cover types was used. This is also the appropriate averaging technique for surface resistance (Blyth 1996); an areal weighted average of $r_{s\,min}$ /β was used. The momentum roughness length in Table 2 are suitable for homogeneous patches. A blending height aggregation scheme (e.g. Mason 1988), which includes enhanced intra-patch roughness, was compared with simple logarithmic averaging but little difference was found and the simpler scheme was used.

## PATTERNS OF SURFACE ENERGY BALANCE FOR 8 OCTOBER 1992

The MMM was run for 8 October 1994. This was at the end of the HAPEX-Sahel observation period and three weeks after the last rains. The resultant pattern of β for the HAPEX-Sahel square for the 8th October (Figure 3) shows values ranging from 0.3 to 0.76 with a mean value of 0.66. The simulation is described in detail in Taylor *et al.* (1996), comparisons with surface and aerological measurements show good agreement. The fields of net radiation, sensible heat flux and latent heat flux calculated by the MMM show a substantial variability (Table 3 and Figure 4). The net radiation has a moderate variability (coefficient of variation [= Standard deviation/average] of 6%), responding to the input albedo field and the calculated field of surface temperature. The latent and sensible heat fluxes have a large variation (coefficient of variation 16 and 21% respectively).

The influence of the different driving variables and input parameters on the variability of the latent heat flux can be explored by re-running the surface scheme

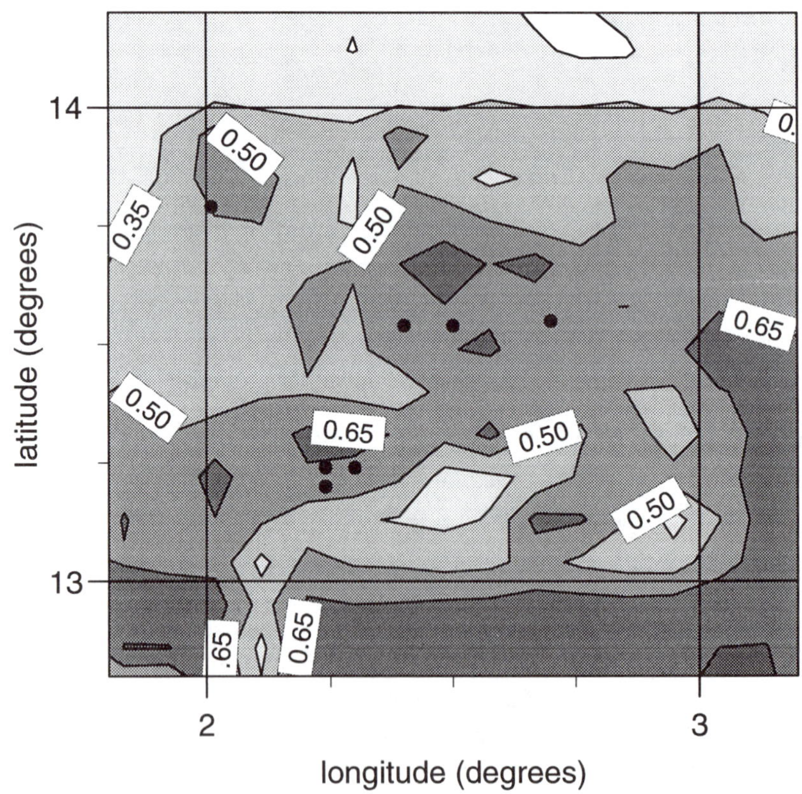


**Figure 3** Pattern of soil water stress factor, β, over the HAPEX-Sahel square for 8 October 1992. Dots indicate the positions of the measurement sites.

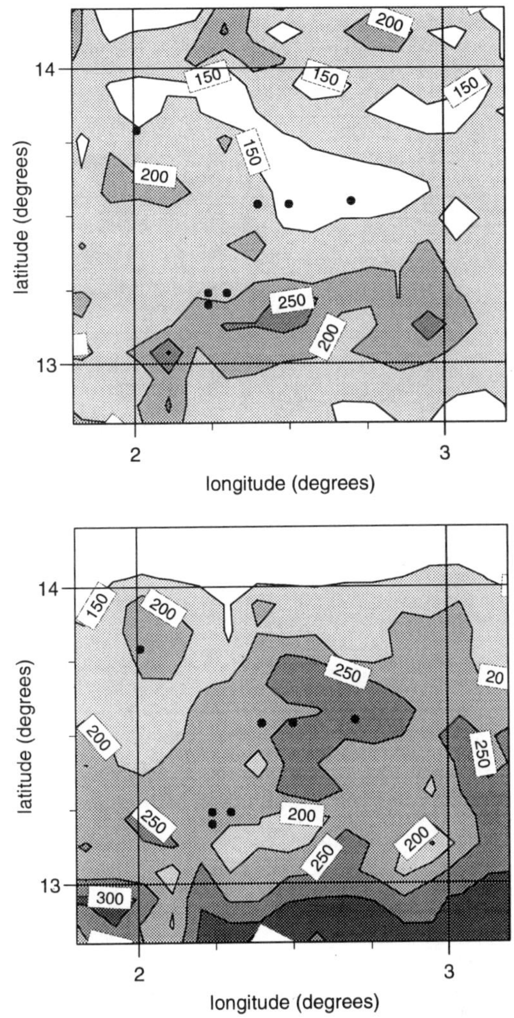


**Figure 4** Pattern of sensible heat flux (top) and latent heat flux (bottom) over HAPEX-Sahel square at midday on 8 October 1992

for each grid cell and keeping one variable (or parameter) constant in turn. This analysis shows that the variations in soil moisture and minimum surface resistance, $r_{s\,min}$, have the greatest influence on the range of latent heat flux (Table 4). Keeping soil moisture storage constant reduces the standard deviation of the latent heat flux from 34 W m$^{-2}$ to 22 W m$^{-2}$ and keeping the minimum surface resistance constant reduces the standard deviation from 34 to 26 W m$^{-2}$. Both these reductions are small compared with keeping both the soil moisture and minimum surface resistance constant which reduces the standard deviation from 34 W m$^{-2}$ to 9.2 W m$^{-2}$. Keeping all the driving variables constant shows that the variability of the driving variables has very little effect on the variability of the evaporation.

**Table 3** Averages and standard deviations (S.D.) over the HAPEX-Sahel square of driving variables, input parameters and surface outputs for land surface scheme, driving variables and outputs are midday values

| | Average | S. D. | Maximum | Minimum |
|---|---|---|---|---|
| *Driving variables (midday)* | | | | |
| Air Temperature, °C | 33.7 | 0.39 | 34.3 | 33.3 |
| Specific humidity, g $kg^{-1}$ | 11.8 | 0.9 | 14.7 | 10.2 |
| SHD, g $kg^{-1}$ | 21.1 | 1.5 | 23.4 | 15.7 |
| Solar radiation, W $m^{-2}$ | 1029 | 2.1 | 1034 | 1025 |
| Long-wave down, W $m^{-2}$ | 434 | 5.4 | 454 | 421 |
| Wind speed, m $s^{-1}$ | 3.3 | 0.98 | 4.9 | 1.4 |
| *Parameters* | | | | |
| Albedo* | 0.360 | 0.011 | 0.395 | 0.306 |
| $r_{s\,min}$, s $m^{-1}$ | 118.5 | 23.5 | 210 | 80 |
| $z_{0M}$, m | 0.27 | 0.12 | 0.45 | 0.01 |
| Soil moisture storage, mm | 66.1 | 13.2 | 94 | 37 |
| *Outputs (midday)* | | | | |
| Net radiation, W $m^{-2}$ | 470 | 29.5 | 553 | 379 |
| Latent heat flux, W $m^{-2}$ | 217 | 34.0 | 229 | 135 |
| Sensible heat flux, W $m^{-2}$ | 178 | 39.0 | 307 | 94 |
| Ground heat flux, W $m^{-2}$ | 82 | 10.4 | 114 | 63 |
| Surface temperature, °C | 48.0 | 3.3 | 60.7 | 41.5 |

* Note an addition of 12% was made to the albedo values to account for the effects of reflection by aerosols in the atmosphere.

**Table 4** The effect on the mean and standard deviation of the latent heat flux (W $m^{-2}$) across the HAPEX-Sahel square keeping certain principle driving variables and parameters constant (at midday)

| | Mean | Standard deviation |
|---|---|---|
| All varying | 217 | 34.1 |
| Soil moisture constant | 218 | 22.3 |
| $r_{s\,min}$ constant | 217 | 25.6 |
| Soil moisture and $r_{s\,min}$ constant | 219 | 9.2 |
| Albedo constant | 216 | 32.5 |
| Driving variables constant | 225 | 35.7 |

## THE CALCULATION OF THE AREAL AVERAGE EVAPORATION

To investigate the effects of averaging methods on the estimation of the surface energy balance the input parameters and driving variables to the land surface model were averaged using the simple averaging rules described by Noilhan and Lacarrère (1995). An inverse average was used for the surface resistance and for the roughness length either a logarithmic or a simple arithmetic average was made. These mean parameters and driving variables were used in the same land surface scheme as the MMM. The diurnal variation in the latent heat flux calculated either by the MMM and averaged over the square or by using averaged parameters, is shown in Figure 5. There is a very small difference between the two methods of calculation: less than 5% for the total evaporation over the day.

This finding is in agreement with that of Noilhan and Lacarrère (1995) for the HAPEX-MOBILHY study area in SW France but contrasts with the more extreme, but hypothetical, simulation described by Harding *et al.* (1996). This latter study was for the same region and date as the current study but a hypothetical pattern of soil moisture was imposed; where the soil was either of dry soil or a surface well supplied with water and hence the stress function, $\beta$, was either zero or one. In this case the depth of the boundary layer varied strongly with soil wetness with resulting strong contrasts in air temperature and humidity (this variation was not observed in the current, more realistic, simulation with smaller contrasts in soil moisture). However, in Harding *et al.* (1996) the biggest difference between the average evaporation and the evaporation calculated with average parameters was found to be due to the influence of the non-linearity of the transfer equations. In contrast, in

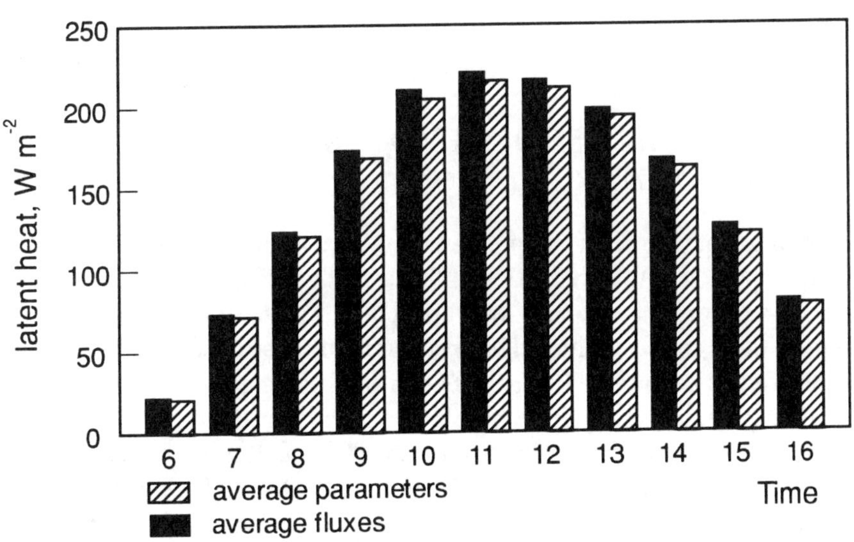


**Figure 5** Diurnal variation of mean latent heat over HAPEX-Sahel square: calculated as an average of the fluxes (hached) and from average parameters (solid)

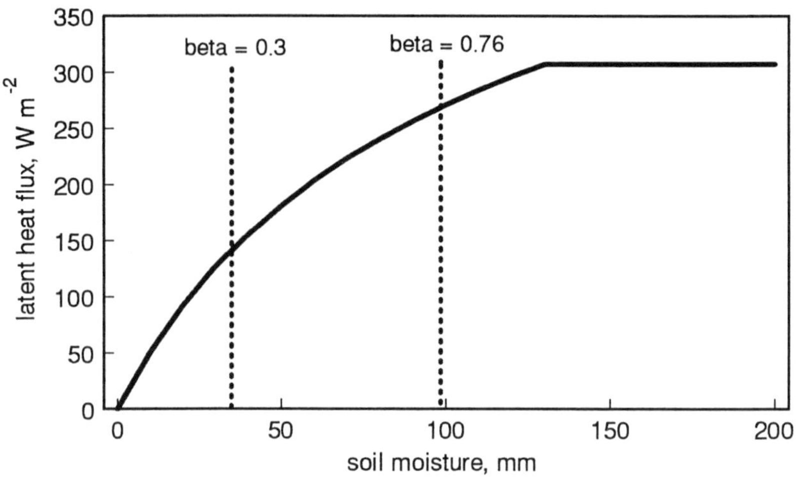


**Figure 6** Functional dependence of latent heat flux on soil moisture storage within the rooting zone in the land surface model, calculated using the mean surface paramenters and meteorological variables for the HAPEX-Sahel square for 8 October.

the current study the range of soil moisture is such that the evaporation is essentially a linear function of soil moisture, with $\beta$ in the range 0.3 to 0.76 (Figure 6) and the effect of the averaging techniques is small.

It is still possible that at different times of the year the calculation of the latent heat flux with average parameters may fail to represent the true average evaporation. For example, in the wet season the soil moisture contents are likely to be in a range in which the evaporation is not linear. Wet season simulations for HAPEX-Sahel have yet to be undertaken.

## DISCUSSION OF THE USE OF REMOTE SENSING FOR SURFACE PARAMETERS

The approach adopted here in the production of the surface fields has been the 'traditional' approach of the production of the land cover map, with a very limited number of land cover classes, and a look-up table to produce the parameter fields. A distributed, daily soil water model has been used to construct fields of available soil moisture over the domain. The lack of good areal coverage of land cover maps and climate measurements coupled with the large variability in this region necessitates the use of remotely-sensed images in this region. Thus satellite-derived land cover and rainfall fields were used extensively in this work. The HAPEX-Sahel dataset provides the opportunity to calibrate a certain number of the land surface parameters in the MMM and the soil moisture model. This was done for five sites covering three of the major land cover classes (savannah, tiger bush and millet). The parameters calibrated were: surface resistance, albedo, roughness and rooting depth.

This approach has the disadvantage that only a limited number of land cover classes can be discriminated at this spatial scale using remotely-sensed data. In addition, there are usually surface observations to calibrate the land surface scheme for only a few if any of the vegetation types. This is particularly a problem in semi-arid regions such as the Sahel, where the agricultural practices tend to result in relatively small areas of tilled land mixed with natural vegetation which often shows gradational rather than abrupt changes. Although there is some scope for improvement in the analysis of remotely sensed data by using spectral mixture modelling (Settle and Drake, 1993, Price and Bausch, 1995), rather than pixel-wise classifiers, there are still problems of assigning calibrated parameters to areas with varying vegetation density. A further complication arises with the definition of soil parameters. A good digital soil map of the region does not exist. Although existing soil maps are being digitised, the soil classifications are usually based on soil series and so need a link to the physical parameters of the land surface scheme. What is really required is an explicit map of soil parameters and there is no prospect of obtaining this from remotely-sensed data in the foreseeable future.

Unfortunately, despite the availability of the HAPEX-Sahel data set, there are as yet no data with which to verify these fields other than at the small number of ground-based measurement sites (most of which have been used for calibration). Remote sensing obviously has the ability to provide average fields for some of the parameters, such as albedo, and output fields, such as sensible heat flux, but these have not yet become available.

It is unlikely that remote sensing will ever be able to provide all the surface fields required for this model directly. Albedo should be possible (Arino *et al.*. 1991, Valiente *et al.*. 1995) but not the really important fields of surface resistance and soil moisture in the rooting zone. The requirement for correct areal fields is so great that consideration may have to be given to changing the formulation of the Soil Vegetation Atmosphere Transfer (SVAT) schemes so that areal fields generated from remotely-sensed data can be input directly. An example may be the use of leaf area, 'greenness' or APAR (=Absorbed Photosynthetically Active Radiation) as an alternative to surface resistance (Choudhury, 1987, Sellers, 1987, Xinmei *et al.*. 1995).

Equally, changes in the surface temperature or the radiative properties of the surface may provide a quantitative estimate of water stress to replace the soil water based approach. However it is unlikely that the whole modelling community will move away from the vegetation class approach completely. Remotely-sensed surface fields are useful for mesoscale models which run for a specific day for a specific area but are not suited for GCM runs which are used to predict the effects of future landcover or $CO_2$ scenarios. In these latter cases remotely-sensed fields should, however, have a role to play in verifying the surface fields of input parameters and output variables.

## CONCLUSIONS

A case study has been described of the implementation of a MMM to a region in west Africa centred on the HAPEX-Sahel square. Considerable emphasis has been placed on the provision of realistic surface parameter fields. It can be shown that it is essential to get these fields right if a realistic simulation is to be achieved. The model output shows a considerable variability in the surface fluxes (for example, the latent heat flux ranges from 135 to 229 W $m^{-2}$). The primary cause of this variability is the variation in soil moisture content in the rooting zone; minimum surface resistance also makes a significant contribution and a smaller contribution from albedo. The variation in the driving variables (air temperature, humidity, wind speed and downwelling radiation) have only a very small effect on the variability. The output of the model was used to test simple aggregation rules to calculate the areal average surface fluxes. It is shown that simple averaging of the surface parameters can provide an accurate representation of the areal average surface fluxes. Thus the real problem remains of how to estimate accurately the mean fields of surface parameters and soil moisture.

Unfortunately, it is unlikely that remote sensing will be able to provide the parameters required for the surface descriptions for this type of model directly, although it is one of the only technologies which can provide information at the required scale. It must be concluded that the SVAT models will have to be reformulated to enable remotely-sensed fields to be used more directly.

## ACKNOWLEDGEMENTS

We would like to thank all those involved in the field observations in HAPEX-Sahel without whom this modelling activity would not be possible. We would particularly like to thank the EPSAT team from ORSTOM, Niamey, for developing and providing a network of rainfall observations unique in the Sahelian region and CNRM, Toulouse, for providing the surface flux measurements from the East Central Site. Thanks are also due to the UK Meteorological Office for making the Unified model code available together with advice on its implementation. This work was funded under the UK TIGER programme.

## REFERENCES

Allen, S.J., Wallace, J.S., Gash, J.H.C. and Sivakumar, M.V.K. 1994. Measurements of albedo variation over natural vegetation in the Sahel. *Int. J. Climatol.*, **14**, 625-636.

Anderson, L. and Harding, R. J. 1991. Soil moisture deficit simulations with models of varying complexity for forest and grassland sites in Sweden and the UK. *Wat. Res. Man.*, **5**, 25-46.

Arino, O., G. Dedieu, and P.Y. Deschamps,1991. Accuracy of satellite land surface reflectance determination. *J. Appl. Meteorol.*, **30**, 960-972.

Beljaars, A.C.M. and Holtslag, A.A.M. 1991. Flux parameterization over land surfaces for atmospheric models. *J. Appl. Meterol.*, **30**, 327-341.

Beljaars, A.C.M., Viterbo, P., Miller, M.J. and Betts, A.K. 1996. The anomalous rainfall over the USA during 1993: sensitivity to land surface parameterization. *Mon. Wea. Rev.*, **124**, 362-383.

Blyth, E.M. 1996. Representing heterogeneity at the southern super site with average surface parameters. *J. Hydrol HAPEX-Sahel Special Issue*. In press.

Blyth, E.M. and Dolman, A.J. 1995. The roughness length for heat of sparse vegetation. *J. Appl. Meteorol.*, **34**, 583-585.

Blyth, E.M. and Harding, R.J. 1995. Application of aggregation models to surface heat flux from Sahelian tiger bush. *Agri. For. Meteorol.*, **72**, 213-235.

Blyth, E.M., Noilhan, J. and Dolman, A.J. 1994. The effect of forest on mesoscale rainfall: an example from HAPEX-MOBILHY. *J. Appl. Meteorol.*, **33**, 445-454.

Calder I. R., Harding R. J. and Rosier P. 1983. An objective assessment of soil moisture deficit models. *J. Hydrol.*, **60**, 329-355.

Choudhury, B.J. 1987. Relationships between vegetation indices, radiation absorption and net photosynthesis evaluated by a sensitivity analysis. *Remote Sensing Environ.* **22**, 209-233.

Dolman, A.J., Lloyd, C.R., and Culf, A.D. 1993. Aerodynamic roughness of an area of natural open forest in the Sahel. *Ann. Geophys.*, **10**, *930-934* .

Gash, J.H.C., Kabat, P., Monteny, B.A., Amadou, M., Bessemoulin, P., Billing, H., Blyth, E.M., de Bruin, H.A.R., Elbers, J.A., Friborg, T., Harrison, G., Holwill, C.J., Lloyd, C.R., Lhomme, J-P., Moncreiff, J.B., Puech, D., Sogaard, H., Taupin, J. D., Tuzet, A. and Verhoef, A. 1996. The variability of evaporation during the HAPEX-Sahel intensive observation period. *J.Hydrol. HAPEX-Sahel special issue. In press*

Goutorbe, J.-P., Lebel, T., Tinga, A., Bessemoulin, P., Brouwer, J., Dolman, A.J., Engman, E.T., Gash, J.H.C., Hoepffner, M., Kabat, P., Kerr, Y.H., Monteny, B., Prince, S., Saïd, F., Sellers, P. and Wallace, J.S., 1994. HAPEX-Sahel: a large-scale study of land-atmosphere interactions in the semi-arid tropics. *Ann. Geophys.* **12**,53-64.

Harding, R.J., Blyth, E.M. and Taylor, C.M. 1996. Issues in the aggregation of surface fluxes from a heteorogeneous landscape: from sparse canopies up to the GCM grid scale. SEB Seminar Series, Cambridge University Press.

Huntingford, C., Allen, S.J. and Harding, R.J. 1995. An intercomparison of single and dual-source vegetation-atmosphere transfer models applied to transpiration from Sahelian savannah. *Bound.-Layer Meteor.*, **74**, 397-418.

Lebel, T., Taupin, J.D. and D'Amato, N. 1996. Rainfall monitoring during HAPEX-Sahel: 1. General rainfall conditions and climatology. *J. Hydrol. HAPEX-Sahel Special Issue.* In press.

Leemans, R. and Cramer, W.P. 1991. The IIASA data base for mean monthly values of temperature, precipitation and cloudiness on a global terrestrial grid. IIASA R-91-18.

Lloyd, C.R., Gash, J.H.C. and Sivakumar, M.V.K. 1992. Derivation of the aerodynamic roughness parameters for a Sahelian savannah site using the eddy correlation technique. *Bound.-Layer Meteor.*, **58**, 261-271.

Mahli, Y. 1996. The behaviour of roughness length for temperature over heteorogenous terrain. *Bound.-Layer Meterol.* In press.

Mason, P.J. 1988. The formation of areally-averaged roughness lengths. *Q. J. Roy. Met.. Soc.* **114**, 399-420.

Milford, J.R. and Dugdale, G. 1992. Estimation of rainfall using gestationary satellite data in applications of remote sensing in agriculture. *Proceedings of the 48th Easter School in Agricultural Science, University of Nottingham. Butterworth, London.*

Mitchell, K.E. 1994. GCIP initiatives in operational mesoscale modeling and data assimilation at NMC. *Proc. of AMS Fifth Conference on Global Change Studies, Nashville TN.*

Noilhan, J. and Lacarrère, P. 1995. GCM grid-scale evaporation from mesoscale modeling. *J. Clim.* **8**, 206-223.

Pielke, R.A., Dalu, G.A., Snook, J.S., Lee, T.J. and Kittel, T.G.F. 1991. Nonlinear influence of mesoscale land use on weather and climate. *J. Clim.* **4**, 1053-1069.

Price, J.C. and Bausch, W.C. 1995. Leaf area index estimation from visible and near-infrared reflectance data. *Remote Sensing Environ.*, **52**: 55-65.

Sellers, P.J. 1987. Canopy reflectance, photosynthesis and transpiration. II. The role of biophysics in the linearity of their interdependance. *Remote Sensing Environ.*, **21**, 143-183.

Settle, J.J. and Briggs, S.A. 1987. Fast maximum likelihood classification of remotely sensed imagery. *Int. J. Remote Sensing*, **8**, 723-734.

Settle, J.J. and Drake, N.A. 1993. Linear mixing and the estimation of ground cover proportions. *Int. J. Remote Sensing*, **14**, 1159-1177.

Taylor, C.M., Harding, R.J., Thorpe, A.J. and Bessemoulin, P. 1996. A mesoscale simulation of land surface heterogeneity from HAPEX-Sahel. *J. Hydrol. HAPEX-Sahel Special Issue*. In press.

Valiente, J.A., M. Nunez, E. Lopez-Baeza, and Moreno, J.F. 1995. Narrow-band to broad-band conversion for Meteosat-visible channel and broad-band albedo using both AVHRR-1 and AVHRR-2 channels. *Int. J. Remote Sensing*, **16**, 1147-1166.

Warrilow, D.A., Sangster, A.B. and Slingo, A. 1986. Modelling of land surface processes and their influence on European climate. Meteorological Office, London Road, Bracknell, Berkshire, RG12 2SZ. Met. 0. 20 DCTN 38.

Willmott, C.J., Rowe, C.M. and Mintz, Y. 1985. Climatology of the terrestrial seasonal water cycle. *J. Climatol.*, **5**, 589-606.

Wilson, M.F. and Henderson-Sellers, A. 1985. A global archive of land cover and soils data for use in general circulation models. *J. Climatol.*, **5**, 119-144.

Xinmei, H., Lyons, T.J., Smith, R.C.G. and Hacker, J.M. 1995. Estimation of land surface parameters using satellite data. *Hydrological Processes*, **9**, 631-643.

# 5 Surface radiation budget: scaling up from local observations

**J.R. GARRATT and A.J. PRATA**
*CSIRO Division of Atmospheric Research, Aspendale, Australia*

## INTRODUCTION

The importance of the surface radiation budget (SRB) lies in the significance of radiation to the total energy exchange between the atmosphere and surface, and in the significance of both net radiation and surface temperature to the evaporation from land. The SRB has not been as amenable to remote measurement as the radiation budget at the top-of-the-atmosphere (TOA), so it is less understood. In addition, the lack of comprehensive, long-term observations has prevented a thorough validation of the SRB, both globally and regionally, as simulated in climate models.

There is no doubt that satellite data can play a major role in studies and model validation of the SRB. But these satellite data themselves require careful validation and calibration at selected sites ("ground truth"). We are faced with a scaling up problem, from essentially "point" measurements (noting that ground-based upward looking instruments "see" a horizontal scale of tens of km, ground-based downward instruments only a few metres depending on the height of measurement) to the satellite pixel and beyond. To illustrate the problems we focus on two issues:

(i) Validation of satellite radiances at the pixel scale - an example is given of a site in Australia designed to tackle this problem, providing SRB data at a land scale of 1 km, comparable with the resolution of many of the currently used satellite sensors.

(ii) Validation of land surface radiation fluxes in satellite-based global data sets and global climate models (GCMs) at the grid scale (hundreds of kilometres). We consider direct observations at single stations made over long periods of time, from weeks to many years. These are observations generally available for climate studies and model validation.

*Scaling up in hydrology using remote sensing*. Edited by J.B. Stewart, E.T. Engman, R.A. Feddes and Y. Kerr

## THEORETICAL BACKGROUND AND SOME BASIC RELATIONS

From an atmospheric perspective, the local or regional evaporation is closely linked to the net radiation through the surface energy balance (SEB):

$$R_N - G_o = H + \lambda E \tag{1}$$

where $\lambda$ is the latent heat of vaporization, $R_N$ is the net radiative flux density to the surface (net radiation), H is the sensible heat flux, E is evaporation and $G_o$ is the heat flux into the soil at the surface. This is satisfied both instantaneously and for long-term averages, and both locally and regionally. The net radiation itself is made up of a number of flux components, as seen in the SRB equation

$$R_N = (1 - \alpha)\, S_d + \varepsilon L_d - L_u \tag{2}$$

where the downwelling shortwave flux $S_d$, the downwelling longwave flux $L_d$, the upwelling longwave flux $L_u$, the broadband surface albedo $\alpha$ and surface emissivity $\varepsilon$ are taken here as positive quantities. With suitable modelling of terms in Equation 1, solutions of the SEB equation define the surface radiative (land) temperature, $T_o$, which determines the upwelling longwave flux via $L_u = \varepsilon\sigma T_o^4$, where $\sigma$ is the Stefan-Boltzmann constant.

Surface-layer turbulent theory can be used to derive bulk relations describing the turbulent fluxes H and E in terms of air-surface differences, though there is an inherent problem in defining (and observing) the surface values of temperature, and particularly humidity. In addition, the application of these relations over complex or heterogeneous surfaces has yet to be fully demonstrated, and spatial averaging of the fluxes when they are parameterised in terms of surface temperature, surface roughness lengths and canopy (surface) resistances is problematic.

The nature of the parameterisations and the non-linear nature of the transfer processes mean that only with surface temperature, as well as the fluxes themselves, can some form of linear averaging be done to give the effective larger-scale (area) average. The magnitude and dominant scales of spatial variations across an area of interest will influence the level of errors involved in the averaging process (upscaling). For the energy (H and E) and radiation fluxes (including net radiation) the area-average flux $\langle F \rangle$ is given by

$$\langle F \rangle = \Sigma a_i F_i \tag{3}$$

where $F_i$ is the local time-averaged flux, $a_i$ is the fraction of the total area taken up by surface type i and i goes to infinity. For small i (and $i > 1$), the question of spatial heterogeneity and representativeness is of paramount importance. Equation 3 and the bulk transfer relation for sensible heat flux can be used to infer the effective area-average surface temperature $\langle T_o \rangle$ in terms of the component temperatures $T_{oi}$. With

$<F> = (<T_o> - T_a)/<r_a>$ and $F_i = (T_{oi} - T_a)/r_{ai}$, where $T_a$ is the air temperature assumed constant in space and $<r_a>$ is the area- average aerodynamic resistance given, in terms of the component resistances $r_{ai}$, by $<r_a^{-1}> = \Sigma a_i / r_a^i$, then it readily follows that

$$<T_o> = \Sigma \gamma_i \, T_{oi} \tag{4}$$

where $\gamma_i = a_i <r_a>/r_{ai}$ (e.g. see Garratt 1978; Sun and Mahrt 1995). Note that $r_a$ is the resistance required to match the heat flux and observed surface temperature through the bulk relation. Further, only when $r_{ai}$ is the same for all surface types is <To> given by the simple area-weighted average, i.e. $<T_o> = \Sigma a_i T_{oi}$.

## OBSERVATIONS OF MEAN MONTHLY FLUXES

For many purposes, climate studies require mean monthly fields (either in the form of simple flux values or as a diurnal variation of fluxes). Long-term observations of surface energy and radiation fluxes over land are now available in a variety of forms, which we summarise below.

*Climatological*

Henning (1989) gives maps of net radiation, sensible heat flux and evaporation for selected months for all the continents whilst a global data set based on Willmott *et al.* (1985) provides monthly evaporation in grid point form at 1° intervals. Both sets are based on mean monthly values of reported station climatological variables recorded over many years (for the period 1931-60 in the case of Henning), and a range of empirical relations which relate the observed variables to the above three quantities. Values of the fluxes derived from these maps represent the most comprehensive data source for comparison with, and validation of, model-derived components of the surface energy balance.

*Historical stations*

Mean monthly values of energy and radiation fluxes measured at a number of continental observatories or stations for varying lengths of time are available in the Global Energy Balance Archive (GEBA) held in Zurich (Ohmura and Gilgen, 1991). Rossow and Zhang (1995) for example use the mean monthly radiation flux data for comparison with their satellite-based estimates of downwelling short and longwave fluxes at the ground.

Plans are well advanced for implementation of the World Climate Research Programme's Baseline Surface Radiation Network (BSRN). The main thrust involves a commitment to uniformly high quality global observations of SRB component fluxes, with supplementation of the present surface network where possible, development of radiation instrumentation and archiving of the observations with quality control procedures, through the already existing GEBA facilities. In addition, the USA has instigated an auxiliary network for specialised radiation measurements over several years. In Australia, a small network of stations making comprehensive measurements of all SRB component fluxes and land-surface temperature is being installed by the Commonwealth Scientific and Industrial Research Organisation (CSIRO).

*Satellite-based products*

Global data sets are now available (some in the public domain) of all-sky and clear-sky TOA and surface radiation fluxes at 2.5° resolution for a number of years (e.g. Darnell *et al.*, 1992; Zhang *et al.*, 1995). The data are based on the International Satellite Cloud Climatology Project (ISCCP) data (available at three-hourly intervals) which include cloud statistics, atmospheric temperature and humidity profiles, surface temperature and ozone amount. The use of these data in radiative transfer calculations enables monthly mean radiation fluxes to be calculated.

In summary, a variety of direct observations or observationally-based products is available covering a range of space scales (local station to satellite pixel (1 to 10 km) to 2.5° grid averages) and time scales (three- hourly to long-term monthly averages). The space distribution of the station observations is very limited, whilst the satellite-based products available as global distributions lack comprehensive validation and calibration.

## VALIDATION OF SATELLITE OBSERVATIONS AT THE 1 KM SPATIAL SCALE

*Satellite measurements*

Generally, satellite observations of surface properties have a spatial scale of tens of metres to tens of kilometres. To use these data quantitatively, the measurements must be calibrated accurately and the derived geophysical variables (e.g. surface temperature) must be validated. Our discussion here will focus on the problem of geophysical validation of satellite-derived quantities as this pertains more directly to the SRB and is critically linked to the spatial variability, hence scaling up, problem.

For reasons of practicality most satellite-derived measurements are validated against point data, which is a clearly deficient method for a variable such as surface temperature. A better approach is to obtain sufficient samples within the field-of-view of the instrument so that the variability is captured at an appropriate spatial scale. The First ISLSCP (International Satellite Land Surface Climatology Project) Field Experiment (FIFE) (Sellers *et al.*, 1992) used this approach.

*Hay (Australia) site*

Since mid-1992 the CSIRO has operated a field site for the purpose of providing vicarious calibration and validation data for satellite-derived measurements of land surface variables. Anticipating the need to capture spatial variability, a number of ground-based instruments have been laid out at a large, flat site on the Hay plain, in southern New South Wales, Australia (the site is centred at geographic coordinates: 34.392°S 145.305°E). To minimise difficulties associated with terrain height and atmospheric variability, the site was chosen to be in a very uniform, tree-less region with low atmospheric water vapour loadings, low aerosol contents and a low probability of cloudiness. The ground-based instrumentation consists of upward-, and downward-looking pyranometers and pyrgeometers, surface-mounted temperature sensors, humidity and wind speed sensors spread out over an area of

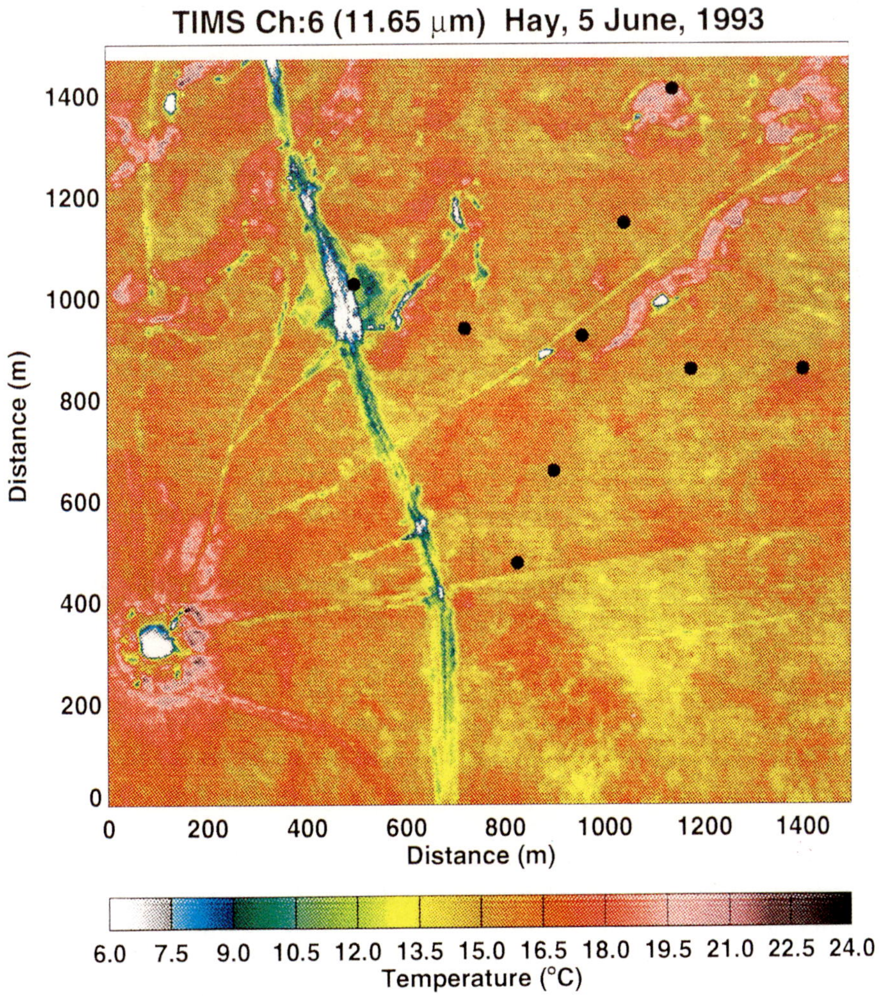


**Figure 1** Daytime TIMS thermal image over south-east Australia for the Hay site at 5m pixel scale. The temperatures have been atmospherically corrected, but not corrected for emissivity effects. The circles show the locations of the instruments at the site. Variations in shading are mainly the result of vegetation-cover variations (with yellow areas representing little or no vegetation).

approximately 1x1 $km^2$, within a region that is uniform at a spatial scale of about 30x30 $km^2$. Data are logged at approximately 2-5 minute intervals. The Hay field site has provided these measurements continuously from August 1992 to the present, with occasional interruptions due to system upgrades and re-calibration exercises. Further details concerning the Hay field site may be found in Prata (1994a).

Extensive satellite data sets from the Advanced Very High Resolution Radiometer (AVHRR) and Along Track Scanning Radiometer (ATSR) are available for this site, and in addition there are airborne scanner data (at spatial scales of a few metres). Figure 1 shows a temperature image of the Hay field site obtained from the airborne Thermal Infrared Multispectral Scanner (TIMS) (Palluconi and Meeks, 1995) together with the locations of instrument sites. The pixel size of the image is about 5 m and the entire image fits within a single pixel of the AVHRR. The temperatures have been corrected for atmospheric effects, but not for emissivity effects and so the image is not a true surface temperature map. The spatial variability is high — part of this variability is due to variability in the land-surface emissivity. Notwithstanding this inherent variability, the validation of land-surface temperature can be achieved by averaging sufficient *in situ* point measurements, at least at the Hay site. Prata (1994b) has shown that 1 km-scale land-surface temperature (i.e. derived from AVHRR or ATSR satellite measurements) is highly correlated ($r^2 > 0.96$) with the linear average of the point measurements made at Hay.

*Surface temperature variability*

Land surfaces tend to be extremely heterogeneous. A typical area may contain many land components such as grass, trees and shrubs, soil and rock. The temperature can vary by 10 K or more in just a few metres and by 50 K over the daily cycle. Extreme maxima may reach 90°C (Garratt, 1992). Arguably the most significant of the problems associated with surface temperature ground-truthing is the problem of spatial variability.

On a horizontal scale of tens of metres or less, greatest variations are often associated with incomplete canopy cover. Garratt (1978; his Table 2) showed daytime temperature differences of 20 K between transpiring leaves and (warmer) bare soil and described a method of combining the individual temperatures (soil, grass and leaves) to yield an effective area average applicable to the heat flux determination. More recently Sun and Mahrt (1995) described temperature variations between sunny and shady parts of an open forest of order 5-10 K, and showed how the relationship between the heat flux and an average air-surface temperature difference is sensitive to this microscale variability in surface temperature.

At scales of 1 km or less, variations often depend on a range of both natural and man-made surface types. For example, Prata (1994b) shows daytime standard deviations over a 1 $km^2$ region of crops in south-east Australia close to 4 K. Even areas apparently uniform to the eye can be remarkably heterogeneous so far as thermal properties are concerned (see Figure 1). Data from the Hay field site for January 1994 are used to illustrate the variability (quantifiable, for example, as the

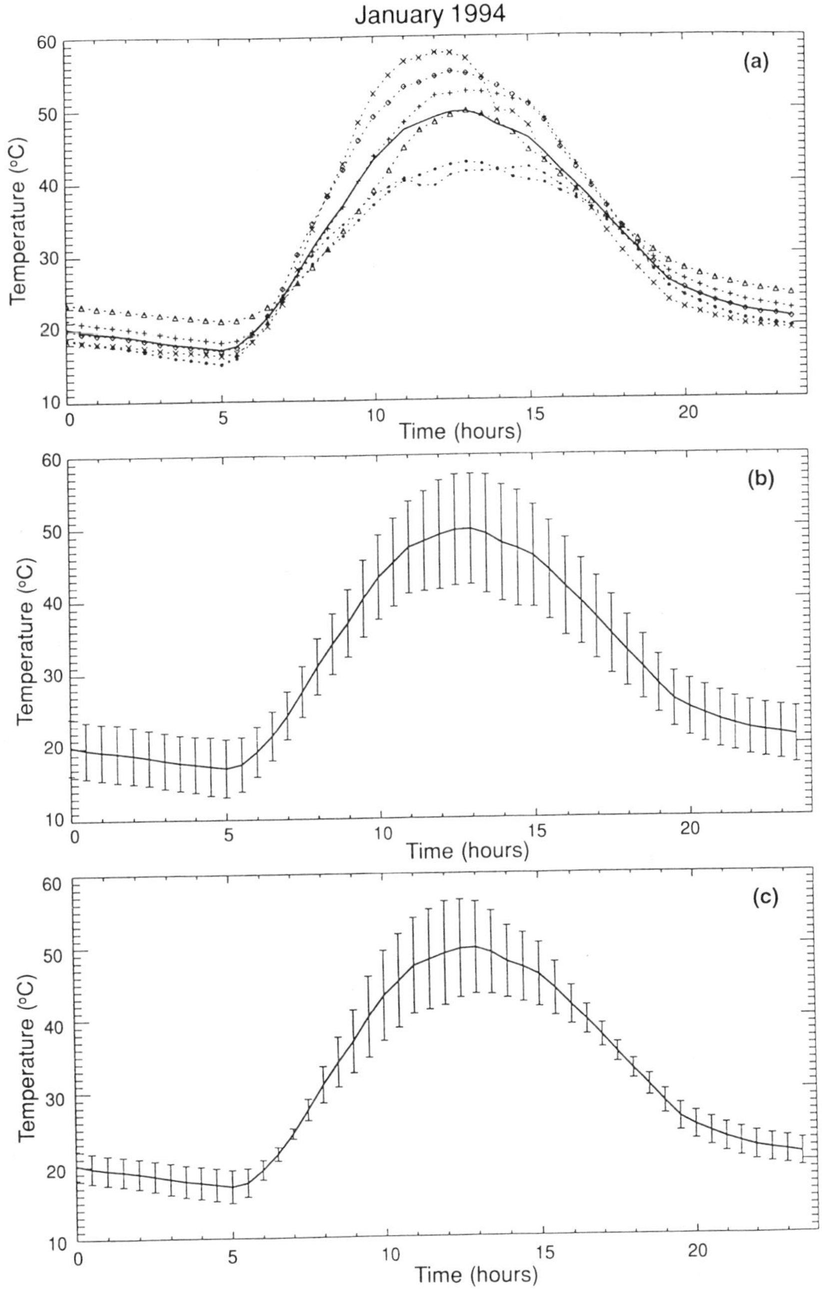


**Figure 2** Monthly mean diurnal cycles of surface temperature at the Hay site in January 1994, for six stations, (a) shows for each station the 30-day mean temperature at 30-minute intervals, (b) shows 6-station, 30-day mean temperature, together with the temporal rms variations (rms of the 30- minute mean value at the specified time of day about the 30-day mean); (c) as in (b), but with the spatial rms variations.

standard deviation $\sigma_T$ of the individual observations about the time or area mean) in surface temperature in Figure 2. The top panel (Figure 2a) shows the monthly-averaged diurnal cycle in surface temperature for six locations (dashed and dotted lines) within the 1x1 $km^2$ site and the average of these (solid line). Spatial variability is high (third panel Figure 2c) with $\sigma_T$ exceeding 7 K at the height of the afternoon in mid-summer. Variability is lowest just after sunrise and just before sunset having a $\sigma_T$ value of about 2 K. These spatial variabilities may be compared with the temporal variability, with a $\sigma_T$ of more than 7 K at 1300 local time and about 3 K at 0700 local time (Figure 2b). The spatial and temporal variabilities, as a function of the time of day, are strongly correlated at these scales since surface temperature response is strongly correlated with insolation and the magnitude of the variability is linked to times of maximum and minimum insolation.

Quantification of the spatial variability in the monthly mean is given in Table 1, and for each of the six stations the time variability (in the form of the standard deviation) is also shown. The variability at each station is due to the relatively large amplitude of the diurnal cycle of surface temperature and its day-to-day variation, while the variability in the ensemble mean has an additional component due to spatial variations in temperature. Since the diurnal cycle is large at this site, the temporal contribution dominates over that due to spatial variability. However, this will not necessarily be the case for land sites with a smaller continental influence (e.g. maritime).

**Table 1** Monthly mean values and standard deviations of the individual (30-minute average) surface temperature at six different locations at the Hay field site for January 1993 and January 1994. The ensemble means and standard deviations are also shown.

| | Jan 93 | | Jan 94 | |
|---|---|---|---|---|
| Site | T (°C) | $\sigma_T$(°C) | T (°C) | $\sigma_T$(°C) |
| 1 | 28.3 | 10.1 | 32.2 | 12.4 |
| 2 | 29.3 | 9.1 | 32.7 | 13.9 |
| 3 | 28.8 | 9.4 | 31.8 | 9.7 |
| 4 | 29.4 | 10.0 | 28.2 | 9.7 |
| 5 | 28.3 | 8.9 | 27.9 | 14.8 |
| 6 | 27.0 | 7.5 | 31.6 | 14.8 |
| Mean | 28.5 | 9.1 | 30.7 | 12.6 |

*Longwave flux variability*

At spatial scales of several kilometres or less, large horizontal variations in energy and outgoing radiation fluxes occur whenever concomitant variations in soil moisture, and hence surface temperature, are present (e.g. in irrigated regions). Very little spatial variability at scales of a few kilometres is expected in the downwards flux at the surface since this quantity is strongly linked to the near-surface air temperature and vapour pressure, neither of which has significant variability at 1 km scales. The upward energy and longwave fluxes can be expected to exhibit high spatial variability since they are closely linked to the surface (radiative) temperature.

Small-scale spatial variations in radiative fluxes can be inferred using measurements from an array of instruments. National networks, or smaller scale arrays as part of special field experiments, may provide such data but issues related to instrument errors and calibration drifts must necessarily be tackled. Thus Field *et al.* (1992) described such problems in the measurement of net radiation at FIFE, where seven different designs of radiometer by five different manufacturers were used. Daytime corrected net radiation values contained systematic rms instrumental differences of 20 W $m^{-2}$.

At Hay, measurements of both the downwelling and upwelling components of the longwave radiation flux are made. Figure 3 shows the monthly averaged downwelling flux (a), and the upwelling longwave flux calculated from surface temperature (b and c), and their variabilities (temporal in (a) and (b), spatial in (c)). The spatial variability in the upward longwave flux (quantified as the standard deviation) is about 5 $Wm^{-2}$ at 0700 and about 50 $Wm^{-2}$ at 1300 local times.

There is a small error incurred when the average (temporal or spatial) of the upwelling longwave flux is computed using the mean of the surface-temperature measurements rather than by using the mean of the fluxes corresponding to each surface-temperature measurement (call this the actual flux). This error is related to the spatial variability of surface temperature. The difference ($\Delta F$) between the actual flux and the flux calculated by averaging the fluxes derived from the mean surface temperature can be shown to be approximately

$$\Delta F = 0.5(\delta^2 F/\delta T^2)\, \sigma_T^2 \qquad (5)$$

where the second derivative is calculated at the mean (spatial) temperature. For $F = \sigma T^4$, it follows that $F = 6\sigma T^2\, \sigma_T^2$, so that for Hay measurements in January the maximum difference ($\Delta F$) is about 3 W $m^{-2}$. A similar analysis can be done for the temporal variability where the error for estimating monthly mean fluxes from monthly mean surface temperatures is of similar size (see Garratt, 1995).

*Emissivity problems*

In trying to infer the upwelling longwave flux from the surface temperature (as opposed to direct measurements) variations in surface emissivity must also be taken into account (cf. Equation 2). There are virtually no coherent data available for land surface emissivity; either for its magnitude or for its spatial, temporal and spectral variability. It is usual to assume some (ad hoc) constant value for the broadband emissivity of between 0.9 and 1.0 for natural surfaces. This situation is not as bad as it might seem however because the process of scaling up is likely to cause the emissivity to approach a high value and for the variability at the greater spatial scale to be smaller. The reason for this behaviour is that the small-scale features in the landscape — leaves, bushes, trees, etc. — behave like cavities, trapping radiation

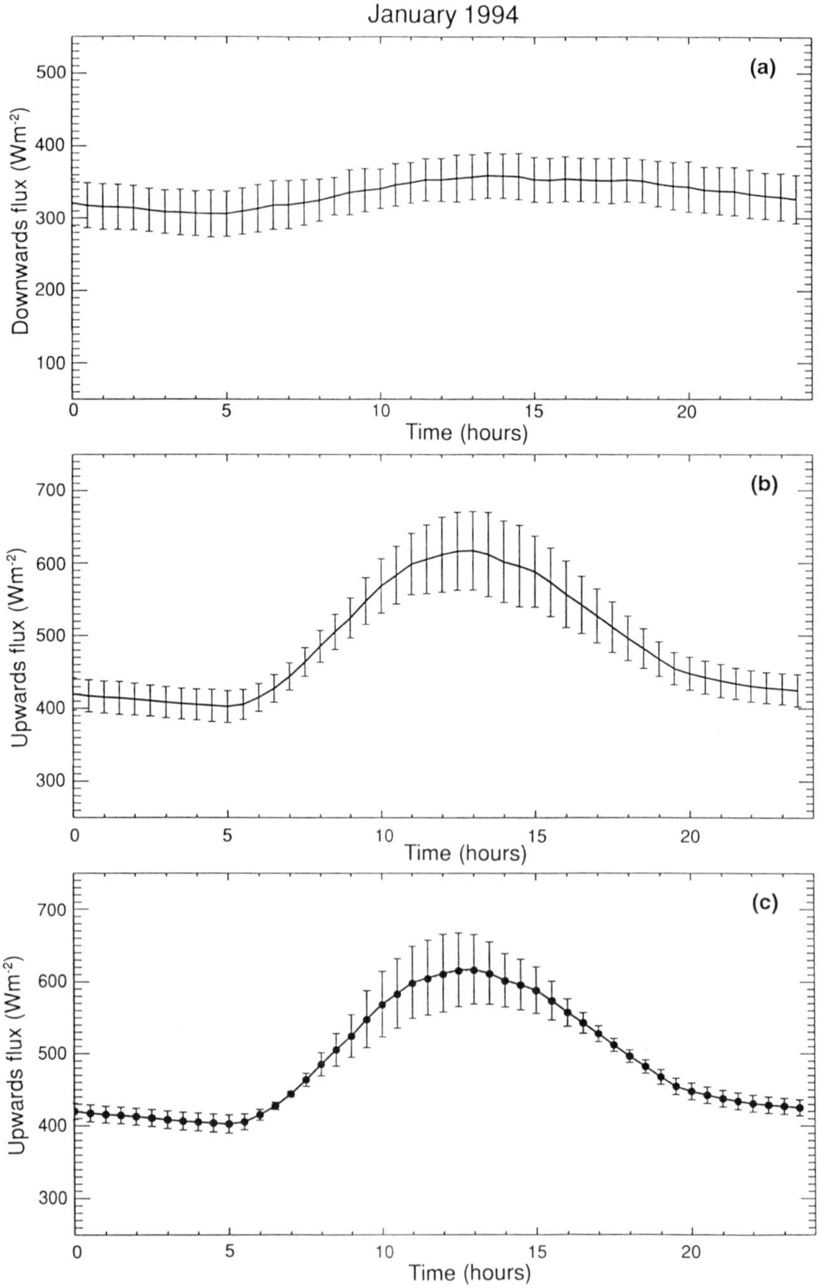


**Figure 3** Monthly mean diurnal cycles of (a) observed downwards flux at the central Hay site, together with the temporal rms variations (rms of the 30-minute mean value at the specified time of day about the 30-day mean); (b) calculated upwards flux averaged for six stations, together with the temporal rms variations; (c) calculated upwards flux averaged for six stations, together with the spatial rms variations. In (b) and (c) the solid line represents the flux calculated from instantaneous temperatures. In (c), the closed circles represent the flux calculated from the monthly mean surface temperature at the specified time of day.

and leading to greater efficiency in absorption and emission of longwave radiation. There is also some evidence that this cavity- like behaviour smoothes out spectral emissivity variation. At spatial scales where bushes and trees can be resolved the spatial emissivity variation is likely to be very high. Some of the emissivity variation is manifest in the thermal (temperature) image shown in Figure 1.

## VALIDATION OF SRB AT GRID SCALES OF HUNDREDS OF KILOMETRES

Model validation of the SRB in climate models has become increasingly reliable in recent years through the availability of long-term station observations of radiation fluxes, net radiation and screen-air temperature (which can be used as a surrogate for the upwelling longwave flux). For example, Garratt (1994, 1995), Wild et al. (1995) and Garratt and Prata (1996) described comparisons between climate-model derived mean monthly and annual fluxes with analogous continental station values of the fluxes. The comparisons were based on interpolation of the model grid-point values to the station locations. Future, more comprehensive, validation of models will soon be possible using the new global datasets of grid-averaged fluxes based on satellite-derived cloud statistics (e.g. Zhang *et al.,* 1995) but, as with climate models, these too require validation with direct surface measurements (e.g. Rossow and Zhang, 1995).

In any of these validation studies, differences between station and model (or satellite-based) estimates are influenced by:

(i) Model biases due to incorrect representation of physical processes (this is what we are trying to identify);

(ii) Observational biases due to instrumental systematic errors;

(iii) Sub-grid variations in the real world, so that one station is not representative of a grid-scale area. Comparisons of closely located stations helps to quantify this non-representativeness but such comparisons are also affected by instrumental biases.

If instrumental biases are known, some inferences on model biases may then be possible.

Components of the SRB which are particularly sensitive to sub-grid spatial variability are the upwelling shortwave flux, which is primarily affected by variations in surface albedo, and the upwelling longwave flux, which is primarily determined by the land-surface temperature. In contrast, the downwelling shortwave flux is relatively insensitive to surface effects but, at a given location, time of day and season, is strongly affected by cloudiness. We choose this component to illustrate the approach by which sub-grid effects can be minimised and, consequently, validation made more reliable.

We first take a small-scale spatial example from Rossow and Zhang (1995; see their Table 6). In their paper, the primary focus is on the validation of calculated radiative fluxes, both TOA and surface, contained within a four-year global data set at a resolution of 280 km and three hours (such a data set is most likely to be used in the form of mean monthly fluxes). Part of their validation involved satellite-based fluxes calculated at 10 pixel locations using observations from 17 closely-spaced ground stations within a region of 100x100 km$^2$. Data used in the analysis comprised 30-minute averages from observations made over a three-week period. Table 2 summarises their results.

**Table 2** Comparison of satellite-based calculated downwelling shortwave flux with direct surface measurements from a 17-station array within a 100x100 km$^2$ region in the USA. Observations are averaged for 30- minute periods to provide the basic data set for analysis. From Table 6 of Rossow and Zhang (1995). Statistics are for 4 types of comparison:

1. At individual times for each station.
2. At individual times for the array average.
3. Daily mean values for each station.
4. Daily and array averages.

| Comparison | | 1 | 2 | 3 | 4 |
|---|---|---|---|---|---|
| Mean difference (Wm$^{-2}$) | $\Delta$ | 14 | 14 | 9 | 8 |
| Standard deviation (Wm-2) | $\sigma_F$ | 57 | 35 | 16 | 9 |
| Number of data | N | 1658 | 104 | 34 | 21 |

In brief, the mean differences (biases) vary little, but standard deviations are greatly reduced, as the scales of averaging (time and space) are increased. Comparisons 1 and 2 both use the raw 30-minute averaged data, and it is clear that use of these short time averages leads to unacceptably high standard deviations and meaningless estimates of the bias. Only when daily averages are used (Comparisons 3 and 4), and thus the strong diurnal effects removed, does the standard deviation diminish to acceptable values. The results reflect the fact that much of the flux variability due to sun position and cloudiness occurs at scales of less than one day and 100 km. Of course, significant contributions are still to be found at larger scales (see also Rossow and Zhang, 1995).

In a second, larger-scale, example, we take a region of approximately 1500x1500 km$^2$ in the USA containing 35 ground stations, whose long-term observations of downwelling shortwave flux are available through the GEBA.

These data were used by Garratt (1994) in the validation of 20 grid-point values of monthly fluxes from a number of climate models (see Garratt, 1994 — reference his USA region). Table 3 summarises the results for one such model and, as in the previous smaller-scale example, standard deviations decrease significantly as the scale of averaging increases. For these climatological station data, one month apparently provides a sufficiently long averaging period with which to perform meaningful comparisons (both with climate model and satellite-based global gridded mean monthly fluxes), but spatial averaging is critical.

**Table 3** Comparison of GCM (CSIRO9) values of downwelling shortwave flux with direct long-term measurements for 35 climatological stations within a 1500x1500 $km^2$ region in the USA. Observations used are long-term monthly mean values available in the Global Energy Balance Archive (Ohmura and Gilgen, 1991). See also Garratt (1994).

Statistics are for 3 types of comparison:

1. For single months and single stations.
2. For annual mean and single stations.
3. For single months and the 35-station average.

| Comparison | | 1 | 2 | 3 |
|---|---|---|---|---|
| Mean difference (W $m^{-2}$) | $\Delta$ | 22 | 22 | 22 |
| Standard deviation (W $m^{-2}$) | $\sigma_F$ | 35 | 14 | 4 |
| Number of data | N | 420 | 35 | 12 |

It is clear from Tables 2 and 3 that comparisons of single-station fluxes with nearest pixel or grid-point values are least reliable, and both time and space averaging help reduce the uncertainties. At the pixel scale (1 to 10 km), time averaging out to periods of at least one day is critical for comparison of station and satellite-based fluxes. Diurnal effects are thus removed. At the model grid scale, or that of the new satellite-based global data sets (several hundreds of km), space averaging out to scales if the order of 1000 km (involving tens of grids points and stations) is critical, with one month providing a long enough averaging period for reliable comparisons. Use of annual means of course removes the complications of seasonal effects.

## CONCLUDING REMARKS

We have looked at three practical problems where the upscaling issue is relevant. Consideration has been given to:

(i) Local ground-truth measurements of land temperature and longwave radiative fluxes (spatial scale of 1 km) used in the calibration and validation of pixel scale (1 to 10 km) satellite observations. The time averages are typically minutes to a month;

(ii) Observations of downwelling shortwave flux from an array of stations (spatial scale 100 km) used in the validation of the new global data sets of satellite-based surface radiation fluxes available at 2.5° resolution (Rossow and Zhang, 1995). The time averages are one hour to several weeks;

(iii)Observations of downwelling shortwave radiative fluxes (from extended time series) from an array of stations (spatial scale of 1000 km) used in the validation of the SRB at the grid scale in climate models (grid scales of several hundred km). Time averages are typically monthly to annual. Such measurements can also be used to validate the grid-scale fluxes referred to in (ii) above.

Typical magnitudes and scales of spatial variability are described that are pertinent to all three problems, and practical solutions to minimising the effects of sub-grid variability discussed (spatial averaging). In all three cases, where space scales range from 1 to 1000 km, some level of spatial averaging is critical. When observations at sub-diurnal time scales are used, some level of time averaging, at least to daily means, is highly desirable. Such time and space averaging will help ensure meaningful validation of land temperature and surface radiative fluxes derived from satellite-based calculations or from climate model simulations.

## REFERENCES

Darnell, W.L., Staylor, W.F., Gupta, S.K., Ritchey, N.A. and Wilber, A.C.1992. Seasonal variation of surface radiation budget derived from International Satellite Cloud Climatology Project C1 data, *J. Geophys. Res.*, **97**(D14), 15741-15760.

Field, R.T., Fritschen, L.J., Kanemasu, E.T. Smith, E.A., Stewart, J.B. Verma, S.B. and Kustas, W.P. 1992. Calibration, comparison, and correction of net radiation instruments used during FIFE. *J. Geophys. Res.*, **97**(D17), 18,681-18,695.

Garratt, J.R. 1978. Transfer characteristics for a heterogeneous surface of large aerodynamic roughness. *Quart. J. Roy. Meteor. Soc.*, **104**, 491-502.

Garratt, J.R. 1992. Extreme maximum land surface temperatures. *J. Appl.Meteor.*, **31**, 1096-1105.

Garratt, J.R. 1994. Incoming shortwave fluxes at the surface - a comparison of GCM results with observations. *J. Climate*, **7**, 72-80.

Garratt, J.R.. 1995. Observed screen (air) and GCM surface/screen temperatures - implications for outgoing longwave fluxes at the surface. *J. Climate*, **8**, 1360-1368.

Garratt, J.R. and Prata, A.J. 1996. Downwelling longwave fluxes at continental surfaces - a comparison of observations with GCM simulations and implications for the global land-surface radiation budget. *J. Climate*, **9**, 646-655.

Henning, D. 1989.*Atlas of the Surface Heat Balance of the Continents*. Gebtruder Borntraeger, Berlin, 402pp.

Ohmura, A. and Gilgen, H. 1991. Global Energy Balance Archive GEBA. Report 2: The GEBA Database: Interactive Applications, Retrieving Data (Heft 44). Geographisches Institut, ETH Zurich. 66pp. [available from Prof. A. Ohmura, Geographisches Institut, ETH Zurich, Switzerland]

Palluconi, F.D. and Meeks, G.R. 1995. Thermal Infrared Multispectral Scanner (TIMS): An investigator's guide to TIMS data, JPL Publication 85-32, Jet Propulsion Laboratory, Calif. Inst. Technology, Pasadena, CA, 22pp.

Prata, A.J. 1994a. Validation data for land surface temperature determination from satellites, CSIRO, Division of Atmospheric Research Technical Paper No. 33, 36pp.

Prata, A.J. 1994b. Land surface temperatures derived from the AVHRR and ATSR 2, Experimental and validation of AVHRR algorithms. *J. Geophys. Res.*, **99**, 13025-13058.

Rossow, W.B. and Zhang, Y.-C. 1995. Calculation of surface and top of atmosphere radiative fluxes from physical quantities based on ISCCP data sets. 2. Validation and first results. *J. Geophys.Res.*, **100**, D1, 1167-1197.

Sellers, P.J., Hall, F.G. Asrar, G. Strebel, D.G. and Murphy, R.E. 1992. An overview of the First International Satellite Land Surface Climatology Project (ISLSCP) Field Experiment (FIFE), J. *Geophys. Res.*, **97**, D17, 18345- 18371.

Sun, J. and Mahrt, L. 1995. Relationship of surface heat flux to microscale temperature variations: Application to BOREAS. *Bound. Layer Meteor.*, 76, 291-301.

Wild, M., Ohmura, A., Gilgen, H. and Roeckner, E. 1995. Validation of general circulation model radiative fluxes using surface observations. *J. Climate*, **8**, 1309-1324.

Willmott, C.J., Rowe, C.M. and Mintz, Y. 1985. Climatology of the terrestrial seasonal water cycle. *J. Climatol.*, **5**, 589-606.

Zhang, Y.-C., Rossow, W.B. and Lacis, A.A. 1995. Calculation of surface and top of atmosphere radiative fluxes from physical quantities based on ISCCP data sets. 1. Method and sensitivity to input data uncertainties. *J. Geophys.Res.*, **100**, D1, 1149-1165.

# 6 Estimation of surface resistance and Priestley-Taylor α-parameter at different scales

W.G.M. BASTIAANSSEN[1], H. PELGRUM[1], M. MENENTI[1] and R.A. FEDDES[2]
[1] *DLO-The Winand Staring Centre, Wageningen, The Netherlands*
[2] *Agricultural University of Wageningen, The Netherlands*

## INTRODUCTION

Hydrological processes need to be described at a wide variety of spatial and temporal scales: field scale water balance studies can be conducted for assessing crop water use, surface runoff or the design of sub-surface drainage systems whereas regional scale hydrological studies are often related to areal water resource management. In the last two decades, a description of the meso-scale hydrological processes for atmospheric circulation studies has received growing attention. Land surface evaporation is an important component in all these studies and exhibits a significant spatial variability over heterogeneous land surfaces. The evaporation behaviour depends on (i) soil water content, (ii) the air humidity, (iii) the momentum flux, (iv) the surface temperature, (v) the air temperature, (vi) the surface roughness and (vii) the net available energy, being all essential components of the Penman-Monteith equation. The current paper emphasizes the hydrological control on evaporation at different spatial and temporal scales and demonstrates how remote sensing measurements can contribute to understanding that.

It is quite customary to describe the effects of soil water content on evaporation by means of a resistance or conductance term (Monteith, 1965; Jarvis, 1976). Canopy resistance of homogeneous vegetation stands has been measured for several biome types (e.g. *pine forest*: Stewart, 1988; *grass prairie*: Stewart and Gay, 1989; *millet*: Wallace *et al*., 1993; *irrigated barley*: Brunet *et al*., 1994). Soil-Vegetation-Atmosphere-Transfer (SVAT) models are developed to predict the canopy evaporation by means of a network of evaporation resistances (e.g. Choudhury and Monteith, 1988). Simulation results with SVAT models match the observations reasonably well when the resistance to evaporation is calibrated locally against soil water content (e.g. *oak forest*: Ogink-Hendriks, 1994) or minimum surface resistance (*Sahelian tiger bush*: Blyth and Harding, 1995).

*Scaling up in hydrology using remote sensing*. Edited by J.B. Stewart, E.T. Engman, R.A. Feddes and Y. Kerr

Resistances for bare soil evaporation have also been frequently expressed as a function of soil water content (e.g. van de Griend and Owe, 1994; Wallace, 1995). Drawbacks of both canopy and bare soil resistance are that most relationships between resistance and soil water content have an empirical character and that they are validated for small sized relatively homogeneous patches. Several, if not most, atmospheric studies require the surface energy partitioning to be known at much coarser scales and for lack of better information, resistance formulations developed for homogeneous patches are also applied on a coarser scale for which they were not designed. The implications of such handling are not fully understood because large-area resistances are difficult to assess in an independent way: heterogeneous land surfaces are characterized by a large spatial variation of surface resistances and their distribution as well as their aggregated value cannot be straightforwardly assessed. Although much emphasis has been given recently to computing with coupled SVAT-Planetary Boundary Layer models the average latent heat flux from composite terrain (e.g. *grass+forest*: Blyth *et al.*, 1993; *desert+crop+lake+forest*: Raupach, 1995) and these exercises are necessary to understand the mixing and averaging of parameters of the soil-vegetation-atmosphere continuum, these studies fall short with respect to a realistic representation of the natural variations of evaporation resistances. Bastiaanssen *et al.* (1994) showed by means of remote sensing measurements for a Spanish rainfed and irrigated agricultural area that the shape of the probability density function changes significantly with the land surface conditions. Aggregation models should therefore incorporate at least the probability density function (pdf) of surface resistance which differs for each region.

This chapter describes a technique to derive area-effective bulk surface resistance and Priestley-Taylor $\alpha$-parameter using SEBAL in association with a set of aggregation rules which can be used to validate the large area surface resistance and evaporation predicted by SVAT models and bucket type of land surface models (Manabe, 1969). The results of a proposed aggregation rule to bridge the mismatch between pixel scale and the region as well as the daytime behaviour of the aggregated values will be given at different spatial and temporal scales. The effects of measurement resolution (high *versus* low resolution) on area-average and the probability density function of the bulk surface resistance and Priestley-Taylor $\alpha$-parameter will be described.

## SEBAL PROCEDURE

The Surface Energy Balance Algorithm for Land, SEBAL (Bastiaanssen *et al.*, 1994; Bastiaanssen, 1995), solves the instantaneous surface energy balance pixel-by-pixel and needs surface hemispherical reflectance, surface temperature and vegetation index from remotely operating radiometers. Since we are concentrating on how scales, heterogeneity and spatial resolution are related, the description of the SEBAL procedure to derive fluxes from remote sensing observations is confined to a minimum. New aspects of SEBAL as compared to existing flux algorithms, for instance Seguin and Itier (1983) and Moran *et al.* (1994) are that (i) fewer micro-meteorological observations are needed (heat fluxes, net radiation, air temperature, windspeed and vapour pressure deficit are not necessary), (ii) a land classification is not longer required, (iii) the soil heat flux in sparse canopies is accounted for and (iv) the coupling between surface temperature, air temperature and the surface roughness length for heat is physically sound.

Net radiation is obtained from distributed surface hemispherical reflectance $r_0$ and surface temperature $T_0$ data in combination with variable zenith angles for the determination of the clear sky incoming shortwave solar radiation $K_d$. The incoming longwave radiation $L_d$ is held constant over the area. Soil heat flux is obtained from a new empirical soil heat flux/net radiation fraction which includes (i) the vegetation index to describe attenuation effects, (ii) the surface temperature to describe the soil heat flux for sparse canopy condition and the phase difference between soil heat flux and net radiation due to a delayed warming of the soil and (iii) surface hemispherical reflectance to describe the at-surface soil water content when the soil is bare. The area-average momentum flux is computed from the area-effective aerodynamic resistance for dry land surface elements which in turn can be determined from the linearised slope between surface hemispherical reflectance and surface temperature (see Menenti *et al.*, 1989 for more details). The momentum $\tau$ and sensible heat flux H are based on the resistance version of the flux-profile relationships. For each pixel, (which may encompass a mixture of landscape elements), a one-layer resistance scheme for $\tau$ and H will be applied.

The vertical near-surface air temperature difference is obtained by an inversion of the equation for sensible heat transfer from sub-areas with specific land surface conditions: entirely wet (zero sensible heat flux) and entirely dry (zero latent heat flux). These sub-areas can be allocated on the basis of the surface temperature and surface hemispherical reflectance of the pixels covering these sub-areas: pixels with a surface hemispherical reflectance less than 10% and a low surface temperature are considered to fulfil the condition of zero sensible heat flux. Pixels which have a surface hemispherical reflectance more than 30% and having a negative slope between surface hemispherical reflectance and surface temperature are considered to have zero latent heat flux. The advantage of this procedure is that sensible heat fluxes will be always scaled between zero and the net available energy and the need for air temperature records does no longer apply. Hence the vertical difference in near-surface air temperature is obtained at sub-areas where sensible heat flux is assumed to be either zero or maximum, i.e. equal to the net available energy and the roughness for heat is predefined. The areal interpolation of the near-surface air temperature difference, which differs significantly for the two chosen sub-areas, is done with a surface temperature map. Finally, latent heat flux $\lambda E$ is obtained as the residual of the land surface energy balance.

Ground measurements of surface temperature $T_0$, shortwave atmospheric transmittance $\tau_{sw}$, and surface roughness $z_0$ are required as input data for SEBAL. Collateral information on wind speed are not needed since the areal effective momentum flux is obtained from the negative slope between surface temperature and surface hemispherical reflectance. The output of SEBAL is summarized in Table 1. Some of the listed variables are directly obtained from empirical relationships, e.g. surface emissivity, $\varepsilon_0$, and surface roughness for momentum transport, $z_{0m}$, while others are obtained by iterative computation procedures such as friction velocity, $u_*$, and Monin Obukhov length, L. For a comprehensive description on the procedures to estimate the parameters listed in Table 1, refer to Bastiaanssen *et al.* (1994) and Bastiaanssen (1995).

**Table 1** Pixel-dependent output of the SEBAL flux algorithm driven by remote sensing measurements of surface hemispherical reflectance, surface temperature and vegetation index

| Symbol | Parameter | Symbol | Variable |
|---|---|---|---|
| $r_0$ | Surface albedo | L | Monin-Obukhov length |
| $T_0$ | Surface temperature | $\psi_h$ | Stability correction |
| $\varepsilon_0$ | Surface emissivity | $r_{ah}$ | Air resistance to heat |
| $K_d$ | Solar radiation | $\delta T_{a\text{-}sur}$ | Vertical air temperature difference |
| $G_0$ | Soil heat flux | H | Sensible heat flux |
| $z_{0h}$ | Surface roughness for heat | $\lambda E$ | Latent heat flux |
| $u_*$ | Friction velocity | | |

## REMOTE SENSING IMAGES USED WITHIN EFEDA

The EFEDA project on the determination of areal surface energy balances and executed in Central Spain (Bolle *et al.*, 1993) comprised three supersites: Tomelloso, Barrax and Rada de Haro (Figure 1). A NASA C130 airborne mission with the NS001 radiometer was performed over these sites on June 29, 1991. The overpass time was 10:21 GMT and the NS001 pixels were 18.5 m by 18.5 m. The surface energy balance of the Tomelloso and Barrax supersites has been determined with SEBAL using the NS001 spectral radiances. Tomelloso is homogeneous with respect to land use and consists of rainfed vineyards and natural woodlands. Due to irrigation facilities, the Barrax site has heterogeneous land use and hydrological conditions. The irrigated crops were mainly maize, wheat, barley and alfalfa. The latent heat fluxes obtained with SEBAL and validated with ground fluxes measured by eddy correlation and Bowen-ratio energy balance techniques (Bastiaanssen *et al.*, 1996) are depicted in Figure 2.

Hourly METEOSAT images of 10:00 and 11:00 GMT on June 29, 1991, have also been processed with SEBAL covering the entire Iberian Peninsula. The Normalized Difference Vegetation Index (NDVI) data necessary for the operation of SEBAL have been taken from the visible and near-infrared channels of the AVHRR radiometer onboard the NOAA satellites. The METEOSAT pixels were re-sampled to 3.9 km by 4.1 km. To enable a multi-resolution study of surface fluxes, results of the NS001 and METEOSAT data were intercompared. The overlay of the many

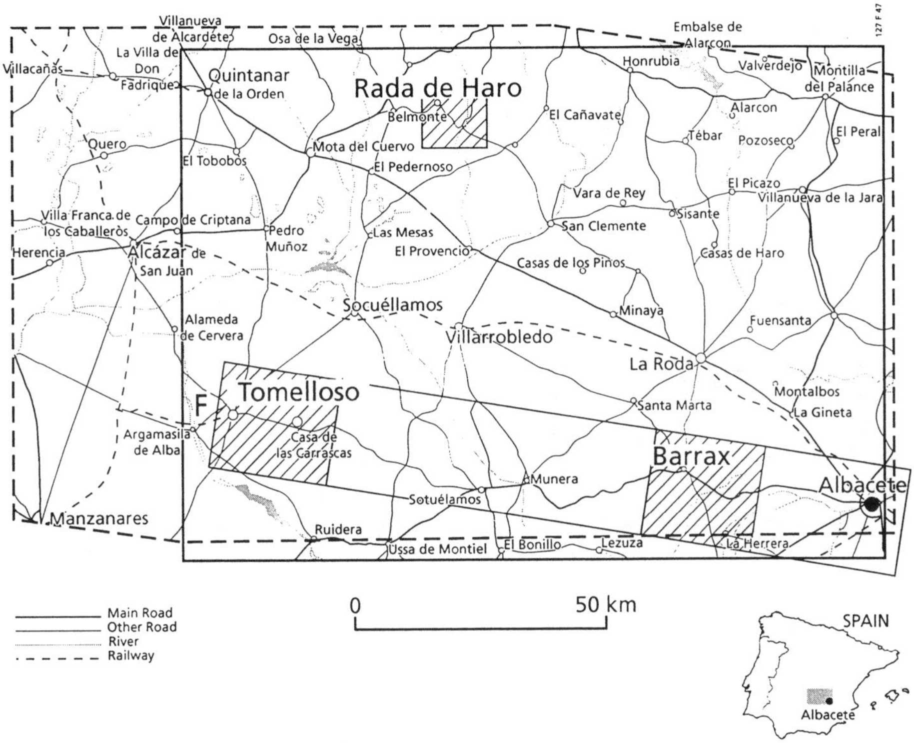


**Figure 1** Location of the EFEDA grid in Castilla la Mancha, Central Spain. The Tomelloso, Barrax and Rada de Haro supersites are indicated. The hatched squares represent the locations shown in Figure 3. The East-West leg flown by the NS001 aircraft on 29 June 1991 is described by E and F.

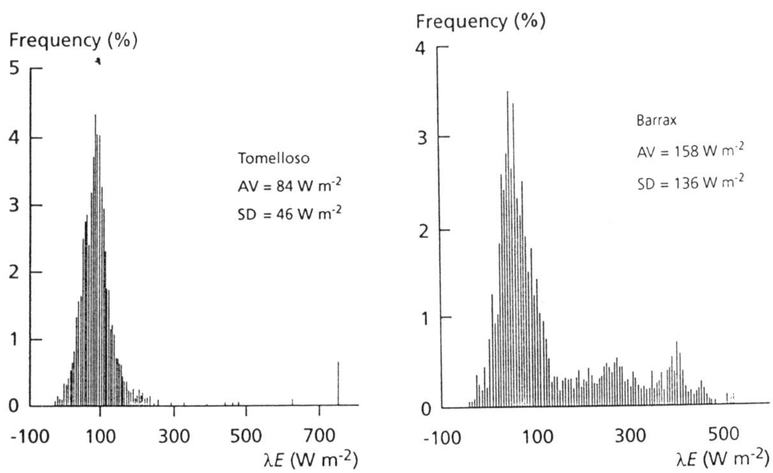


**Figure 2** Spatial variation of the latent heat flux at the Tomelloso and Barrax supersites during the NS001 flight on 29 June 1991.

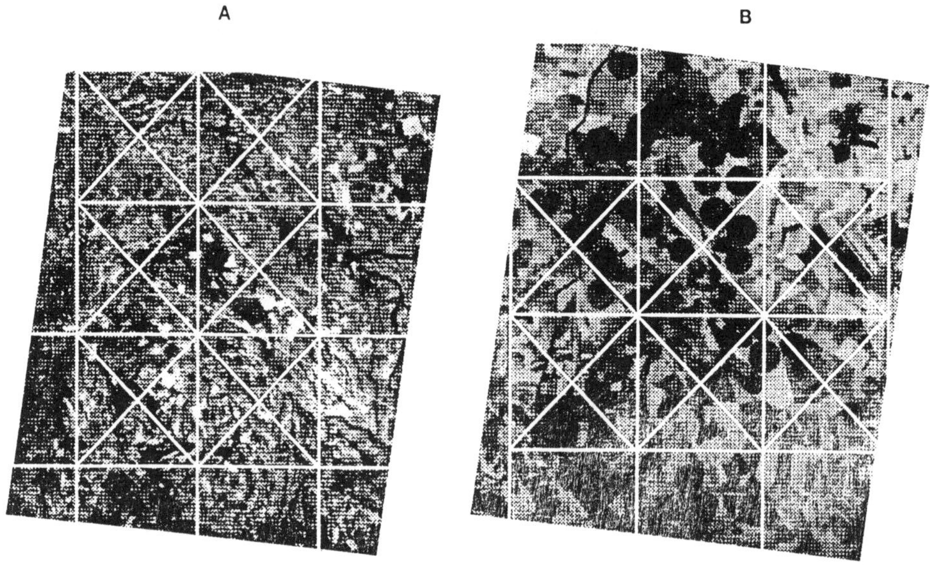


**Figure 3** Overlay of the METEOSAT and NS001 images (Part A) Tomelloso and (Part B) Barrax

small NS001 pixels with the few large METEOSAT pixels of the Tomelloso and Barrax supersites is illustrated in Figure 3. Only the twelve METEOSAT pixels marked by a cross in Figure 3 will be used in the further analysis. The validation of sensible and latent heat fluxes obtained from the METEOSAT and AVHRR measurements at a spatial scale of 100 km was established with aircraft flux measurements (Jochum, 1993). A detailed description of the validation study is given by Bastiaanssen *et al.* (1996) and will not be repeated here.

## SURFACE ENERGY PARTITIONING

Bulk surface resistance $r_s$ or Priestley-Taylor $\alpha$-parameter couple the water and energy budget and are therefore termed *moisture indicators*. These variables can be inferred from $\lambda E$ to arrive at pixelwise values:

$$r_s = \frac{\rho_a c_p}{\gamma \lambda E} [e_{sat}(T_0) - e] - r_{av} \qquad (\mathrm{s\ m^{-1}}) \qquad (1)$$

and

$$\alpha = \frac{\lambda E}{R_n - G_0} \frac{s_a + \gamma}{s_a} \qquad (-) \qquad (2)$$

where $\lambda E$ is obtained as the residual of the surface energy budget and $R_n$, $G_0$ and H are obtained by SEBAL:

$$\lambda E = R_n - G_0 - H \qquad (\mathrm{W\ m^{-2}}) \qquad (3)$$

where $R_n$ (W m$^{-2}$) is the net radiation, $G_0$ (W m$^{-2}$) is the soil heat flux, $s_a$ (mbar K$^{-1}$) is the slope of the saturated vapour pressure curve, g (mbar K$^{-1}$) is the psychrometric constant, $r_a c_p$ (J m$^{-3}$ K$^{-1}$) is the air heat capacity, $e_{sat}$ (mbar) is the saturated vapour pressure, e (mbar) is the actual vapour pressure at the evaporating surface and $r_{av}$ (s m$^{-1}$) is the aerodynamic resistance to vapour transfer. Combining Equations 1 and 2 demonstrates analytically that a is inversely proportional to $r_s$:

$$\alpha = \frac{1}{r_s + r_{av}} \left\{ \frac{\rho_a c_p \, [e_{sat}(T_0) - e] \, [s_a + \gamma]}{s_a \, \gamma \, (R_n - G_0)} \right\} \qquad (-) \qquad (4)$$

Earlier work of McNaughton & Spriggs (1989) based on Planetary Boundary Layer (PBL) model simulations indicated a systematic trend between α and $r_s$. Monteith (1994) suggested a relationship between α and $r_s^{-1}$ to be simpler than between α and $r_s$. This hypothesis was tested in Figure 4 for several spatial scales and resolutions using $r_s$ and α derived from Equations 1 and 2 respectively. Because of the asymptotic shape of the $\alpha(r_s^{-1})$ relationship, the surface conductance $r_s^{-1}$ is mathematically easier to fit to α than is $r_s$ (this appears also from Equation 4). Figure 4 shows that α is rather insensitive to fractional changes in $r_s^{-1}$ when $r_s^{-1} > 0.04$ m s$^{-1}$, over ranges of resolution from 18.5 to 4000 m and scales of 18 km to 1000 km. The controlling range of evaporation lies essentially in Priestley-Taylor terms between $\alpha = 0.1$ and 1.0 which according to Figure 4 is equivalent to a range of $r_s = 25$ to 1000 s m$^{-1}$. Hence, resistance values of $r_s < 25$ s m$^{-1}$ and $r_s > 1000$ s m$^{-1}$ hardly affect the evaporation—at micro-scale (part A and B), meso-scale (part C) or macro-scale (part D).

Equation 4 explains the wide scatter of values for Barrax in part B of Figure 4, a consequence of strong spatial variations in $R_n$, $G_0$, $r_{av}$, $e_{sat}(T_0)$ and $s_a$ at a scale of 18.5 m. This phenomenon did not appear in the 'homogeneous' Tomelloso supersite. The low resolution of METEOSAT pixels is held responsible for the low scatter in the $\alpha(r_s^{-1})$ relationship observed for EFEDA and the Iberian Peninsula: variations at spatial scales smaller than 4 km are not seen by the remotely sensed observations. Some other data from the FIFE (Kansas, USA), MOBILHY (Southwest France) and ARME (Amazonia, Brazil) large-scale field experiments (Shuttleworth, 1993) are included in the graph for the Iberian Peninsula to demonstrate a common trend among a and $r_s$ for a wide variety of surface and climate types. The latter implies that a is a simple and appropriate candidate to describe surface energy partitioning at a scale of 4 km or larger. At smaller scales, anomalies between α and $r_s^{-1}$ may exist because of the significant spatial variation of the controlling variables in Equation 4, and then $r_s^{-1}$ is technically preferred to describe the surface energy partitioning.

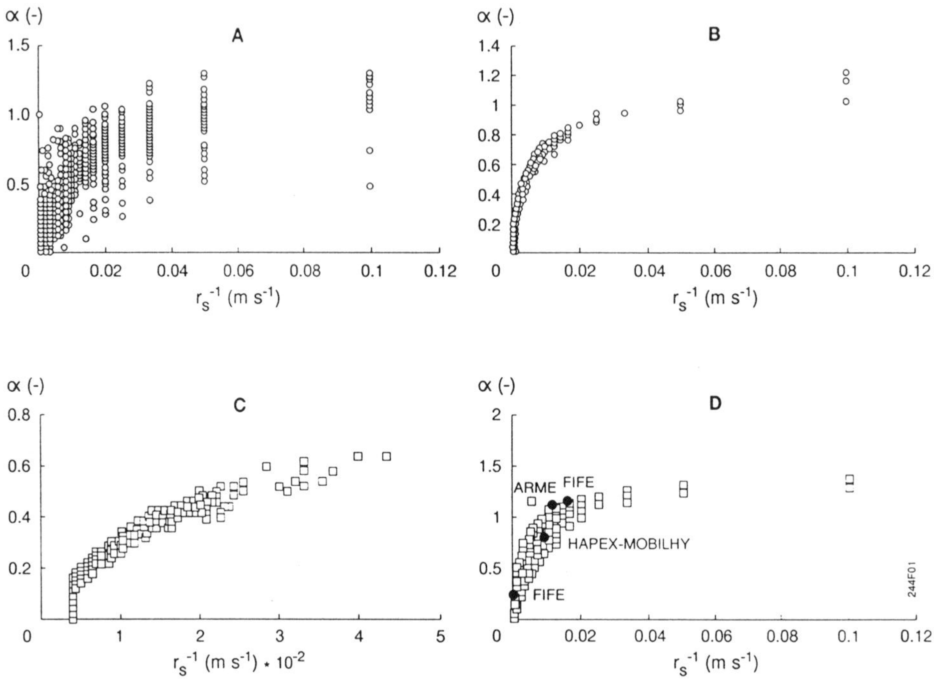


**Figure 4** Relationship between bulk surface conductance $r_s^{-1}$ and Priestley & Taylor α-parameter. (Part A) 18.5 m NS001-Tomelloso, (Part B) 18.5 m NS001-Barrax, (Part C) 4 km METEOSAT-EFEDA and (Part D) 4 km METEOSAT Iberian Peninsula. Also shown are some data points for FIFE (Kansas, USA), HAPEX-MOBILHY (Southwest France) and ARME (Amazonia, Brasil) taken from Shuttleworth (1993)

## SCALE DEPENDENCY OF SURFACE ENERGY PARTITIONING

The scale at which surface energy partitioning becomes scale-invariant can be studied with a semi-variogram. A semi-variogram is a plot that expresses the spatial dependence of any parameter value on the parameter at a neighbouring location. The correlation length of a surface energy partitioning semi-variogram describes the scale at which the semi-variance of the energy partitioning is maximized: sampling at any larger size does not affect the quantification. If the scale at which the SVAT model applies is much larger than the correlation length scale of the hydrological processes, which is pertinent for surface flux measurements at a heterogeneous land surface, in-situ observations at a fraction of the correlation length cannot be used to validate large area fluxes predicted by models.

The validated sensible H and latent heat flux λE maps being based on the SEBAL algorithm and NS001 spectral radiances can be explored to compile semi-variograms of evaporative fraction, λE/(λE+H). A number of rows have been selected (three north-south; three west-east) for the Barrax and Tomelloso supersites. An example of one variogram for Barrax is given in Figure 5. Table 2 summarizes all sills (i.e.

maximum semi-variance) and ranges (i.e. correlation lengths). On the basis of Table 2, it may be concluded that the semi-variance at sill-level in Barrax is with $1.9 \times 10^{-2}$ larger than for Tomelloso ($1.4 \times 10^{-2}$) which implies that Barrax contains more heterogeneity in surface energy partitioning (see also Figure 2). Surprisingly, the average correlation length for Barrax (3293 m) is smaller than that for Tomelloso (3900 m). This implies that, although the spatial variation is more diverse, the energy partitioning in Barrax is more geometrically even than for Tomelloso, i.e. wet and dry patches in Barrax are regularly spread across the image. The only patch of the Tomelloso supersite that is irrigated is in the Northern part, some kilometres away from the vineyard area in the centre of the image, and this irregularity in geometry enhances the range.

The correlation length provides a suitable basis to understand the effect of shifting from high to low resolution images on the areal distribution of bulk surface resistance and Priestley-Taylor α-parameter (see Figure 6).

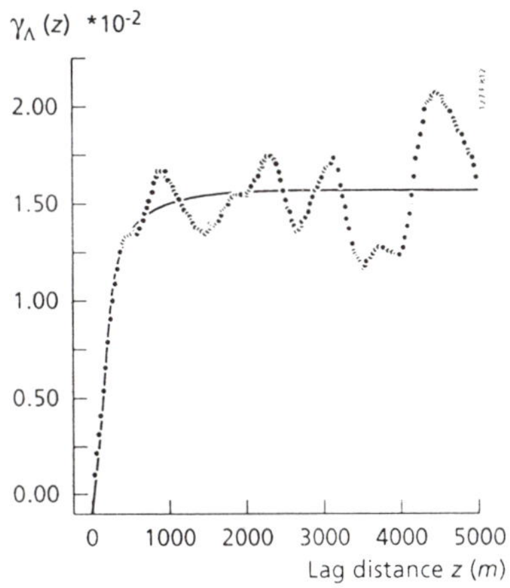


**Figure 5** Semi-variogram of evaporative fraction ($\lambda E/(\lambda E+H)$) for a transect across the Barrax NS001 sub-scene

**Table 2** Semi-variogram statistics on evaporative fraction, $\lambda E/(\lambda E+H)$. The highest resolution was 18 m.

| Transect | Sill | Range (m) | Transect | Sill | Range (m) |
|---|---|---|---|---|---|
| Tomelloso-1 | $1.8 \times 10^{-2}$ | 2500 | Barrax-1 | $1.8 \times 10^{-2}$ | 4000 |
| Tomelloso-2 | $1.2 \times 10^{-2}$ | 3000 | Barrax-2 | $2.5 \times 10^{-2}$ | 1400 |
| Tomelloso-3 | $1.3 \times 10^{-2}$ | 4000 | Barrax-3 | $1.5 \times 10^{-2}$ | 5000 |
| Tomelloso-4 | $1.2 \times 10^{-2}$ | 4500 | Barrax-4 | $1.2 \times 10^{-2}$ | 2000 |
| Tomelloso-5 | $1.5 \times 10^{-2}$ | 5500 | Barrax-5 | $1.6 \times 10^{-2}$ | 2000 |
| Tomelloso-6 | $1.2 \times 10^{-2}$ | 4000 | Barrax-6 | $3.0 \times 10^{-2}$ | 5000 |
| Average | $1.4 \times 10^{-2}$ | 3900 | Average | $1.9 \times 10^{-2}$ | 3293 |

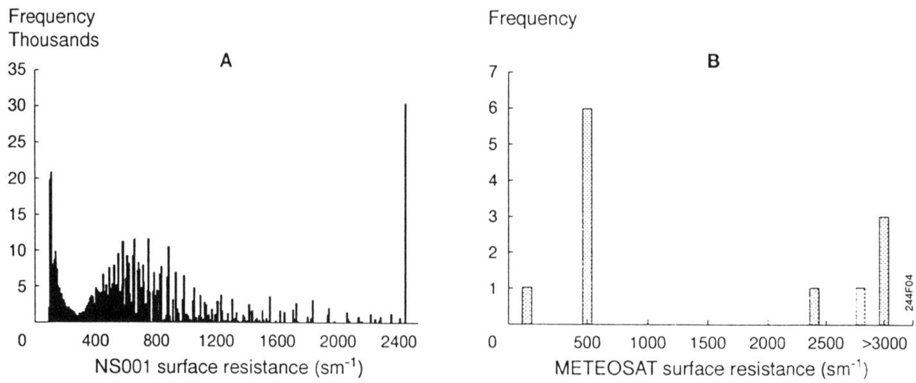


**Figure 6** Frequency distributions of bulk surface resistance for an area of 200 km$^2$ being based on (Part A) 18.5 m NS001 and (Part B) 4 km METEOSAT pixels

Although the frequency distribution of the METEOSAT pixels coincides with the classes of the highest frequencies on the $r_s^{NS001}$ distribution, it is obvious that the overall shape is not similar: the class of 750 to 2250 s m$^{-1}$ is entirely absent on the $r_s^{METEOSAT}$ histogram because (i) part A contains 560000 pixels whereas part B contains 12 pixels and (ii) the size of a METEOSAT pixel at 4 km is larger than the correlation length of 3.3 to 3.9 km at which scale heterogeneity no longer changes with scale.

## AGGREGATING MOISTURE INDICATORS BY FLUX AVERAGING AND INVERSION

The main arguments for aggregating $\alpha$ and $r_s$ from pixel scale to coarser scales are (i) the feasibility of comparing the results of remote sensing images with different resolution and (ii) to bridge the gap in scales between pixel-based information and the mesoscale at which SVAT models and other types of land surface models are operational. Surface fluxes average linearly over uniform sites and heterogeneous land surfaces with a disorganized structure (Shuttleworth, 1988). The mathematical averaging of fluxes is straightforward to apply for remote sensing images since the areal size of each pixel is known and 100% of the area is sampled. In the formulation presented hereafter, the pixel dependency of the fluxes is indicated by means of the (x,y)-notation and the arithmetic mean procedure is given by chevron brackets.

$$<R_n> = \frac{1}{n} \sum R_n(x,y) \qquad (\mathrm{W\ m^{-2}}) \qquad (5)$$

$$<K_d> = \frac{1}{n} \sum K_d(x,y) \qquad (\mathrm{W\,m^{-2}}) \qquad (6)$$

$$<H> = \frac{1}{n} \sum H(x,y) \qquad (\mathrm{W\,m^{-2}}) \qquad (7)$$

$$<\lambda E> = \frac{1}{n} \sum \lambda E(x,y) \qquad (\mathrm{W\,m^{-2}}) \qquad (8)$$

Knowledge of the area-averaged surface fluxes can be deployed to find in an inverse manner the area-effective values of the involved spectrum of hydrometeorological parameters:

$$r_0^{eff} = \frac{1}{n <K_d>} \sum [r_0(x,y)\, K_d(x,y)] \qquad (-) \qquad (9)$$

$$L_u^{eff} = <K_d> - r_0^{eff} <K_d> + L_d - <R_n> \qquad (\mathrm{W\,m^{-2}}) \qquad (10)$$

$$T_0^{R\text{-}eff} = \left(\frac{L_u^{eff}}{\sigma}\right)^{0.25} \qquad (\mathrm{K}) \qquad (11)$$

$$\varepsilon_0^{eff} = \{\varepsilon_0(x,y)\}_{Med} \qquad (-) \qquad (12)$$

$$T_0^{eff} = \frac{T_0^{R\text{-}eff}}{(\varepsilon_0^{eff})^{0.25}} \qquad (\mathrm{K}) \qquad (13)$$

where $T_0^R$ (K) is the radiation temperature, $\varepsilon_0$ (–) is the surface emissivity in the thermal infrared range and L is the incoming longwave radiation. If the value for L under clear skies and at a regional scale is kept spatially constant, the area-averaged radiation balance should remain preserved, such as defined in Equation 10. The latter aspect was ignored by for instance Lhomme (1992) and Braden (1995). Lhomme (1992) computed $T_0^{eff}$ from area-average surface fluxes and Braden (1995) used Stefan Boltzmann law to linearly average the emitted longwave fluxes. The median value for $\varepsilon_0$ was considered in Equation 12 because its pdf may have strongly uneven distributions. Under these conditions, the arithmetic mean is less advisable and a median average is more correct to describe an eventual log-normal distribution.

An essential component of the SEBAL procedure is the estimation of the vertical near-surface air temperature differences, $\delta T_{a\text{-}sur}$ on the basis of $T_0(x,y)$ according to:

$$\delta T_{a\text{-}sur}(x,y) = c_1 + c_2 T_0(x,y) \qquad (K) \qquad (14)$$

where $c_1$ and $c_2$ are area and time dependent calibration coefficients and $\delta T_{a\text{-}sur}$ is obtained from the specific sub-areas with H=0 and $\lambda E$=0. Equation 14 can be used to obtain the pdf of $\delta T_{a\text{-}sur}$. Likewise for $e_0$, it is suggested to obtain $\delta T_{a\text{-}sur}$ by means of a median procedure:

$$\delta T_{a\text{-}sur}^{eff} = \{\delta T_{a\text{-}sur}(x,y)\}_{Med} \qquad (K) \qquad (15)$$

Equation 14 can be used to infer local estimations of air temperature $T_a(x,y)$ according to Equation 16, which is necessary to solve $e_{sat}^{eff}$ and $s_a^{eff}$ in Equations 17 and 18 respectively.

$$T_a(x,y) = T_0(x,y) - \delta T_{a\text{-}sur}(x,y) \qquad (K) \qquad (16)$$

$$e_{sat}^{eff} = \{e_{sat}\, f(T_a)(x,y)\}_{Med} \qquad (mbar) \qquad (17)$$

$$s_a^{eff} = \{s_a\, f(T_a)(x,y)\}_{Med} \qquad (mbar\ K^{-1}) \qquad (18)$$

The area-effective resistance for heat, $r_{ah\text{-}sur}^{eff}$, and vapour transport, $r_s^{eff}$, can be finally obtained as:

$$r_{ah\text{-}sur}^{eff} = \rho_a c_p \frac{\delta T_{a\text{-}sur}^{eff}}{<H>} \qquad (s\ m^{-1}) \qquad (19)$$

$$r_s^{eff} = \rho_a c_p \left(\frac{e_{sat}^{eff} - e}{\gamma <\lambda E>}\right) - r_{ah\text{-}sur}^{eff} \qquad (s\ m^{-1}) \qquad (20)$$

$$\alpha^{eff} = \frac{<\lambda E>}{<R_n> - <G_0>} \; \frac{s_a^{eff} + \gamma}{s_a^{eff}} \qquad (-) \qquad (21)$$

Note that $r_{av\text{-}sur}^{eff}$ which should have appeared in Equation 20 is replaced by $r_{ah\text{-}sur}^{eff}$ to solve $r_s^{eff}$.

## AGGREGATION RULES AND RESOLUTION EFFECTS ON AREA-REPRESENTATIVE BULK SURFACE RESISTANCE AND α-PARAMETER

Equations 5 to 21 have been applied with the SEBAL variables specified in Table 1 to obtain $r_s^{eff}$ and $\alpha^{eff}$ for 46700 NS001 pixels to match one METEOSAT pixel. This yields 12 pairs of $r_s^{eff}$ and $r_s^{METEOSAT}$ at the same spatial scale of 3.9 km by 4.1 km

(Tables 3 and 4). In already published aggregation studies (e.g. Blyth *et al.*, 1993; Shuttleworth, 1993), potential solutions on the basis of surface resistance in parallel and in series or the mean of these two approaches were used. These methods have been used with the $r_s(x,y)$ data and the results are also presented in Table 3 as $r_s^{com}$.

The aggregated values estimated from the NS001 measurements, $r_s^{eff}$, are perceived as the 'truth', which is not unlikely because the pdf of $r_s(x,y)$ could be well described with a 18.5 m sampling resolution and Equations 5 to 21 are physically sound, assuming that laws valid at microscale are also applicable at mesoscale. Obviously, the values of the parallel sum for Tomelloso lie rather close to the aggregated value, $r_s^{eff}$. Although TOM5 resulted in $r_s^{eff} = 2015$ s m$^{-1}$ and $r_s^{par.sum} = 1412$ s m$^{-1}$, Figure 3 has shown that in this range of $r_s$ (>1000 s m$^{-1}$), α and λE are hardly affected. Hence, the concept of parallel sum works out quite well for a homogeneous rainfed agricultural region with small spatial variations in λE and $r_s$. The arithmetic mean bulk surface resistance $r_s^{mean}$ systematically underestimates $r_s^{eff}$ and is therefore a less favourable solution. The option with a combined value $r_s^{com}$ shows more deviations from $r_s^{eff}$ than from $r_s^{par.sum}$, but again, since $r_s^{com}$ >1000 s m$^{-1}$, this affects the prediction of <λE> only marginally.

The parallel sum approach showed however a poor performance for the Barrax area and the values of $r_s^{par.sum}$ are substantially lower than for $r_s^{eff}$. It should be remarked that $r_s^{com}$, as suggested by Blyth *et al.* (1993), gives a strong improvement in the case of the heterogeneous Barrax area, compared to the value of $r_s^{par.sum}$. A

**Table 3** Area-representative and statistical measures of the bulk surface resistance, $r_s$, of an one-layer parameterization for evaporation. TOM refers to the Tomelloso pixels and BAR to the Barrax pixels. $r_s^{mean}$ represents the setting of all resistances in series, $r_s^{par.sum}$ in parallel and $r_s^{com}$ reflects the combined series-parallel connection.

| Site | $r_s$ METEOSAT | $r_s^{eff}$ NS001 Eq. 20 | $r_s^{min}$ NS001 | $r_s^{max}$ NS001 | $r_s^{mean}$ NS001 | $r_s^{median}$ NS001 | $r_s^{par.sum}$ NS001 | $r_s^{com}$ NS001 |
|---|---|---|---|---|---|---|---|---|
| TOM1 | – | 1225 | 0 | 2500 | 615 | 560 | 1083 | 849 |
| TOM2 | 29793 | 1244 | 0 | 2500 | 600 | 550 | 1161 | 881 |
| TOM3 | 95208 | 1493 | 0 | 2500 | 953 | 770 | 1272 | 1113 |
| TOM4 | 46130 | 1647 | 0 | 2500 | 887 | 700 | 1605 | 1246 |
| TOM5 | 2380 | 2015 | 50 | 2500 | 1034 | 830 | 1412 | 1223 |
| TOM6 | 2739 | 1630 | 110 | 2500 | 842 | 690 | 1629 | 1236 |
| BAR1 | 418 | 443 | 10 | 2500 | 676 | 380 | 224 | 450 |
| BAR2 | 464 | 256 | 0 | 2500 | 429 | 140 | 171 | 300 |
| BAR3 | 481 | 853 | 10 | 2500 | 761 | 620 | 570 | 666 |
| BAR4 | 454 | 411 | 0 | 2500 | 706 | 300 | 198 | 452 |
| BAR5 | 435 | 382 | 0 | 2500 | 340 | 290 | 193 | 267 |
| BAR6 | 453 | 682 | 0 | 2500 | 598 | 480 | 429 | 514 |

tentative conclusion is that the parallel sum is suitable for land surfaces with minor variations in $\lambda E$ while the combined parallel-series connection is suited for land with major spatial variability in $\lambda E$. The statistics of $\alpha$ are given in Table 4.

**Table 4** Area-representative and statistical measures of the Priestley-Taylor $\alpha$-parameter

| Site | $\alpha$ METEOSAT | $\alpha^{eff}$ NS001 Equation 21 | $\alpha^{min}$ NS001 | $\alpha^{max}$ NS001 | $\alpha^{mean}$ NS001 | $\alpha^{median}$ NS001 |
|---|---|---|---|---|---|---|
| TOM1 | - | 0.30 | 0 | 1.36 | 0.30 | 0.28 |
| TOM2 | 0.01 | 0.30 | 0 | 1.16 | 0.30 | 0.28 |
| TOM3 | 0.00 | 0.26 | 0 | 1.38 | 0.26 | 0.22 |
| TOM4 | 0.01 | 0.24 | 0 | 1.10 | 0.24 | 0.24 |
| TOM5 | 0.15 | 0.21 | 0 | 0.86 | 0.21 | 0.22 |
| TOM6 | 0.13 | 0.25 | 0 | 0.68 | 0.25 | 0.24 |
| BAR1 | 0.52 | 0.43 | 0 | 1.42 | 0.40 | 0.24 |
| BAR2 | 0.43 | 0.54 | 0 | 1.44 | 0.50 | 0.48 |
| BAR3 | 0.44 | 0.26 | 0 | 1.24 | 0.24 | 0.18 |
| BAR4 | 0.49 | 0.45 | 0 | 1.42 | 0.42 | 0.30 |
| BAR5 | 0.43 | 0.49 | 0 | 1.44 | 0.47 | 0.32 |
| BAR6 | 0.46 | 0.31 | 0 | 1.30 | 0.29 | 0.22 |

Although not scientifically justified, $\alpha^{mean}$ and $\alpha^{eff}$ agree well. For the situation in Tomelloso, no difference could be noted. For Barrax, the differences are not significant. Hence, the area aggregation of $\alpha$-values is considerably easier than for $r_s$ and it is recommended to validate mesoscale energy partitioning predicted by SVAT models by $\alpha$ rather than by $r_s$.

To gain knowledge on resolution effects of $r_s^{eff}$ and $\alpha^{eff}$, results from NS001 and METEOSAT were also compared. The effects of sampling resolution (18.5 vs. 4000 m) are also included in Tables 3 and 4. As a consequence of having a pixel size larger than the correlation length, the inter-pixel variations of $r_s^{METEOSAT}$ and $\alpha^{METEOSAT}$ are small. The comparison between $r_s^{eff}$ and $r_s^{METEOSAT}$ shows variable agreement. Discrepancies between $r_s^{eff}$ and $r_s^{METEOSAT}$ can be attributed to:

- non-linearity between fluxes and surface variables (surface temperature, surface hemispherical reflectance, vegetation index) in a heterogeneous landscape;
- aggregation rules imposed by Equations 5 to 21;
- the difference in size of the sub-areas allocated by NS001 and METEOSAT pixels for zero sensible and zero latent heat flux.

For comparing the resistance with mesoscale SVAT models, it is fairer to compare the bulk behaviour of all six METEOSAT pixels rather than single ones such as done in Tables 3 and 4 (see Table 5). The entire aggregation procedure for the given fluxes has for this purpose been repeated (Equations 5 to 21).

**Table 5** Resolution effects on area-representative bulk surface resistance and α-parameter for areas with a size of 96 km$^2$

| | $r_s^{eff}$ NS001 | $\alpha^{eff}$ NS001 | $r_s^{eff}$ METEOSAT | $\alpha^{eff}$ METEOSAT |
|---|---|---|---|---|
| Tomelloso | 1506 | 0.26 | 629 | 0.05 |
| Barrax | 471 | 0.42 | 450 | 0.46 |

The area coverage of 6 METEOSAT pixels equals 96 km$^2$ and $r_s^{eff}$ and $\alpha^{eff}$ thus reflect the hydrological situation of heterogeneous land surfaces. At this scale, the resistance can no longer be interpreted bio-physically and reflects a range of vegetation cover, plant species, soil types and soil water content. Table 5 gives the impression that even a small number of six METEOSAT pixels is sufficient to obtain a realistic value for $r_s^{eff}$ for heterogeneous landscapes such as Barrax. Unfortunately this conclusion cannot be supported by the findings for Tomelloso, and a larger set of METEOSAT pixels seems to be necessary.

## DAYTIME EVOLUTION OF AREA-REPRESENTATIVE SURFACE RESISTANCE AND α-PARAMETER

With a Bowen-ratio surface energy balance system, Bastiaanssen (1995) showed that the $H/R_n$ ratio is fairly constant throughout the day for wet surfaces in the Western Desert of Egypt. It was interpreted that the turbulent heat transfer into the atmosphere responds directly to surface warming which has an almost negligible time lag behind the incoming solar radiation. Hence, the daytime evolution of $\delta T_a^{sur}$ and $K_d$ is similar and this is the clue that H implies a constant fraction of $R_n$. To study the $H/R_n$ behaviour at the mesoscale, $H/R_n$ was calculated for each METEOSAT pixel on 29 June. The scaled-up version, $\langle H\rangle/\langle R_n\rangle$, shows something important: between 0900 and 1600 hours GMT; $\langle H\rangle/\langle R_n\rangle$ is also temporally stable for the entire EFEDA area (100 km×100 km) and even for the whole Iberian Peninsula (1000 km×1000 km). The $\langle H\rangle/\langle R_n\rangle$ value for Iberia is much smaller than for the EFEDA area because large parts of the Mediterranean Sea and Atlantic Ocean are included in the image. The stability of $H/R_n$ in association with $\Lambda=\lambda E/(H+\lambda E)$ explains the temporal stability of Λ noted by Shuttleworth *et al.* (1989) and Brutsaert and Sugita (1992).

Since the EFEDA grid shown in Figure 1 has a scale typical of one cell of an

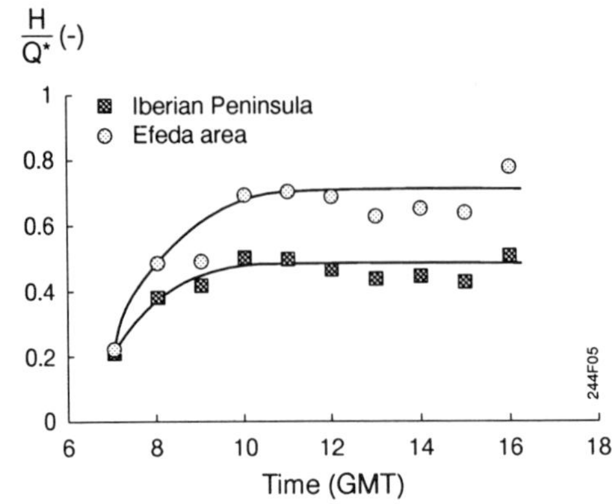


**Figure 7** Diurnal variation of the area-representative sensible heat flux/net radiation fraction $<H>/<R_n>$ for the EFEDA grid at a 100 km by 100 km scale and the Iberian Peninsula at a 1000 km by 1000 km grid on 29 June 1991.

atmospheric model, Figures 7 and 8 can be used to validate the large area heat fluxes, bulk surface resistance and Priestley-Taylor $\alpha$-parameter from heterogeneous land surfaces. Although all hourly SEBAL calculations have been calculated independently, they show a strong coherence: $r_s$ varies between 1200 and 2000 s $m^{-1}$ for the entire EFEDA grid. Comparing this range with the $\alpha$ ($r_s^{-1}$) relationship illustrated in Figure 4 shows that $\alpha$-fluctuations are limited in this range of $r_s$, which can be confirmed from Figure 8B. These values are appropriate for the summer season in a Mediterranean climate and are supported by the EFEDA field campaign. It is remarkable that the daytime $r_s$ behaviour is not comparable with the behaviour of the resistance at a patch scale. Hence, the implementation of resistances developed for the microscale in atmospheric models should be done with great care, because the value changes with scale and the daytime evolution can differ significantly.

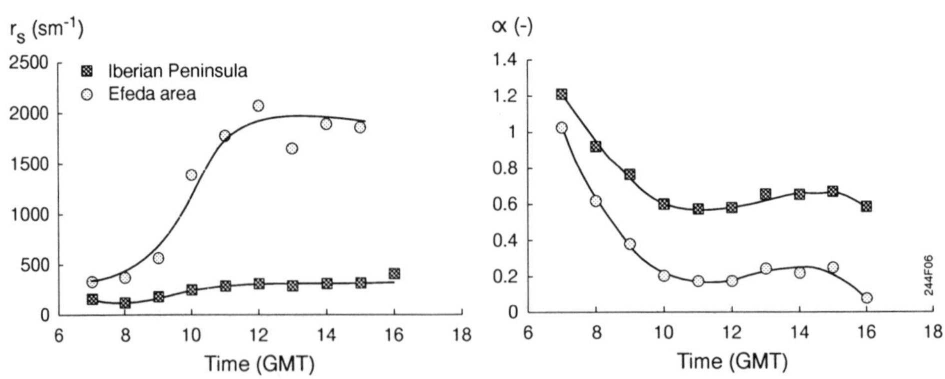


**Figure 8** Diurnal variation of the area-representative values for the EFEDA grid at a 100 km by 100 km scale and for the Iberian Peninsula at a 1000 km by 1000 km scale on 29 June 1991 for (part A) bulk surface resistance $r_s^{eff}$ and (Part B) Priestley-Taylor $\alpha$-parameter $\alpha^{eff}$.

## CONCLUSIONS

- The spatial variation of the bulk surface resistance and Priestley-Taylor α-parameter are significant;
- The relationship between α and $r_s^{-1}$ is scale invariant and the controlling range of resistance in evaporation lies between $25 < r_s < 1000$ s m$^{-1}$;
- Distributed SVAT models with pdf's for $r_s$ should be based on high resolution satellite image such as Landsat Thematic Mapper because the correlation length for energy partitioning in Central Spain is approximately 3 to 4 km;
- The concept of correlation lengths suggests that field measurements at the patch scale should not be applied to validate mesoscale SVAT models;
- With remote sensing techniques and aggregation rules, area-representative moisture indicators for vast heterogeneous land surfaces can be obtained and applied to validate mesoscale SVAT models and other types of hydrological models;
- The Priestley-Taylor α-parameter is easier to aggregate than the bulk surface resistance and suitable for estimating the surface energy partitioning at large scales;
- The error in $r_s^{eff}$ estimated by a single METEOSAT pixel is significant. The error reduces substantially if a small number of pixels, say 20, are averaged;
- The applicability of simple aggregation rules such as the combined parallel sum and series connections should be tested with more detailed descriptions of the natural variation of surface conditions in combination with PBL models. There is an indication that the combined resistance works well;
- The physical basis behind stomatally and soil physically controlled evaporation resistances decrease with increasing scale because the quantity changes with scale and the daytime behaviour is totally different;
- The $\langle H \rangle / \langle R_n \rangle$ ratio remains stable for the entire Iberian Peninsula including its coastal regions during daytime, suggesting that the evaporative fraction ($\lambda E/(\lambda E+H)$) can be explored to temporally integrate instantaneous surface fluxes;

## ACKNOWLEDGEMENT

The authors are grateful to two anonymous reviewers for providing constructive criticism on an earlier version of this paper. The help of Dr John Stewart (Institute of Hydrology) with the final editing of the mathematical equations is appreciated.

## REFERENCES

Bastiaanssen, W.G.M., Hoekman, D.H. and Roebeling, R.A. 1994. A methodology for the assessment of surface resistance and soil water storage variability at mesoscale based on remote sensing measurements, a case study with HAPEX-EFEDA data, *IAHS Special Publication* No. **2**.

Bastiaanssen, W.G.M. 1995. Regionalization of surface flux densities and moisture indicators in composite terrain (PhD thesis, *Agricultural University of Wageningen*), Report **109**, DLO-Winand Staring Centre, Wageningen, The Netherlands.

Bastiaanssen, W.G.M., Pelgrum, H., Wang, J., Ma, Y., Roerink, G.J., Moreno, J.F. and van der Wal, T. 1996. A remote sensing surface energy balance algorithm for land (SEBAL): part 2: Validation (*submitted to Rem. Sens. Environ.*).

Blyth, E.M., Dolman, A.J. and Wood, N. 1993. Effective resistance to sensible and latent heat flux in heterogeneous terrain, *Q.J.R. Met. Soc.* **119**, 423-442.

Blyth, E.M. and Harding, R.J. 1995. Application of aggregation models to surface heat flux from the Sahelian tiger bush. *Agric. For. Meteorol.* **72**, 213-235.

Bolle, H.J. *et al.* 1993. EFEDA: European field experiments in a desertification-threatened area. *Ann. Geophys.* **11**, 173-189.

Braden, H. 1995. Energy fluxes from heterogeneous terrain: averaging input parameters of the Penman-Monteith formula. *Agric. For. Meteorol.* **75**, 121-133.

Brunet, Y., Itier, B., McAneney, J. and Lagouarde, J.P. 1994. Downwind evolution of scalar fluxes and surface resistance under conditions of local advection. Part II: measurements over barley. *Agric. For. Meteorol.* **71**, 227-245.

Choudhury, B.J. and Monteith, J.L. 1988. A four layer model for the heat budget of homogeneous land surfaces. *Q.J.R. Met. Soc.* **114**, 373-398.

Griend, van de, A.A. and Owe, M. 1994. Bare soil surface resistance to evaporation by vapour diffusion under semiarid conditions. *Wat. Resour. Res.* **30**, 181-188.

Jarvis, P.G. 1976. The interpretation of the variation in leaf water potential and stomatal conductance found in canopies in the field. *Phil. Trans. R. Soc. Lond. B* **273**, 593-610.

Jochum, A.M. 1993. Evaporation and energy fluxes during EFEDA: horizontal variability and area averaging, *IAHS Publication No. 212*, 373-380.

Lhomme, J.P. 1992. Energy balance of heterogeneous terrain, averaging the controlling parameters, *Agric. For. Meteor.* **61**, 11-21.

Manabe, S. 1969. Climate of the ocean circulation, the atmospheric circulation and the hydrology of the earth's surface. *Mon. Weath. Rev.* **97**, 739-774.

McNaughton, K.G. and Spriggs, T.W. 1989. A mixed layer model for regional evaporation, *Bound.-Layer Meteorol.* **34**, 243-262.

Menenti, M., Bastiaanssen, W.G.M., van Eick, D. and Abdel Karim, M.H. 1989. Linear relationships between surface reflectance and temperature and their application to map actual evaporation of groundwater. *Advances in Space Research* **9**, 165-176.

Monteith, J.L. 1965. Evaporation and the environment. In: The state and movement of water in living organisms. *XIXth Symp., Soc. for Exp. Biol., Swansea*, Cambridge University Press, 205-234.

Monteith, J.L. 1994. Fifty years of potential evaporation. *Proc. Conf. Trinity College Dublin*, 29-45.

Moran, M.S., Clark, T.H., Inoue, Y. and Vidal, A. 1994. Estimating crop water deficit using the relation between surface-air temperature and spectral vegetation index. *Rem. Sens. Environ.* **49**, 246-263.

Ogink-Hendriks, M.J. 1995. Modelling surface conductance and transpiration of an oak forest in the Netherlands, *Agric. For. Meteorol.* **74**, 99-118.

Priestley, C.H.B. and Taylor, R.J. 1972. On the assessment of surface heat flux and evapotranspiration using large scale parameters. *Mon. Weath. Rev.* **100**, 81-92.

Raupach, M.R. 1995. Vegetation-atmosphere interaction and surface conductances at leaf, canopy and regional scales. *Agric. For. Meteorol.* **73**, 151-179.

Seguin, B. and Itier, B. 1983. Using midday surface temperature to estimate daily evaporation from satellite thermal IR data. *Int. J. Rem. Sens.* **4**, 371-383

Shuttleworth, W.J. 1988. Macrohydrology, the new challenge for process hydrology. *J. Hydrol.* **100**, 31-56.

Shuttleworth, W.J., Gurney, R.J., Hsu, A.Y. and Ormsby, J.P. 1989. FIFE: The variation in energy partitioning of surface flux sites. *IAHS Publ. No. 186*, 67-74.

Shuttleworth, W.J. 1993. The soil-vegetation-atmosphere interface. (Eds.) E. Raschke & D. Jacob, *NATO-ASI series*, **15**. Springer Verlag, 323-364.

Stewart, J.B. 1988. Modelling surface conductance of pine forest, *Agric. For. Meteorol.* **43**, 19-35.

Stewart, J.B and Gay, L.W. 1989. Preliminary modelling of transpiration from the FIFE site in Kansas, *Agric. For. Meteorol.* **48**, 305-315.

Wallace, J.S., Lloyd, C.R. and Sivakumar, M.V.K. 1993. Measurements of soil, plant and total evaporation from millet in Niger. *Agric. For. Meteorol.* **63**, 149-170.

Wallace, J.S. 1995. Calculating evaporation: resistance to factors. *Agric. For. Meteorol.* **73**, 353-366.

# 7 Spatial structure and scaling of surface fluxes in a Great Basin ecosystem

**L.E. HIPPS, D. OR and C.M.U. NEALE**
*Utah State University, Logan, USA*

## INTRODUCTION

Arid and semi-arid lands are a large feature of the planet and are inherently inhomogeneous at a range of scales. As a result, they present special difficulties for studying surface-atmosphere interactions and their dependence upon scale. The nature of the spatial variability of such surfaces suggests the utility of incorporating remote sensing information where possible. Recent studies have explored several approaches to estimate surface fluxes which used remote sensing estimates of critical surface variables such as surface temperature, vegetation indices, and available energy. For semi-arid regions these studies include Kustas *et al.* (1994) and Bastiaanssen (1995).

The Great Basin region in North America is a very extensive region of semi-arid lands. The unique climate of cold winter temperatures, dominance of precipitation as snow, and dry summers, combine with the closed-basin hydrology result in shrub-dominated vegetation distinct from warm desert regions (Caldwell, 1985). The development of approaches to characterize the surface fluxes of heat and water and their spatial distribution is necessary to understand the hydrology, ecology, and climate of the region.

*Unique nature of the surface*

These basin surfaces consist of two components: plants that are irregularly distributed on a usually-dry soil background. A very large difference will exist between the surface conductance of dry soil and plants. Much greater in fact than variations among plants. Similarly, values of net radiation will be large for green plants as compared to those for bare soil because of the lower albedo and temperatures of the plant foliage. The large disparity in these properties define a nearly-binary system comprised of green patches of high conductance and net radiation, overlying a background of low conductance and net radiation. The large convective heat fluxes typical of these lands ensure a strong coupling between surface and atmosphere, thereby minimizing horizontal gradients of air temperatures and humidities.

*Scaling up in hydrology using remote sensing.* Edited by J.B. Stewart, E.T. Engman, R.A. Feddes and Y. Kerr

Our studies (Hipps *et al.*, 1995; Neale *et al.*, 1995) indicate that vegetation in this ecosystem responds very rapidly to changes in soil water availability with commensurate changes in green foliage. Because of the great disparity in radiance properties between shrubs and the underlying soil, and the close coupling between water use and green foliage area, the ability to estimate the changes in the water vapour fluxes from remote sensing data seems a reasonable prospect. If the horizontal distribution of vegetation can be detected and quantified, this system provides opportunities to develop flux aggregation approaches. The spatial scales of vegetation suggest that high spatial resolution remote sensing data will be required.

## OBJECTIVES

Here we outline a proposed approach combining high spatial resolution remote sensing data and surface flux information, to study fluxes at various scales and their governing properties on a Great Basin landscape. Specifically the goals are:

- Examine the ability of high resolution remote sensing information to quantify the cover and spatial distribution of surface vegetation.
- Propose a procedure to combine high resolution remote sensing information with the unique nature of this surface to derive scale-dependent effective parameters for areal average surface fluxes.
- Develop a small perturbation analysis to relate spatial variations of evaporation to those of the governing variables for the landscape.

## THEORETICAL CONSIDERATIONS

*Aggregation and up-scaling*

Lhomme *et al.* (1994) identified two approaches to modeling surface fluxes over large scale heterogeneous terrain: (i) formulate the physics of the process at the scale of interest; or (ii) scaling up a "small-scale-physics" model. For many practical applications the latter approach appears to provide reasonable approximations. However, it should be recognized that due to the nonlinear nature of surface interactions, the aggregated values of the controlling surface parameters are not their simple areal averages. The prevalence of a strongly convective lower atmosphere homogenizes the spatial distribution of saturation deficit, rendering the transpiration primarily a function of the ratio of saturation deficit to stomatal resistance. Thus, spatial variations in water vapour exchanges are likely to be strongly related to the spatial distribution of plants.

*Effective parameters of Penman-Monteith equation for heterogeneous surface*

Consider that a heterogeneous landscape may be represented as a collection of n surface patches, of which the patch i occupies an area fraction $a_i$ , has available energy $R_{ni}$ - $G_i$ , and aerodynamic and surface resistances $r_{ai}$ and $r_{ci}$, respectively. $R_n$ is net radiation and G is soil heat flux. Raupach (1991) argues that because the atmospheric boundary layer (ABL) in such regions is usually convective and well-mixed, and does not respond to individual surface patches, all patches experience a similar saturation deficit (D). The most influential surface parameters governing the variation in water fluxes over many areas in the Great Basin are therefore: the area, available energy (Rn -G) , $r_a$ , and $r_c$ for each patch.

Lhomme *et al.*, (1994) presented a procedure for estimating the up-scaled or effective parameters. The method will be illustrated for surface parameters governing water fluxes according to Penman-Monteith equation. In general, the aggregation and preservation rules for surface fluxes may be expressed as:

$$<\phi> = \sum_i a_i \phi_i \; ; \quad \phi = H, LE, (R_n - G) \tag{1}$$

where H and LE are the sensible and latent heat fluxes. Aggregation of fluxes based on summation of local fluxes should equal the flux computed using the (yet undetermined) effective parameters at coarser scales. The flux can be expressed in terms of the appropriate equation containing the surface properties which govern the process. This is expressed in general terms as:

$$\phi = \sum_{k=1}^{m} Y_k(x_j) \; ; \quad x_j = [\alpha, r_a, r_c, G, T_s] \tag{2}$$

where the flux is a summation of *m* terms ($Y_k$), each a function of surface properties $x_j$. Here $\alpha$ is surface albedo. Preservation of the quantity $\phi$ requires that:

$$<\phi> = \sum_i a_i \phi_i = \sum_i a_i \sum_{k=1}^{m} Y_k(x_j^{\,i}) = \sum_{k=1}^{m} \sum_i a_i Y_k(x_j^{\,i}) \tag{3}$$

The same quantity in (3) written in terms of effective parameters is:

$$<\phi> = \sum_{k=1}^{m} Y_k(<x_j>) \tag{4}$$

Matching the right-hand side of Equations 3 and 4 yields a practical definition for effective parameters. Note that in this scheme the spatial arrangement of the patches was completely ignored.

*Small perturbation analysis of Penman-Monteith equation*

The effective parameter approach ignores the spatial distribution of patches within an area of interest. Statistical theories for spatial averaging in heterogeneous systems require spatial covariances of the properties of interest to be defined (Journel and Huijbregts, 1978). Estimation of flux spatial covariances is likely to be prohibitively expensive due to the extensive and dense measurement network required. As an alternative, we propose a small-perturbation analysis of relatively simple-to-define surface parameters which govern the surface flux. The idea is to use a physically-based model such as the Penman-Monteith equation for water flux to determine how small perturbations of estimable surface parameters induce spatial perturbation in surface water flux. These perturbations will later be used to estimate (to a first order) the spatial structure of surface water flux in an area.

The governing equation in this analysis is the Penman-Monteith equation for evaporation (E):

$$\lambda E = \frac{s(R_n - G) + \rho c_p \dfrac{(q_s - q_a)}{r_a}}{s + \gamma\left(1 + \dfrac{r_c}{r_a}\right)} \tag{5}$$

where $q_a$ and $q_s$ are the specific humidity of air and its saturated value, respectively, $s = dq_s/dT$, $\rho$ is air density, $c_p$ is specific heat capacity of air, $\gamma$ is defined as $c_p/L$, where L is latent heat of vaporisation. Our first assumption is that the main source of randomness in E is due to spatial variations in A=(Rn-G) and in $r_c$ (Raupach, 1991). The local values of these random space functions are comprised of a (local) mean value and a random fluctuation:

$$A = m_A + A' \quad ; \quad \langle A \rangle = m_A \quad ; \quad \langle A' \rangle = 0$$
$$r_c = m_{rc} + r_c' \quad ; \quad \langle r_c \rangle = m_{rc} \quad ; \quad \langle r_c' \rangle = 0 \tag{6}$$

where angle brackets denote the expected value. The two-point covariances and cross-covariances of the parameters are specified as products of the respective fluctuations:

$$C_A(x,x') = \langle A'(x) A'(x') \rangle$$
$$C_{r_c}(x,x') = \langle r_c'(x) r_c'(x') \rangle \tag{7}$$
$$C_{Ar_c}(x,x') = \langle A'(x) r_c'(x') \rangle$$

and **are assumed stationary**, i.e. they are functions of the separation distances $h = |x - x'|$ only, where x=(x,y) is the spatial coordinate in the plane.

The first step in the perturbation analysis is to substitute the parameters A and $r_c$ (Equation 6) into Equation (5). Next, the Penman-Monteith equation is expanded into a Taylor series about the random fluctuations (A' and $r'_c$). When variations in A and $r_c$ are relatively small, first order analysis is sufficient, i.e. E may be expressed by retaining terms up to first order only:

$$E_1 = \frac{s\langle A\rangle + \rho c_p \dfrac{(q_s - q_a)}{r_a}}{s + \gamma\left(1 + \dfrac{\langle r_c\rangle}{r_a}\right)} + A'M + r_c'N \tag{8}$$

with

$$M = \frac{s}{\gamma\left(1 + \dfrac{\langle r_c\rangle}{r_a}\right) + s} \tag{9}$$

and

$$N = \frac{-\gamma\left(\dfrac{\rho c_p D}{r_a} + \langle A\rangle s\right)}{r_a\left[\gamma\left(1 + \dfrac{\langle r_c\rangle}{r_a}\right) + s\right]^2} \tag{10}$$

Where D is defined as $q_s$-$q_a$. Taking the expectation of $E_1$ in Equation 8 yields the expected value <E> as the first term on the right-hand side in Equation 8. The random fluctuation in E, E', is expressed as a weighted sum of fluctuations in A and $r_c$:

$$E' = A'M - r_c'N \tag{11}$$

the spatial covariance of E evaluated at x and x' is simply the expected value of the product E'(x)E'(x'):

$$\begin{aligned} C_E(x,x') &= M(x)M(x')C_A(x,x') + N(x)N(x')C_{r_c}(x,x') \\ &\quad - C_{Ar_c}(x,x')[M(x)N(x') + M(x')N(x)] \end{aligned} \tag{12}$$

For x=x' $C_E(0)$ reduces to the point variance of E, $\sigma^2_E$:

$$\sigma_E^2 = M^2\sigma_A^2 + N^2\sigma_{r_c}^2 - 2\langle MN\, C^2_{Ar_c}(0) \tag{13}$$

To enhance the accuracy of the estimates a second order expansion of E, $E_2$, is given as:

$$E_2 = E_1 - \frac{r_c' A' M^2 \gamma}{s\, r_a} - \frac{(r_c')^2 N M \gamma}{s\, r_a} \tag{14}$$

However, no major changes in the spatial covariance of E are expected, since the fluctuations are assumed to be small which limits the contribution of second order terms (in the numerators) to E.

## METHODS

The field studies were conducted in Goshute Valley located about 20 miles west of Wendover, Nevada. This valley varies in width between 15 and 20 miles, and is nearly 60 miles long, at an elevation of 1750 m above mean sea level.

*Surface flux measurements*

A set of six Bowen ratio stations were sited at various positions in the valley. At each station the air temperature and humidities were measured at two heights, with measurements of net radiation and soil heat flux. These data allowed the calculation of sensible and latent heat fluxes. The stations were run during the entire growing season of 1993 and 1994. Soil moisture was monitored throughout the active season at each site using a combination of Time Domain Reflectometry and neutron probe techniques.

Each summer, several intensive field studies were conducted during four- to seven-day periods. During these periods additional sensible heat and water vapour flux measurements were made at four locations using eddy correlation stations. These stations each consisted of a one-dimensional sonic anemometer and a Krypton hygrometer, which were sampled at 10 Hz to yield the eddy covariances.

*Remote sensing data and analyses*

The Utah State University aircraft flew over regions corresponding to each of the surface flux stations, collecting radiometric datasets with the USU airborne video/radiometer system (Neale, 1992; Neale and Crowther, 1994). This system provided high resolution digital narrow band multispectral imagery centred in the green (0.55 μm ), red (0.65 μm) and near-infrared (0.85 μm). The visible bandwidths were 10 nm , while the near-ir bandwidth was 20 nm. An Exotech radiometer with Thematic Mapper bands TM1 (0.45 - 0.52 μm), TM2 (0.52 - 0.60 μm), TM3 (0.63 - 0.69 μm) and TM4 (0.76 - 0.9 μm) was mounted alongside the cameras and aligned to view a footprint of 20 x 20 pixels in the centre of the video imagery. The correction for irradiance non-uniformities and radiometric calibration of the video imagery using the Exotech radiometer is described by Crowther (1992) and Neale and Crowther (1994). Correction for bi-directional effects in the imagery are described by Qi *et al.* (1995). In addition to the shortwave video imagery, an Inframetrics 760 thermal infrared scanner was used to obtain thermal imagery in the 10 - 12.5 μm band.

The resulting remote sensing data set consists of high resolution multispectral digital video imagery of six different sites in Goshute Valley with pixel sizes varying from 0.17 m to 1.5 metres in the shortwave bands and 0.35 m to 3.0 metres in the thermal band.

## RESULTS

*Discrimination of vegetation and soil*

The ability of the high spatial resolution remote sensing data to quantify the spatial structure of vegetation has been verified. Figure 1 exhibits composite representations of three wavebands corresponding to green, red, and near infrared for two sites in the valley. The spatial resolution of the images is 0.17 m. There are three points illustrated here:

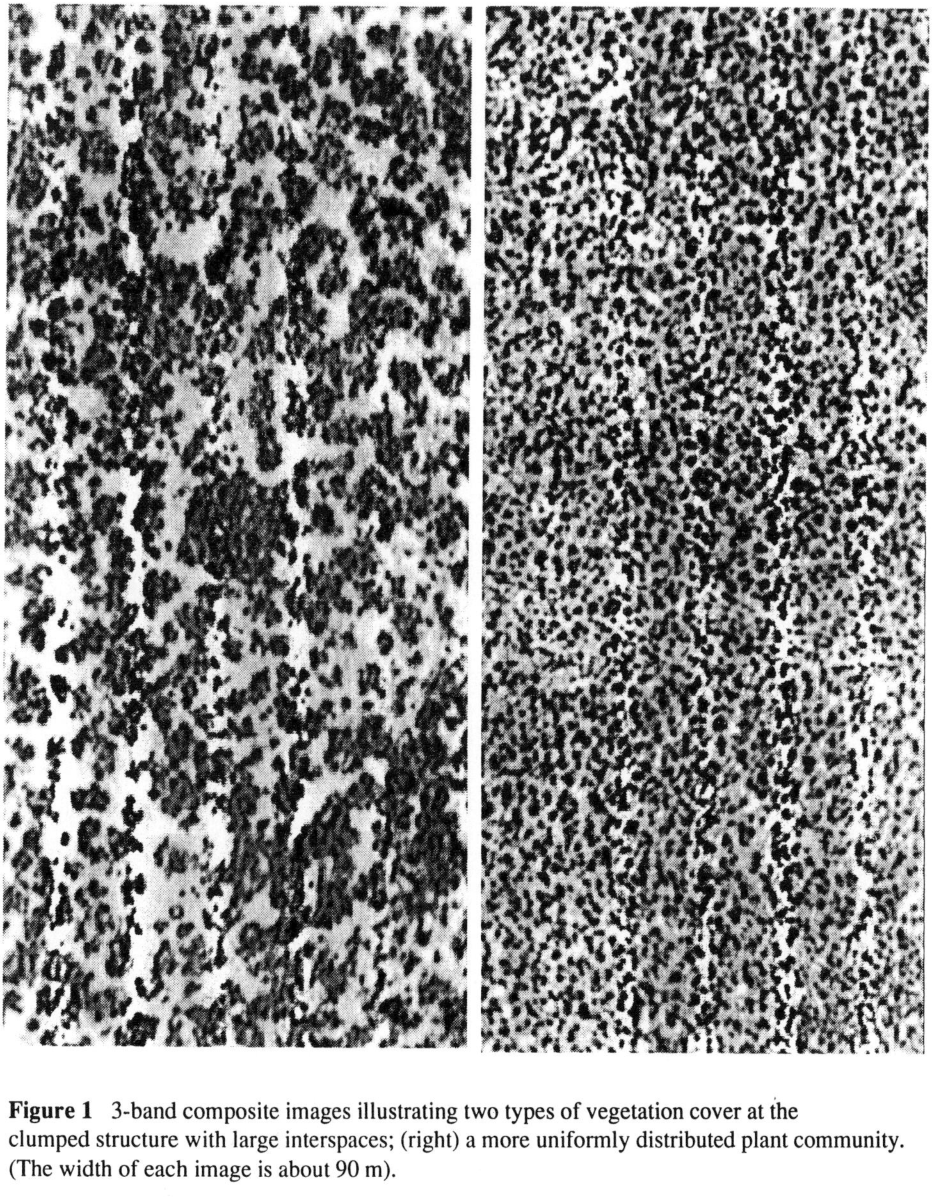

**Figure 1** 3-band composite images illustrating two types of vegetation cover at the clumped structure with large interspaces; (right) a more uniformly distributed plant community. (The width of each image is about 90 m).

- The binary nature (plant versus bare soil) of the system
- Changes in the structure of heterogeneity from location to location
- The very small scales of heterogeneity present on the landscape

By examining actual values of radiance for each of the wavebands in each pixel of the surface, it was possible to classify each pixel as either vegetation, dark aeolian deposited soil, saline soil in interspace, or playa, using maximum likelihood criteria. This was possible because of the large difference in radiance between the components.

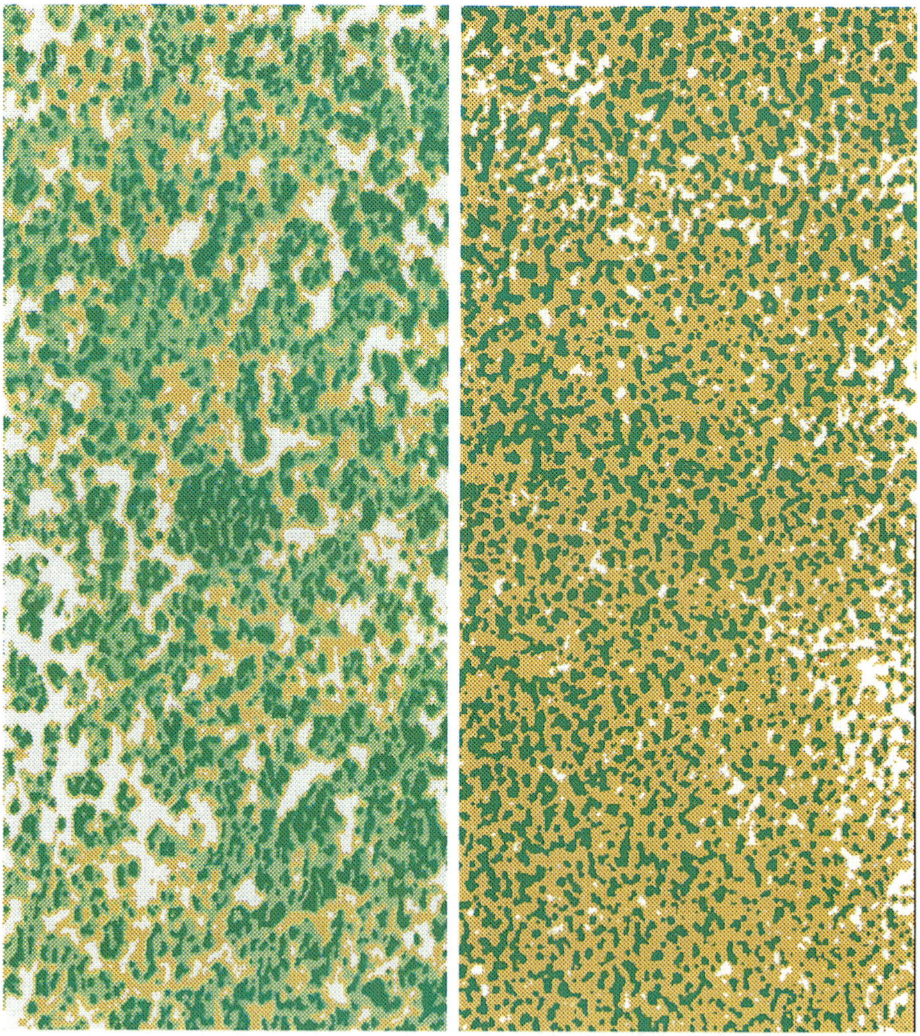

**Figure 2** False colour classification of the sites in Fig. 1 based on radiance properties. Plants are assigned dark-green, other colors are light green – dark aeolian deposited soil, yellow – saline soil in interspace, white – playa.

Figure 2 is a false colour map of the resultant classification. It demonstrates the ability to construct a two-dimensional map of plants and soil on the surface, which is critical to the stated objectives. The resulting estimate of fractional area of plant cover for site (a) is 24.5%, and for site (b) is 32.6%. These are in good agreement with ground-based estimates made by measuring foliage contacts with a vertical probe at equally spaced locations along several long transects.

*Spatial scales of vegetation*

The contribution of various spatial scales to the distribution of plants on the surface was quantified by calculating spatial power spectra for the radiance values at the surface. Figures 3 and 4 exhibit one-dimensional power spectra for the red waveband at the same two locations considered in Figure 1. The red band was used for convenience, as it was more sensitive to the presence of vegetation than the other individual bands. Two different pixel sizes at each location were used in the analyses. These spectra contain information about the spatial distribution of vegetation. Site BR1 shows a peak at about 4.0 m for the high resolution data, shifting to about 8.3 m at the lower resolution data. On the other hand, the BR3 location exhibits a peak at about 10 m for the high resolution pixels, shifting to about

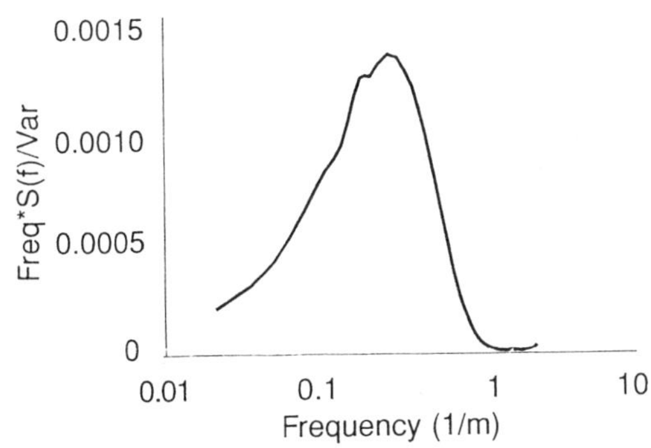


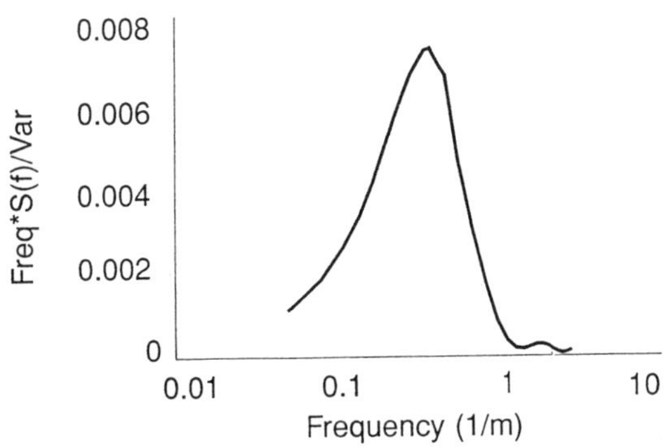


**Figure 3** Spatial power spectra of red radiances for site BR1, corresponding to the left side of Figures 1 and 2. Top is highest resolution of 0.17 m and bottom is lower resolution data of 0.80 m.

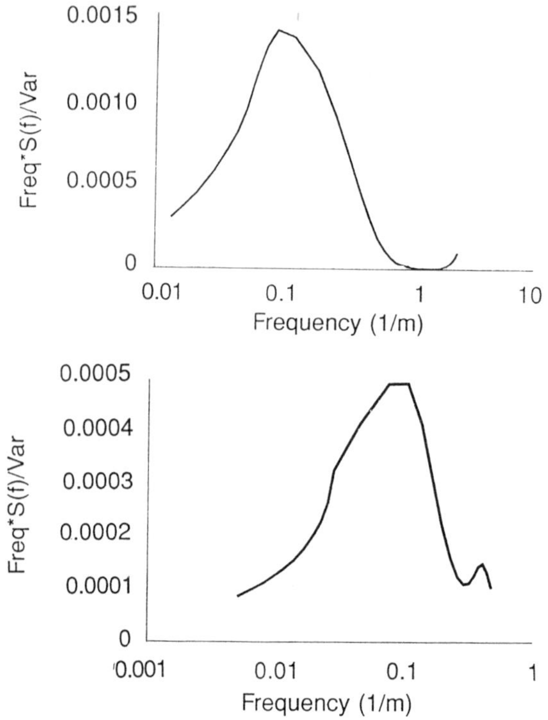


**Figure 4** Spatial power spectra of red radiances for site BR3, corresponding to the right side of Figures 1 and 2. Top is highest resolution of 0.17 m, and bottom is lower resolution data of 1.0 m.

12 m for the more coarse pixels. The shape of the spectrum has also changed for the coarse pixels. The higher frequencies associated with the peak observed in BR3 is related to the clumpiness of that surface, as is evident in Figure 1. These results tell us several things:

- The distribution of plants in a particular location on the landscape is dominated by a well-defined spatial scale.
- The peak and shape of the spatial power spectra are dependent upon the spatial resolution of the surface data.
- The main scales of variability are considerably smaller than those perceived by typical remote sensing platforms.

These results re-enforce those reported by Hipps *et al.* (1995) who also noted that the characteristics of the calculated spatial structure of the surface are dependent upon the spatial resolution with which the surface is observed. Analyses such as those shown in Figures 3 and 4 are important in that they tell us at which scales the spatial structure of vegetation is largely contained for any site. Such scales of variation of the vegetation will greatly affect the spatial structure of the governing variables and surface fluxes. Hence, the above results will guide us in the choice of spatial scales to perform analyses of effective parameters and perturbation.

*Constructing spatial fields of available energy*

Recall that the remote sensing data yields spatial distributions of surface temperature and reflectance in red, green, and near infrared bands. These were combined with other information to estimate net radiation on a pixel by pixel basis as described by Abdalla *et al.* (1996). Briefly, this involved:

- estimating downward and upward solar radiation from radiance values in the discrete wavebands, using relations cited in the literature;
- determining incident longwave from the sky from published empirical expressions;
- combining surface radiometric temperatures with measured emissivities for plants and soil to yield infrared radiation emitted from the surface.

The resulting estimates of net radiation were compared with measured values at the surface stations. The agreement was very good, with a RMSE of 7 W $m^{-2}$, which translated into a mean per cent difference of 1.2 %. So these data can be used to accurately map the 2-D field of net radiation.

Soil heat flux can be estimated from net radiation and remotely-sensed data, as shown by Kustas and Daughtry (1990) and Bastiaanssen (1995). These approaches will be investigated and compared with the measurements available at each surface station.

*Proposed approach to deriving effective parameters*

We propose to estimate the fractional area of the various patches (i.e. $a_i$ in Equation 1) from remotely-sensed imagery. The problem is greatly simplified by the fact that the number of different patch-types is likely to be small in the nearly-binary system of this landscape. For pixels identified as shrubs, a value of surface conductance or $r_c$ can be estimated from the literature. Values of surface conductance of the soil can be determined from the near-surface soil moisture measurements. Note that the soil and plant conductance values will be different. Saturation deficit values are known from the surface flux stations and varied little in space due to the convective ABL.

The above information allows a calculation of the effective parameters of A and $r_c$ for each defined patch by solving Equation 1. These can be combined with the saturation deficit values to calculate areal average values for evaporation. These would be the remote sensing-based fluxes using effective parameters. To evaluate these, they can be checked against the values at the surface flux station in the same patch.

*Procedure for small perturbation analysis*

As described above, the two-dimensional distributions of $R_n$ and $r_c$ can be determined from the remotely-sensed data. This allows calculation of the covariance functions for these variables. The covariance of E in space can be calculated as defined in Equation 12. This function yields the expected value of the evaporation flux in space at various scales.

Summarizing, the proposed scheme will be implemented as follows:

- estimate the spatial distributions of plants to determine the statistical moments of the parameters;

- Use the statistical moments of the parameters (A and $r_c$) to estimate the moments of E, i.e. $<E>$ and $C_E(x,x')$;

- apply the so called "regularization" schemes of Journal and Huijbregts (1978) to estimate the magnitude of the fluxes **and their error-bounds** at coarser scales.

## CONCLUSIONS

Semi-arid natural ecosystems such as the Great Basin can be considered as a two component system of plants distributed in a complex fashion over a dry soil background. The large disparity in the radiative characteristics between soil and shrubs allows the two-dimensional distribution of vegetation to be quantified using remotely-sensed data. In addition, the large difference in net radiation and surface conductance between plants and soil defines a binary system amenable to several flux aggregation schemes. In order to aggregate from the basic unit (individual shrub), remote sensing information must be available at commensurate scales, i.e. high spatial resolution. In addition, information at such a fundamental scale is essential for reconciling ground-based flux measurements to appropriate surface footprints of remote sensing data.

High spatial resolution remote sensing images collected over the Great Basin site in Nevada allowed spatial power spectra to be calculated. These demonstrated that at local scales the spatial distribution of plants varied from site to site but were concentrated at scales of a few metres.

The proven ability to discriminate the spatial distribution of plants and net radiation suggests that proposed flux aggregation schemes are feasible. The two analyses proposed (effective parameterisation and small perturbation) can use the remote sensing imagery at various scales (Figure 2), coupled with supplemental surface properties and environmental inputs, to aggregate the surface fluxes. Pending reliable surface flux footprint reconstruction (using the same imagery), limited testing of such aggregation schemes could be conducted using the local surface flux measurements.

# REFERENCES

Abdalla, S.H., Neale, C.M.U., Malek, E., Hipps, L.E., and Bingham, G.E. 1996. Estimation of net radiation of sparse vegetation using multispectral video imagery. Proceedings of Hydrology Days Meetings, Fort Collins, CO. American Geophysical Union.

Bastiaanssen, W.G.M. 1995. Regionalization of Surface Flux Densities and Moisture Indicators in Composite Terrain. Ph.D. Thesis Landouwuniversiteit Wageningen, The Netherlands.

Caldwell, M. 1985. Cold Desert. In: Chabot, B.F. and H.A. Mooney (eds), *Physiological Ecology of North American Plant Communities*. Chapman and Hall, New York.

Crowther, B.G. 1992. Radiometric calibration of multispectral video imagery. Unpublished M.S. Thesis. Department of Biological and Irrigation Engineering, Utah State University, Logan, UT 84322-4105.

Hipps, L.E., Or, D., Malek, E. and Neale, C.M.U. 1995. Spatial distribution of latent heat flux over a sparsely vegetated arid region. *Abstracts of American Geophysical Union*, Spring 1995, Baltimore, MD. P. S104.

Journel, A.G. and Huijbregts, Ch. J. 1978. *Mining Geostatistics.* Academic Press, London.

Kustas, W.P. and Daughtry, C.S.T. 1990. Estimation of the soil heat flux/net radiation ratio from spectral data. *Agric. For. Meteorol.*, **49**, 205-223.

Kustas, W.P., Moran, M.S., Humes, K.S., Stannard, D.I., Pinter, P.J., Hipps, L.E., Swiatek, E. and Goodrich, D.C. 1994. Surface energy balance estimates at local and regional scales using optical remote sensing from an aircraft platform and atmospheric data collected over semiarid rangelands. *Wat. Resour. Research*, **30**, 1241-1259.

L'homme, J.P., Chehbouni, A. and Monteny, B.A. 1994. Effective parameters of surface energy balance in heterogeneous landscapes. *Bound.-Layer Meteorol.*, **71**, 297-309.

Neale, C.M.U. 1992. An airborne multispectral video/radiometer remote sensing system for natural resource monitoring. Proceedings of the 13th Biennial Workshop on Color Aerial Photography and Videography in Plant Sciences. ASPRS, Bethesda, MD.

Neale, C.M.U. and Crowther, B.G. 1994. An airborne multispectral video/radiometer remote sensing system: development and calibration. *Remote Sensing of Environment*, 49(3): 187-194.

Neale, C.M.U., Artan, G.A., Hipps, L.E., Or, D., Bingham, G.E. and Abdalla, S. 1995. A study of the effects of scale and pixel resolution: Implications on the use of remotely sensed imagery for energy balance studies. *Abstracts of American Geophysical Union*, Spring 1995, Baltimore, MD. P. S104.

Qi, J., Mitchell, T., Moran, M.S., Pinter Jr., P. J., Clarke, T., Neale, C.M.U., Sundararaman, S. and Ahmed, R. H. 1995. Normalization of bidirectional effect in videographic imagery. Proceedings of the 15th Biennial Workshop in Aerial Photography and Videography for Resource Monitoring. ASPRS, Bethesda, MD.

Raupach, M.R. 1991. Vegetation-atmosphere interaction in homogeneous and heterogeneous terrain: some implications of mixed-layer dynamics. *Vegetatio*, **91**, 105-120.

# 8 Sensible heat flux from remotely-sensed data at different resolutions

**W.P. KUSTAS[1] and K.S. HUMES[2]**
[1]*USDA-ARS Hydrology Lab, Beltsville, USA*
[2]*University of Oklahoma, Norman, USA*

## INTRODUCTION

Over the last several years a major emphasis in hydrology and meteorology has been in the development of methods for determining effective parameters for energy flux calculations over heterogeneous surfaces. A recent overview of the scale issues for surface energy balances in heterogeneous terrain is given by Raupach and Finnigan (1995). In that article they follow the work of Raupach (1995) and McNaughton (1994) by adopting the Penman-Monteith model for computing the surface energy fluxes at various scales. Raupach and Finnigan (1995) argue that by adopting the Penman-Monteith or combination equation, they avoid the requirement of having estimates of the elemental concentrations of the surface required by bulk transfer schemes. They consider this to be a major disadvantage when trying to scale up the fluxes.

However, it can also be argued that by using the Penman-Monteith equation, one loses a direct link to the surface temperature which is a key boundary condition for many land surface-atmosphere models, and a variable that can be measured by satellite remote sensing on a global basis. Indeed, considerable progress has been made in accounting for atmospheric effects and variation in surface emissivity so that the uncertainty in radiometric surface temperature measurements from satellites is within 1 to 2 degrees (Norman *et al.*, 1995a). A problem much less tractable is how to account for the many factors (e.g. fractional vegetation cover, vegetation structure and geometry, radiometer view and solar zenith angle) that influence the radiometric surface temperature but are not necessarily related to the aerodynamic temperature or the temperature which satisfies the surface energy balance (Prevot *et al.*, 1994). Nevertheless, progress is being made on this front using a fairly simple dual-source model that requires few additional parameters compared to single-source or bulk transport models (Norman *et al.*, 1995b). Others have studied the resulting effect of using radiometric surface temperature at different pixel resolutions on the computed surface fluxes (e.g., Hall *et al.*, 1992). However, in Hall *et al.* (1992) and in most other cases a constant surface resistance is assumed over the

*Scaling up in hydrology using remote sensing*. Edited by J.B. Stewart, E.T. Engman, R.A. Feddes and Y. Kerr

entire domain. In this study, the variation in surface resistances of the elements at the 120 m and 1000 m resolution is considered by accounting for spatial variations in the surface roughness and fractional vegetation cover on a pixel by pixel basis. As a first approximation, the radiometric surface temperature is linearly averaged when changing pixel resolutions (Lhomme, 1992), although McNaughton (1994) indicated that this aggregation scheme does not conserve the energy balance equation when upscaling. Lhomme *et al.* (1994) subsequently redefined the aggregated surface temperature to be weighted by the surface conductance of each elemental surface. We do not weight the elemental temperatures by their corresponding conductances as we aggregate this variable in order to assess the degree to which this results in the scheme violating the flux-matching requirement.

## METHODOLOGY

### THE SURFACE ENERGY BALANCE FOR EACH ELEMENTAL AREA

The surface energy balance was evaluated for each elemental area i, (defined by the pixel size of the remotely sensed data), using the dual-source model developed and tested by Norman *et al.* (1995b). A main assumption in the model, which simplifies the parameterisation, is that the soil surface and vegetative canopy fluxes can be taken in parallel (Chebhouni *et al.*, 1994). Norman *et al.* (1995b) also derived a set of equations using the more complex series resistance network (e.g., Shuttleworth and Gurney, 1990), but found the differences in model output between the parallel and series resistance network approach were usually insignificant. With the parallel resistance network, the sensible heat flux from the soil, $H_s$, and the canopy, $H_c$, can be related to the total flux by the following equation:

$$H = H_C + H_S = \rho c_P \, [(\Theta_C - \Theta_A)/r_{AH} + (\Theta_S - \Theta_A)/(r_{AH} + r_S)] \tag{1}$$

where $r_s$ is the resistance to heat flow in the boundary layer immediately above the soil surface, $r_{ah}$ is the resistance to heat transfer from the vegetated layer, $\Theta_a$ is the potential temperature at a reference height in the atmosphere, and $\Theta_s$ and $\Theta_c$ are the soil and canopy potential temperatures, respectively.

The expression for $r_{ah}$ is estimated by:

$$r_{ah} = \{\ln[(z - d_o)/z_{om}] - \Psi_m\}\{\ln[(z - d_o)/z_{oh}] - \Psi_h\}/(k^2 u) \tag{2}$$

where $z_{om}$ is the momentum roughness length, $z_{oh}$ is the roughness length for heat, $d_o$ is the displacement height, $\Psi_m$ and $\Psi_h$ are the stability corrections for momentum and heat using formulations developed by Brutsaert (1992) and z is the level where air temperature, $T_a$, and wind speed, u, are measured. The value of $z_{oh}$ was estimated from the ratio $z_{om}/z_{oh} = 7$ (Garrett and Hicks, 1973) and $r_s$ is computed from the equation developed by Sauer *et al.* (1995):

$$r_s = 1/[a + b\,u_s] \tag{3}$$

where $u_s$ is the windspeed (m $s^{-1}$) about 5 cm above the soil surface, a = 0.004 m $s^{-1}$ and b = 0.012. The value of $u_s$ was estimated with equations of Goudriaan (1977).

Since we normally have a radiometric temperature of the composite surface, $T_r$, two assumptions are used to obtain a solution with Equation 1. The first assumption is that $T_r$ is related to canopy and soil surface temperatures, $T_c$ and $T_s$, by the following expression:

$$T_r = [f_\theta\, T_c^{\,4} + (1 - f_\theta)T_s^{\,4}]^{1/4} \tag{4}$$

where $f_\Theta$ is the fraction of vegetation viewed by the radiometer at zenith view angle $\theta_z$. The parameter $f_\theta$ is related to the Leaf Area Index (LAI) and the zenith view angle $\theta_z$ of the radiometer assuming a random canopy with a spherical leaf angle distribution (see Equation (2) of Norman *et al.*, 1995b).

The second assumption is that the absorbed net radiation by the plant canopy, $R_{n,c}$, is partitioned between $H_c$ and $LE_c$ according to the Priestley-Taylor approximation (Priestley and Taylor, 1972). The justification for this approximation is supported by the findings of Ball *et al.* (1986), which relates plant assimilation to stomatal conductance. In essence, the Ball *et al. 1986* study showed that for ample soil moisture availability, the effects on transpiration from increasing vapour pressure deficit and decreasing stomatal conductance are small because the two nearly balance one another for the range in daytime atmospheric humidities typically observed (i.e. 25 to 75%). With the Priestley-Taylor approximation, the sensible heat flux for the canopy is:

$$H_c = \rho c_p\, (\Theta_c - \Theta_a)/r_{ah} = R_{n,c}[1\text{-}1.3\, f_g\, \Delta/(\gamma + \Delta)] \tag{5}$$

where D is the slope of the saturation vapour pressure versus temperature curve, g is the psychrometric constant, the constant 1.3 is the Priestley-Taylor coefficient and $f_g$ is the fraction of green vegetation. The value of $R_{n,c}$ is computed using LAI and the assumption that the extinction of $R_n$ inside the canopy layer can be approximated using Beer's Law (Ross, 1981), namely,

$$R_{n,c} = R_n\,(1 - \exp(-0.45\,\mathrm{LAI})) \tag{6}$$

where the constant 0.45 is the value of the extinction coefficient midway over the observed range (i.e. 0.3 to 0.6) for vegetation (Ross, 1981).

The soil heat flux is then estimated by taking a fraction of the net radiation at the soil surface:

$$G = 0.35\,(R_n - R_{n,c}) \tag{7}$$

where the constant 0.35 is midway over the range for bare soil (i.e. 0.2 to 0.5) given by Choudhury *et al.* (1987). Comparisons between measured and predicted G using Equations 6 and 7 by Norman *et al.* (1995b) indicated that differences can reach nearly 100% but are usually less than 50%, on average. While adjusting the coefficients in Equations 6 and 7 to obtain better agreement between predicted and measured G can be done, Norman *et al.* (1995b) found that this did not necessarily improve the agreement between measured and predicted H and LE.

With Equations 2 and 5, the canopy temperature can be estimated, which permits the estimation of $T_s$ via Equation 4. $H_s$ can then be estimated via Equations 1, 2, and 3. An additional constraint during daytime conditions is for the soil evaporation and plant transpiration (i.e., $LE_s$ and $LE_c$) to be positive or away from the surface. This constraint is needed when large temperature differences between surface and air exist, for example caused by water-stressed vegetation having elevated canopy temperatures. As a result, transpiration may be reduced by overriding the Priestley-Taylor approximation to satisfy Equations 1 and 4 and the surface energy balance. In other words, the partitioning of $Rn_c$ into $H_c$ and $LE_c$ can reach the condition where $LE_c = 0$ and $H_c = R_{n,c}$. Therefore, the model can accommodate water-limiting conditions where the use of the Priestley-Taylor approach would not be valid. For further details of the logic for obtaining a solution under the scenarios of $LE_s < 0$ and $LE_c < 0$, see Norman *et al.* (1995b).

The dual-source model summarized by Equations 1-7 accommodates the difference between radiometric and aerodynamic temperature. Thus there is no need for an empirical correction, such as adjusting $z_{oh}$ in Equation 2 to account for this difference. In fact, Blyth and Dolman (1995) using a dual-source model demonstrated the dependence of $z_{oh}$ on a number of environmental factors including the humidity deficit, the quantity $R_n - G$, fractional cover and soil and vegetation resistances. The model also takes into account through Equation 4 the radiometer view angle effects on the observed radiometric temperature.

Net radiation, $R_n$, was estimated with the following expression:

$$R_n = (1 - \alpha) R_{solar} + \varepsilon_s \varepsilon_a \sigma T_a^{\ 4} - \varepsilon_s \sigma T_r^{\ 4} \quad (8)$$

where $R_{solar}$ is the incoming solar radiation (assumed uniform over the region), $e_a$ is the apparent sky emissivity estimated with the analytical equation from Brutsaert (1975) which uses screen level air temperature and vapour pressure, $e_s$ is the surface emissivity and a is the surface albedo estimated by weighting the reflectances in the visible (RED) and near-infrared (NIR) from Brest and Goward (1987).

Values of LAI were estimated indirectly by assuming a linear relationship between the fraction of absorbed photosynthetically active radiation (fPAR) and normalized difference vegetation index (NDVI = (NIR-RED)/(NIR+RED)) (Goward and Huemmrich, 1992; Myneni and Williams; 1994). Ground-based estimates of fPAR were obtained from measurements of light transmitted through the plant canopy using the Plant Canopy Analyzer* (Model LAI2000, LI-COR, Inc. Lincoln,

*Company and trade names are given for the benefit of the reader and do not imply any endorsement of the product or company by the USDA

Nebraska) and analyzed by Daughtry *et al.* (1991). The RED and NIR data came from the Landsat TM scene (see below). The regression coefficients between fPAR and NDVI, which are site specific, were computed over several sites within the study area covering a range of vegetation cover conditions (Moran *et al.*, 1994). Then, with the linear equation and NDVI values for each remotely sensed pixel, a corresponding estimate of fPAR was computed. Finally, using an exponential relationship between fPAR and LAI (Asrar *et al.*, 1984; Choudhury, 1987), which for randomly distributed leaves meant that LAI could be computed via (J.M. Norman, 1995, *pers. comm.*)

$$LAI = -2 \ln (1 - fPAR) \cos \theta_{solar} \quad (9)$$

where $\theta_{solar}$ is the solar zenith angle at the time of the remote sensing observations. With an estimate of LAI for each pixel, the fractional cover, $f_c$, was estimated assuming a random canopy with spherical leaf angle distribution (Norman *et al.*, 1995b),

$$f_c = 1 - \exp(-0.5\ LAI) \quad (10)$$

## DEFINING ATMOSPHERIC VARIABLES FOR ELEMENTAL AREAS

Two approaches were used to define the meteorological variables, $\Theta_a$ and u, for each elemental area. One was to assign the average $\Theta_a$ from the network of meteorological stations located within the study area to each pixel, but to adjust the average near-surface windspeed by the ratio of logarithms of momentum roughness of the elemental area i, $z_{om,i}$, to the area-average roughness assigned to the region, $\langle z_{om} \rangle$,

$$u_i = \langle u \rangle\ [\ln((z - d_{o,i})/z_{om,i}) - 1]/[\ln((z - \langle d_o \rangle)/\langle z_{om} \rangle) - 1] \quad (11)$$

where the value of 1 is an order of magnitude estimate of the stability correction (Kustas and Humes, 1996).

The other approach is to use atmospheric variables at a height at which wind and scalars can be treated as horizontally uniform over the region. This height or depth in the atmospheric boundary layer having horizontally uniform atmospheric conditions has been called the blending height, $Z_b$, ( Wieringa, 1986) and methods for estimating $Z_b$ are summarized by Raupach and Finnigan (1995). Nevertheless, estimating $Z_b$ for actual landscapes, especially those that are heterogeneous, is not straightforward. Therefore, mixed-layer wind speed ($U_m$) and mixed-layer potential temperature ($\Theta_m$) were assigned to the height of the top of the surface layer ($Z_{sl}$), assumed to be at 50 m above the surface (Raupach and Finnigan, 1995). The sensitivity of the model results to the magnitude of $Z_{sl}$ was evaluated by assigning a value of 20 m to $Z_{sl}$ from the sounding observations over the study area described in Hipps *et al.* (1994), and a value of 150 m, estimated by taking 10% of the atmospheric boundary layer height, $Z_i$, (Stull, 1988). It was found that changes in the computed fluxes using either $Z_{sl}$ = 20 m or 150 m were less than 10%; therefore, only the results with $Z_{sl}$ = 50 m will be given. This approach is also supported by Mahrt and Sun (1996) who found that using a value of potential temperature at the top of the surface layer, without considering spatial variation of atmospheric flow at this level, typically leads to small errors, except in the case of strong surface heterogeneity together with weak regional scale flow.

## EXPERIMENTAL DATA

The data set used by the model was collected during the Monsoon '90 field experiment conducted in the Walnut Gulch Experimental Watershed (31.5° N 110° W) maintained by the Southwest Watershed Research Center in Tucson, Arizona (Kustas and Goodrich, 1994). The primary campaign of the experiment was conducted during a two-week period in the summer rainy season, a time at which the vegetation was very active. This semi-arid rangeland environment supports desert steppe and grassland communities, both of which are contained in the watershed. There is a gradual transition from shrub-dominated communities at the western, lower end of the watershed to grass-dominated ecosystems at the eastern upland portions (see Figure 1). The vegetation cover is highly variable ranging from 25% to 60%. The elevation in the watershed ranges from about 1300 m at the west end and outlet to 1800 m at the northeast corner. The watershed is mildly hilly with ridge to valley heights on the order of 10 m for the west and central portion of the watershed, and reaching 15-20 m for the eastern half. Typical distances between ridges is approximately 500 m.

Meteorological and energy flux (METFLUX) data were collected at eight sites within the watershed. Their approximate locations are shown in Figure 1. At sites 1-8, H was computed using the variance technique (Tillman, 1972) with measurements at 4 m and LE was solved as a residual with measurements of $R_n$ and G (i.e. LE = $R_n - G - H$). At several sites, the variance method was calibrated and tested with one-dimensional eddy correlation data collected with a 1-D sonic anemometer, fast response thermocouple and Krypton hygrometer located at approximately 2 m above the surface (Stannard *et al.*, 1994; Kustas *et al.*, 1994a). Kustas *et al.* (1994a) found differences with 1-D eddy correlation measurements of H and LE were within 20% on average. At sites 1 and 5-8, H was also estimated using a single-propeller eddy correlation system (OPEC) with sensors at 9 m above the surface (Blanford and Gay, 1992). Estimates of H from the OPEC system were generally 20-25% lower than the 1-D eddy correlation measurement at 2 m, which Stannard *et al.* (1994) attributed to differences in source area footprints resulting from the significantly different measurement heights. Measurements of the meteorological variables u and $T_a$ at sites 1-8 were made at nominal heights of 4.5 and 4 m, respectively. All data were averaged over 20-minute intervals.

The radiosoundings were conducted near site 1 (Figure 1) during the first half of

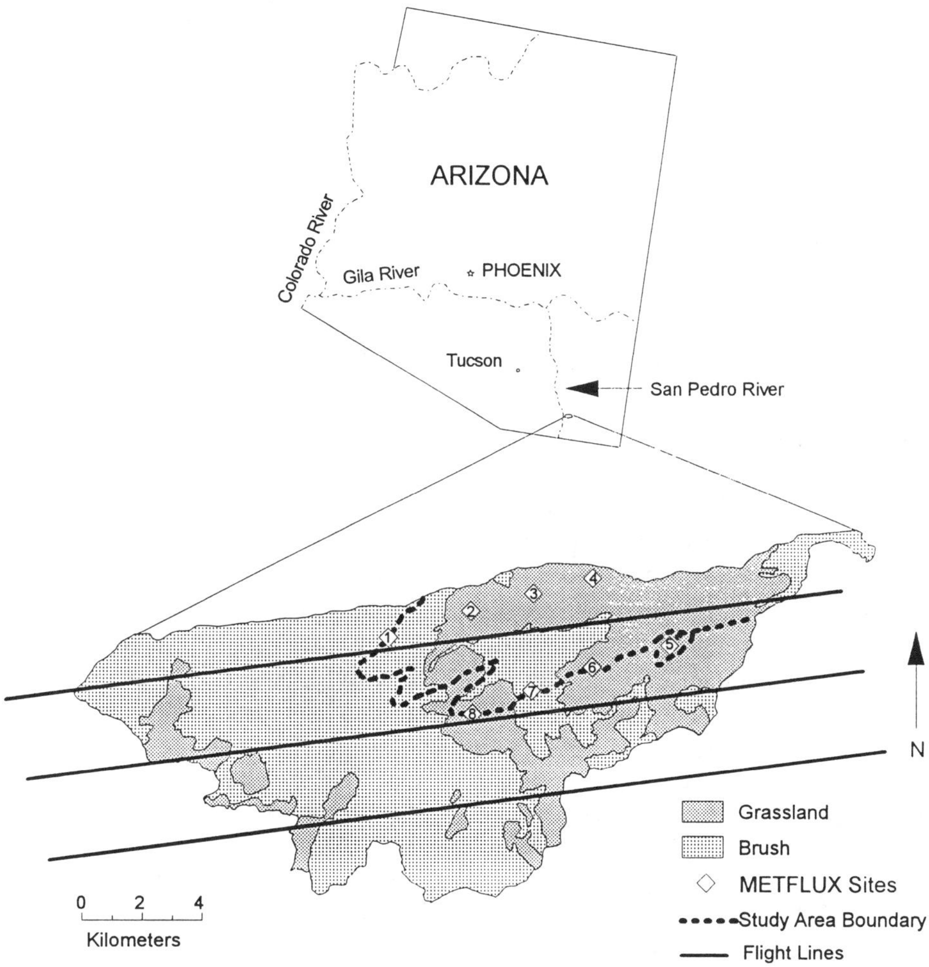


**Figure 1** A schematic plan showing the location of the USDA-ARS Walnut Gulch Experimental Watershed with a map of the areas supporting shrub- and grass-dominated ecosystems. The straight lines represent the transect lines flown by the C-130 aircraft over the study area during the Monsoon '90 field experiment from which the NS001 multispectral data were acquired. The approximate location of each of the METFLUX sites is also illustrated.

the experiment and near the watershed boundary north-west of site 1 for the latter half of the experiment. Measurements were made of dry and wet bulb temperatures, and pressure from the sonde, while elevation and azimuth angles were monitored from a theodolite. These data were used to compute temperature, humidity, wind speed and direction as a function of height. Data from the free soundings were collected up to an elevation of 4-5 km above ground level (agl). On selected days, a series of four or six soundings were made over the course of the daytime period. For a summary of the measurements, see Hipps *et al.* (1994).

The aircraft data used in this analysis were acquired with the NS001 sensor mounted in a NASA C-130 aircraft. The aircraft flew along the transect lines shown in Figure 1. The NS001 instrument has eight bands, seven of which nominally correspond to the seven bands on the Landsat Thematic Mapper instrument. The data used in this analysis were from the thermal band of the sensor (nominal bandpass 10.9-12.3 μm) acquired at approximately 2400 m above the ground surface, corresponding to a pixel size of approximately 6x6 m. For more details see Humes *et al.* (1995).

During the two-week campaign, a number of missions were flown covering a wide range of surface soil moisture conditions from uniformly dry to variably wet conditions. The image data acquired on 9 August, 1990, Day of Year (DOY) 221 at approximately 10:05 a.m. Mountain Standard Time (MST) was selected for this analysis. During the aircraft overpasses, there were clear sky conditions within the study area. The surface soil moisture (0-5 cm) was spatially variable due to several recent rainfall events (Kustas *et al.*, 1993).

Humes *et al.* (1995) evaluated the overall quality of radiometric temperatures derived from the NS001 multispectral scanner data by correcting for atmospheric effects and comparing the resulting temperature with simultaneously-acquired, ground-based remote sensing measurements over two large target areas. The radiative transfer code used for the analysis was Lowtran, Version 7 (Kniezys *et al.*, 1988). The atmospheric corrections computed with Lowtran7 were applied on a pixel-by-pixel basis to the entire image. Adjustments were made to the atmospheric corrections based on the scan angle across the image. The results suggest that the aircraft-based radiometric temperatures generally agree with the ground-based temperature measurements to within approximately 1-2°C.

The Landsat TM scene used to estimate the RED and NIR reflectance factors was acquired during the post monsoon season on 9 September, 1990, DOY 252. Comparison of ground-based observations of NIR and RED reflectance on DOY 221 and 252 by Moran *et al.* (1994) indicated little change in their values, which suggests only minor differences in vegetation conditions would have existed. The procedure for converting the Landsat TM digital counts to values of surface reflectance is summarized by Moran *et al.* (1990). The comparison of the NIR and RED reflectance factors from the Landsat TM image with ground observations showed very good agreement.

The Landsat TM scene was also used in classifying each pixel at the 120 m and 1000 m resolution as either grass-dominated, shrub-dominated, bare soil or riparian vegetation. With each classification, values of canopy height, momentum roughness length and leaf width were assigned (see Table 1). Values of $z_{om}$ and $d_o$ for the different classifications were guided by the analysis of Kustas *et al.* (1994a) and Menenti and Ritchie (1994). Vegetation height estimates were guided by observations (Weltz *et al.*, 1994; Stannard *et al.*, 1994) and the magnitude of $d_o$ assigned to each classification, while leaf width estimates came from M.A. Weltz (1994, *pers. comm.*). For pixels classified as bare soil, a single-source model was run using Equation (2) and $z_{oh}$ estimated assuming a bluff-rough surface which makes $z_{oh}$ a function of the roughness Reynolds number (see Equation (5.29) in Brutsaert,

**Table 1** Surface parameters defined for the four land type classifications determined from Landsat TM data

| Land Type | $z_{om}$ (m) | $d_o$ (m) | $^1h_c$ (m) | $^2w$ (m) |
|---|---|---|---|---|
| Grass-dominated | 0.01 | 0.15 | 0.3 | 0.005 |
| Shrub-dominated | 0.05 | 0.5 | 1 | 0.01 |
| Riparian Vegetation | 0.10 | 1.75 | 2.5 | 0.01 |
| Bare Soil | 0.005 | 0.025 | -- | -- |

$^1h_c$ is the canopy height
$^2w$ is the leaf width

1982).

The 20-minute averages of the near-surface meteorological data, $T_a$ and u, and the incoming solar radiation data from the METFLUX network were used closest in time to the aircraft overpass time (i.e. ~1010 MST). Averages of $U_m$ and $\Theta_m$ derived from the two free soundings bracketing the time of the aircraft overpass were employed in model computations. For the near-surface observations from the METFLUX network, the standard deviation in $T_a$ and u from their mean values of 25.1°C and 5.7 m s$^{-1}$ were 0.63°C and 0.80 m s$^{-1}$. These variations were small enough that attempts by Kustas and Humes (1996) to interpolate and/or extrapolate $T_a$ and u from the METFLUX stations to other portions of the watershed resulted in uncertainties of roughly the same magnitude as these variations. For the mixed-layer variables, the average $\Theta_m$ of 310 K and $U_M$ of 7 m s$^{-1}$ were within ± 0.5K and ± 0.5 m s$^{-1}$ of the values obtained from the two individual soundings bracketing the time of the aircraft overpass.

## RESULTS AND DISCUSSION

### VARIATION IN NDVI AND $T_R$ AT 120 M AND 1000 M PIXEL RESOLUTION

Figure 2 shows the frequency distributions of NDVI using the RED and NIR reflectance factors aggregated to the 120 m and 1000 m pixel resolutions, respectively. For the 120 m resolution, close to 40% of the pixels have a NDVI-value between 0.31 and 0.35, with another 50% of the pixels having NDVI-values equally divided between the range of 0.26-0.30 and 0.36-0.40. The remaining 10% of the pixels are mainly divided between the range of 0.21-0.25 and 0.41-0.45. At the 1000 m resolution, the variation in the magnitude of NDVI reduces with the values falling essentially into three intervals. The largest percentage of pixels at nearly 45% are in the range of 0.31-0.35.

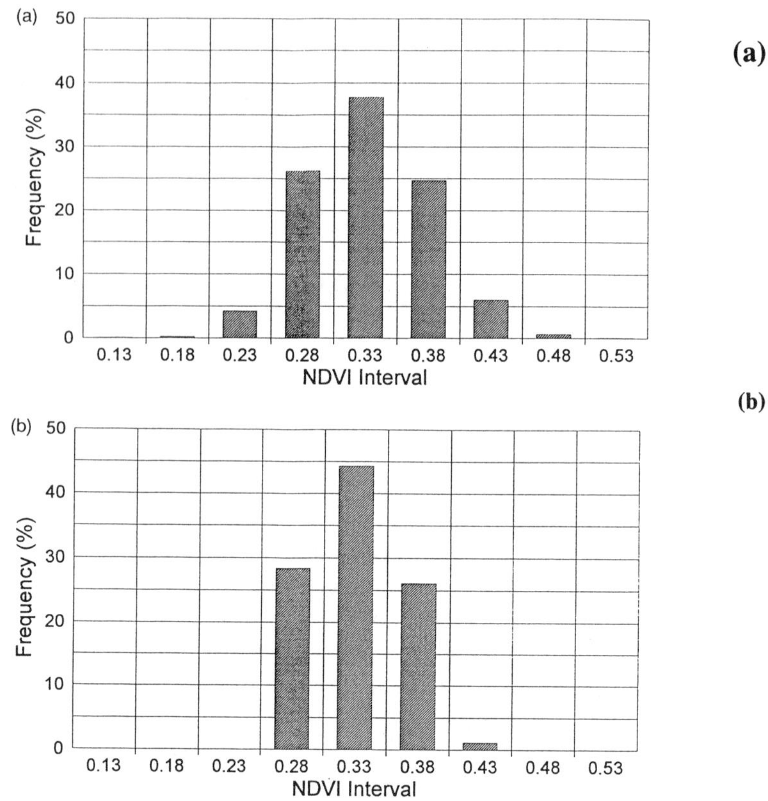


**Figure 2** The frequency histogram for the normalized vegetation index, NDVI, for (a) 120 m and (b) 1000 m pixel resolution data. The NDVI intervals are in 0.02 increments with values along the abscissa representing the midpoint of the interval.

The distributions of the radiometric surface temperature for the 120 m and 1000 m pixels are illustrated in Figure 3. For the 120 m resolution, approximately 75% of the $T_r$-values are between 30° and 34°C with slightly over 15% of the remaining pixels having $T_r$-values in the range of 34°-38°C. The remaining 10% of the $T_r$-values mostly are in the range of 28°-30°C with less than 1% in the ranges of 38°-40°C and 26°-28°C. For the 1000 m data, the variation in $T_r$ reduces such that the values fall into essentially three temperature intervals with nearly 85% of the $T_r$-values being between 30° and 34°C and almost 50% of these being in the range of 30°-32° C. The remaining 15% or so of the pixels have $T_r$-values between 34° and 36°C.

**(a)**

**(b)**

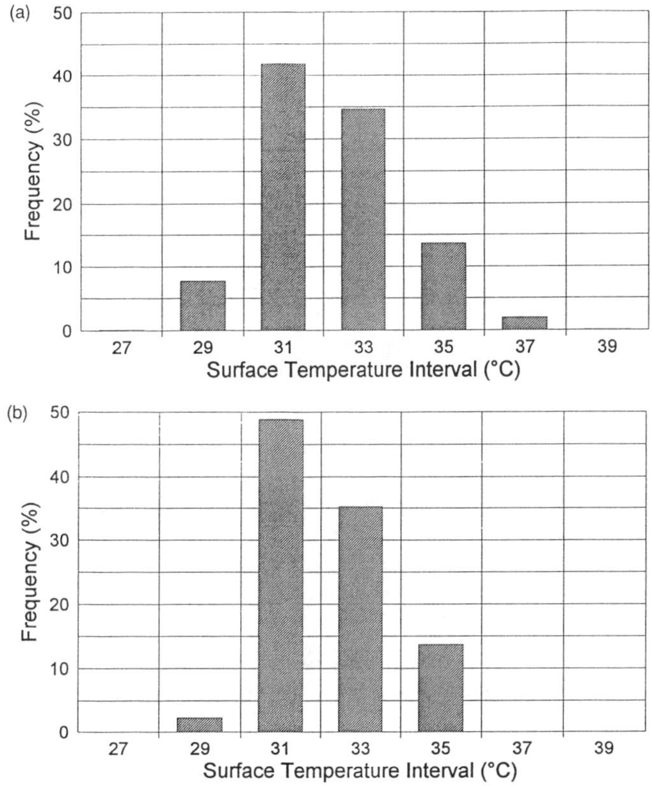


**Figure 3** The frequency histogram for the radiometric surface temperature, $T_r$, for (a) 120 m and (b) 1000 m pixel resolution data. The $T_r$ intervals are in 2° increments with values along the abscissa representing the midpoint of the interval.

## SPATIAL VARIATION IN MODEL OUTPUT OF H

The spatial variation in sensible heat flux, H, determined at the different scales was evaluated in some detail, since model sensitivity to variation in parameters and variables (especially $T_r$) is highest when computing this component of the surface energy balance. The frequency distribution of sensible heat flux, H, predicted by the model using the 120 m pixel data with the near-surface atmospheric forcing variables from the METFLUX network (METNET) or from the mixed-layer variables estimated using the free radiosoundings of the atmospheric boundary layer (ABL), is illustrated in Figure 4. Values of H using either METNET or ABL data range from about 50 W m$^{-2}$ to 250 W m$^{-2}$ with the majority of the values being between 100 and 150 W m$^{-2}$. Compared to the 120 m resolution data, the range in H-values is reduced with the 1000 m pixels (Figure 5), and even more so when the METNET data were used in driving the model. Using the METNET data with

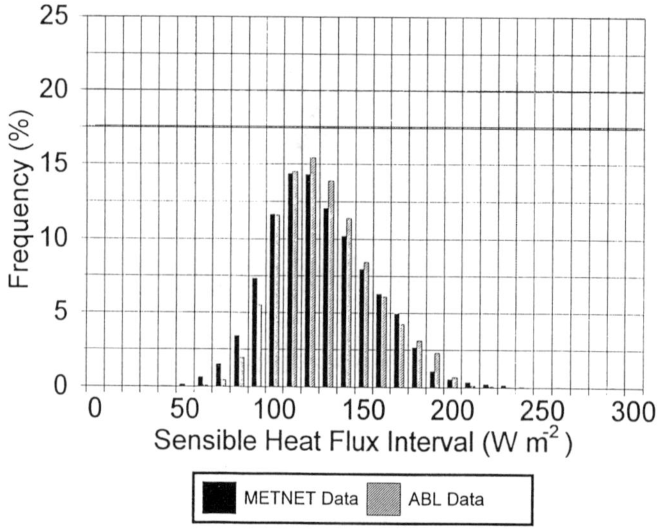


**Figure 4** The frequency histogram for sensible heat flux, H, computed by the model with the 120 m pixel data and using the METNET and ABL data for the atmospheric variables. The H intervals are in 10 W $m^{-2}$ increments with values along the abscissa representing the minimum value for the interval.

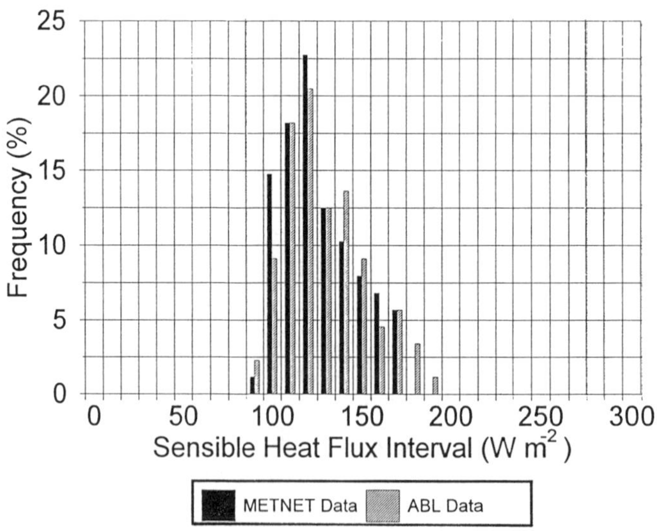


**Figure 5** The frequency histogram for sensible heat flux, H, computed by the model with the 1000 m pixel data and using the METNET and ABL data for the atmospheric variables. The H intervals are in 10 W $m^{-2}$ increments with values along the abscissa representing the minimum value for the interval.

1000 m pixels, the range in H-values is between 90 and 170 W m$^{-2}$ while a few ABL-derived H-values nearly reach 200 W m$^{-2}$. In addition, about 5% more of the pixels have H-values around 120 W m$^{-2}$ using METNET versus the ABL data in driving the model.

The model was also executed using the RED, NIR and $T_r$ values averaged over or aggregated up to the whole image, combined with the ABL data. This gave a pixel size roughly 5000 m x 15000 m, which is slightly larger than the 8000 x 8000 m pixel resolution data available operationally from GOES. The average H computed by the model was 128 W m$^{-2}$. This estimate of the area-average value of H is within 10 W m$^{-2}$ of the area-average values obtained by linearly averaging the H-fluxes computed by the model using the 120 m and 1000 m resolution data (see below).

## COMPARISON OF AREA-AVERAGE SURFACE ENERGY BALANCE COMPONENTS AT DIFFERENT RESOLUTIONS

The comparison of the area-average surface energy balance components using RED and NIR reflectances and $T_r$ aggregated to 120 m, and 1000 m resolution data, and to the whole image (roughly 5000 x 15000 m) is given in Table 2. The results from Table 2 indicate that area-averaged surface energy balance components <$R_n$>, <G>, <H> and <LE> are very similar, regardless of the resolution of the remotely-sensed data being used by the model or whether the METNET or ABL data were used for the atmospheric forcing in the model. Furthermore, the procedure of linearly averaging $T_r$ to the different resolutions does not appear to significantly affect the area-averaged surface energy balance components, nor does it appear to violate to any measurable degree the energy conservation principle (McNaughton, 1994).

The area-average surface fluxes given by the METFLUX network during the time of the aircraft overpass were <$R_n$> = 527 W m$^{-2}$, <G> = 156 W m$^{-2}$, <H> = 151 W m$^{-2}$, and <LE> = 220 W m$^{-2}$. Differences between model-derived (Table 2) and measured <$R_n$>, <G>, <H> and <LE> may suggest that either the measurements

**Table 2** Area-average surface energy balance components using 120 m, 1000 m and 5000 x 15000 m pixel resolutions for RED and NIR reflectance factors and $T_r$

| Resolution (m) | Atmospheric Variables | <$R_n$> (Wm$^{-2}$) | <G> (Wm$^{-2}$) | <H> (Wm$^{-2}$) | <LE> (Wm$^{-2}$) |
|---|---|---|---|---|---|
| 120 | METNET | 549(±18)[a] | 134(±8) | 121(±29) | 294(±44) |
| 120 | ABL | 549(±18) | 134(±8) | 124(±29) | 291(±42) |
| 1000 | METNET | 554(±10) | 135(±4) | 122(±21) | 297(±27) |
| 1000 | ABL | 554(±10) | 135(±4) | 125(±23) | 294(±32) |
| 5000 x 15000 | ABL | 553 | 135 | 128 | 290 |
| METFLUX network | | 527 | 156 | 151 | 220 |

[a] Standard deviation

from the METFLUX network were not representative of the area-average energy fluxes or that values for some of the model variables and/or parameters were incorrectly prescribed. Factors contributing to measurement bias and to model bias that can cause the observed differences in area-average flux estimates are described below.

Assuming the discrepancies are primarily caused by the measurements, a possible explanation of the overestimate of $\langle R_n \rangle$ and underestimate of $\langle G \rangle$ by the model is that the METFLUX measurements were made primarily on ridge tops which contained less vegetation cover and had drier soil moisture conditions than in the nearby ephemeral channels (Stannard *et al.*, 1994). Thus, the measurements of $R_n$ and G from the METFLUX network would have been more representative of the ridge tops and as a result, would tend to yield an underestimate $\langle R_n \rangle$ and an overestimate $\langle G \rangle$ for the study area. Similarly, the estimates of H from the METFLUX sensors at 4 m may have been largely influenced by the drier and more sparsely vegetated ridge tops, so that $\langle H \rangle$ given by the METFLUX network would tend to be higher than the whole study area. In fact, this is supported by the estimate of $\langle H \rangle$ computed with the five OPEC systems for the same period (i.e. $\langle H \rangle$ = 110 W $m^{-2}$) which were sampling over a significantly larger source area including valley bottoms (Stannard *et al.*, 1994). This value of $\langle H \rangle$ is in better agreement with the model estimates in Table 2.

On the other hand, it can also be shown that relatively minor changes in the values of several model input variables can lead to significant changes in the predicted area-average surface energy balance. This results in one reaching a different conclusion from that made assuming a bias in the measured fluxes. For example, "ground-truth" observations suggested different values of surface albedo and radiometric surface temperature from what was used by the model to generate the output in Table 2. For the surface albedo, the 1000 m pixel data for the study area yielded $\alpha = 0.17$, but low-altitude aircraft measurements used by Kustas *et al.* (1994b) indicated an $\alpha = 0.19$. A comparison between ground and aircraft measurements of $T_r$ by Humes *et al.* (1995) indicated that the atmospherically corrected aircraft-derived $T_r$ was still about 1°C less than the ground observations. Discrepancies of around 10% in $\alpha$ and 1°C in $T_r$ are fairly typical for these input variables. If the $\alpha$-values are uniformly increased by 0.02 and the $T_r$-values are increased by 1°C, then the area-averaged fluxes computed by the model using the 1000 m pixel and ABL data change to $\langle R_n \rangle = 532$ W $m^{-2}$, $\langle G \rangle = 130$ W $m^{-2}$, $\langle H \rangle = 141$ W $m^{-2}$, and $\langle LE \rangle = 261$ W $m^{-2}$. Thus, relatively small changes in the values for $\alpha$ and $T_r$ result in significantly better agreement between model-derived $\langle R_n \rangle$, $\langle H \rangle$, and $\langle LE \rangle$ and values given by the METFLUX network. As a result, the earlier statements concerning the possible bias in $\langle R_n \rangle$ and $\langle H \rangle$ with the observations from the METFLUX network may have been overstated.

This analysis indicates that typical levels of uncertainty in the estimate of albedo and radiometric surface temperature can result in variations in model computed $\langle H \rangle$ and $\langle LE \rangle$ of around 10% and a variation in estimated $\langle R_n \rangle$ of nearly 5%. These variations may be large enough to influence the conclusions made when comparing model predictions to measurements. However, one should also consider the fact that the measurements themselves have a level of uncertainty or degree of unexplained variability. For heterogeneous terrain, observations suggest that the uncertainty in $R_n$ is between 5% and 10% and is around 20% for G, H, and LE (e.g., Kustas *et al.*, 1994a; and Stannard *et al.*, 1994). Therefore, differences between modelled and measured $<R_n>$ which are less than 10%, and differences between modelled and measured <G>, <H> and <LE> that are less than 20%, are probably within the level of uncertainty in the measurements.

The range in the area-average Bowen ratio, $<\beta>$ (= <H>/<LE>), is fairly small for the model output listed in Table 2 (i.e. $<\beta> = 0.41$ to 0.44), but increases to 0.41 to 0.54 when $\alpha$- and $T_r$-values are changed. However even with this increase, the range in $<\beta>$-values from the model still fall within the range given by the OPEC and METFLUX systems, namely $<\beta> = 0.42$ to 0.69, respectively. It is also interesting to note that the mean $<\beta>$-value calculated by averaging <H> and <LE> from the OPEC and METFLUX network is equal to the $<\beta>$-value given by the model when $\alpha$- and $T_r$-values are adjusted.

## SUMMARY AND CONCLUSIONS

A two-source model that computes the surface energy balance with primary inputs of radiometric surface temperature and a vegetation index for estimating the fractional vegetation cover or leaf area index was applied to remotely sensed data aggregated to the pixel resolution of Landsat TM (120 m), NOAA AVHRR (1000 m) and to the nominal resolution of data operationally available from GOES (8000 m). The atmospheric data (i.e., wind speed and temperature) to drive the model were supplied by a network of surface meteorological/energy flux METFLUX stations and by free radiosounding observations which provided mixed-layer wind speed and potential temperature. Use of these two sources of atmospheric data resulted in different approaches for computing the surface resistances for each elemental area.

As one might expect, the surface energy balance showed less variability with decreasing resolution in the remotely sensed data. However, the surface fluxes computed with the 120 m and 1000 m pixels produced area-average values of the surface energy balance components, namely $<R_n>$, <G>, <H>, and <LE>, that were very similar to each other and to the model output using the remotely sensed data averaged up to the whole 5000 m x 15000 m area. Furthermore, the model computations were very similar whether the approach using the near-surface meteorological or mixed-layer data was applied to the 120 m and 1000 m pixels.

These results suggest that for the environmental conditions that existed during the period of remote sensing observations over this semi-arid rangeland ecosystem, the use of different resolutions of remotely sensed data given by three operational satellite sensors did not significantly impact the area-average surface energy balance derived from the model. Moreover, computing the coarser resolution

$T_r$-values by linearly averaging the observations did not appear, for all practical purposes, to violate the flux-matching requirement proposed by McNaughton (1994) and Raupach (1995) (i.e. that the elemental energy balance components average linearly). In other words, there should be conservation of the energy balance equation with scale. The spatial resolution of the data did, however, have a noticeable affect on the observed range in the magnitude of the surface fluxes computed by the model.

It was shown that the level of uncertainty in estimating the surface albedo and radiometric surface temperature can result in 5% to 10% variations in model-derived $R_n$, H and LE which can be significant enough to influence the interpretation of the output. However, this is still within the level of uncertainty in the measurement of these energy fluxes. Further sensitivity studies on the effects of uncertainties in the input data on model predictions of the surface energy fluxes will be undertaken in the near future.

If surface conditions such as surface wetness, fractional cover and surface roughness had been more spatially variable, such as in the simulations performed by Lhomme *et al.* (1994), differences in the estimates of $\langle R_n \rangle$, $\langle G \rangle$, $\langle H \rangle$, and $\langle LE \rangle$ using the model computations with 120 m, 1000 m and 8000 m pixels for the region would probably have been greater. In addition, linearly averaging $T_r$ may have also resulted in a significant violation of energy conservation when changing scales. Experimental data containing significant spatial variations in surface conditions need to be collected and analyzed to determine the limitations inherent in the present approach of aggregating the surface energy balance to regional scales.

## ACKNOWLEDGEMENTS

The cooperation and assistance of the USDA-ARS Southwest Watershed Research Center in Tucson, Arizona, and on-site personnel who maintain the Walnut Gulch Experimental Watershed are gratefully acknowledged. The authors also acknowledge David Stannard from the U.S. Geological Survey, Denver, Colorado, and James Blanford, affiliated with the University of Arizona during Monsoon '90, who were mainly responsible for the surface flux data. Dr. Lawrence Hipps and Edward Swiatek collected and processed the radiosonde data. The authors would like to thank Dr. John Norman for his advice on how to estimate LAI from the remotely sensed data. Helpful comments on an earlier version of this manuscript were provided by Drs. Greg Johnson, Larry Mahrt, and Charles Walthall. Funding from NASA Interdisciplinary Research Program in Earth Sciences (NASA Reference Number IDP-88-086) and funds from USDA-ARS Beltsville Area Office provided the necessary financial support to conduct this field study.

## REFERENCES


Asrar, G.,, Fuchs, M., Kanemasu, E.T. and Hatfield, J.L. 1984. Estimating absorbed photosynthetically active radiation and leaf area index from spectral reflectance in wheat. *Agron. J.*, **76**, 300-306.

Ball, J.T., Woodrow, I.E. and Berry, J.A. 1986. A model predicting stomatal conductance and its contribution to control photosynthesis under different environmental conditions. *Progress in Photosynthesis Research*, J. Biggins, (ed). Martinus Nijhoff, Dordrecht, The Netherlands. **44**, 221-224.

Blanford, J.H. and Gay, L.W. 1992. Tests of a robust eddy correlation system for sensible heat flux. *Theor. Appl. Climatol.*, **46**, 53-60.

Blyth, E.M. and Dolman, A.J. 1995. The roughness length for heat of sparse vegetation. *J. Appl. Meteor.*, **34**, 583-585.

Brest, C.L. and Goward, S.N. 1987. Deriving surface albedo measurements from narrow band satellite data. *Int. J. Remote Sens.*, **8**, 351-367.

Brutsaert, W. 1975. On a derivable formula for long-wave radiation from clear skies. *Water Resour. Res.*, **11**, 742-744.

Brutsaert, W.H. 1982. *Evaporation into the atmosphere*. D. Reidel, The Netherlands.

Brutsaert, W. 1992. Stability correction functions in the mean wind speed and temperature in the unstable surface layer. *Geophys. Res. Letters*, **19**, 469-472.

Chebhouni, A., Nichols, W.D., Qi, J., Njoku, E.G., Kerr, Y.H. and Cabot, F. 1994. On the use of radiative surface temperature to estimate sensible heat flux over sparse shrubs in Nevada. *Proceedings of the Sixth International Symposium Physical Measurements and Signatures in Remote Sensing*, 17-21 January, Val d'Isere, France. 777-784.

Choudhury, B.J. 1987. Relationships between vegetation indices, radiation absorption, and net photosynthesis evaluated by a sensitivity analysis. *Remote Sens. Environ.*, **22**, 209-233.

Choudhury, B.J., Idso, S.B. and Reginato, R.J. 1987. Analysis of an empirical model for soil heat flux under a growing wheat crop for estimating evaporation by an infrared-temperature based energy balance equation. *Agric. For. Meteorol.*, **39**, 283-297.

Daughtry, C.S.T., Weltz, M.A., Perry, E.M. and Dulaney, W.P. 1991. Direct and indirect estimates of leaf area index. In: Preprints of the American Meteorological Society *Tenth Conference on Biometeorology and Aerobiology and Special Session on Hydrometeorology*. Salt Lake City UT, pp. 230-233.

Garrett, J.R. and Hicks, B.B. 1973. Momentum, heat and water vapor transfer to and from natural and artificial surfaces. *Quart. J. Roy. Meteorol. Soc.*, **99**, 680-687.

Goudriaan, J. 1977. *Crop Micrometeorology: A simulation study*. Center for Agric. Publ. and Doc., Wagengingen, The Netherlands.

Goward, S.N. and Huemmrich, K. F. 1992. Vegetation canopy PAR absorptance and the normalized difference vegetation index: An assessment using the SAIL model. *Remote Sens. Environ.*, **39**, 110-140.

Hall, F.G., Huemmrich, K.F., Goetz, S.J., Sellers, P.J. and Nickeson, J.E. 1992. Satellite remote sensing of surface energy balance: success, failures and unresolved issues in FIFE. *J. Geophys. Res.*, **97(D17)**, 19,061-19,089.

Hipps, L.E., Swiatek, E. and Kustas, W.P. 1994. Interactions between regional surface fluxes and the ABL over a heterogeneous watershed. *Wat. Resour. Res.*, **30**, 1387-1392.

Humes, K.S., Kustas, W.P., Schmugge, T.J., Moran, M.S. and Hipps, L.E. 1995. Radiometric surface temperature derived from NS001 multispectral scanner data. *Remote Sens. Environ.* (In review)

Kniezys, F.X., Shettle, E.P., Abreu, L.W., Chetwynd, J.H. and Anderson, G.P. 1988. User Guide to Lowtran 7, Air Force Geophysical Laboratory Report No. AFGL-TR-88-0177, Hanscom AFB, MA 01731.

Kustas, W.P. and Goodrich, D.C. 1994. Preface to special issue on MONSOON '90 multidisciplinary experiment. *Wat. Resour. Res.*, **30**, 1211-1225.

Kustas, W. P., Schmugge, T. J., Humes, K. S., Jackson, T. J., Parry, R., Weltz, M. A. and Moran, M. S. 1993. Relationships between evaporative fraction and remotely sensed vegetation index and microwave brightness temperature for semiarid rangelands. *J. Appl. Meteor.*, **32,** 1781-1790.

Kustas, W.P., Blanford, J.H., Stannard, D.I., Daughtry, C.S.T., Nichols, W.D. and Weltz, M.A. 1994a. Local energy flux estimates for unstable conditions using variance data in semiarid rangelands, *Wat. Resour. Res.*, **30**, 1351-1361.

Kustas, W. P., Pinker, R. T., Schmugge, T. J. and Humes, K. S. 1994b. Daytime net radiation estimated for a semiarid rangeland basin from remotely sensed data. *Agric. For. Meteor.*, **71,** 337-357.

Kustas, W.P. and Humes, K.S. 1996. Spatially distributed sensible heat flux over a semiarid watershed Part II: Use of a variable resistance approach with radiometric surface temperature. *J. Appl. Meteorol.* In press.

Lhomme, J.-P. 1992. Energy balance of heterogeneous terrain: averaging the controlling parameters. *Agric. Forest Meteorol.*, **61**, 11-21.

Lhomme, J.-P., Chehbouni, A. and Monteny, B. 1994. Effective parameters of surface energy balance in heterogeneous landscape. *Boundary-Layer Meteorol.*, **71,** 297-309.

Mahrt, L. and Sun, J. 1996. Dependence of surface exchange coefficients on averaging scale and grid size. *Quart. J. Roy. Meteor. Soc.*, **121**, 1835-1852.

McNaughton, K.G. 1994. Effective stomatal and boundary-layer resistances of heterogeneous surfaces. *Plant, Cell Environ.*, **17,** 1061-1068.

Menenti, M. and Ritchie, J.C. 1994. Estimation of effective aerodynamic roughness of Walnut Gulch watershed with laser altimeter measurements. *Wat. Resour. Res.*, **30,** 1329-1337.

Moran, M.S., Jackson, R.D., Hart, G.F., Slater, P.N., Bartell, R.J., Biggar, S.F., Gellman, D.I. and Santer, R.P. 1990. Obtaining surface reflectance factors from atmospheric and view angle corrected SPOT-1 HRV data. *Remote Sens. Environ.*, **32,** 203-214.

Moran, M.S., Clarke, T.R., Kustas, W.P., Weltz, M.A. and Amer, S.A. 1994. Evaluation of hydrologic parameters in semiarid rangeland using remotely sensed spectral data. *Water Resourc. Res.*, **30,** 1287-1297.

Myneni, R.B. and Williams, D.L. 1994. On the relationship between FAPAR and NDVI. *Remote Sens. Environ.*, **49,** 200-211.

Norman, J.M., Divakarla, M. and Goel, N.S. 1995a. Algorithms for extracting information from remote thermal-IR observations of the earth's surface. *Remote Sensing Environ.*, **51,** 157-168.

Norman, J.M., Kustas, W.P. and Humes, K.S. 1995b. A two-source approach for estimating soil and vegetation energy fluxes from observations of directional radiometric surface temperature. *J. Agric. For. Meteorol.*, **77,** 263-293.

Pinker, R.T., Kustas, W.P., Laszlo, I., Moran, M.S. and Huete, A.R. 1994. Basin-scale solar irradiance estimates in semiarid regions using GOES 7. *Water Resour. Res.*, **30,** 1375-1386.

Prevot, L., Brunet, Y., Paw U, K.T. and Seguin, B. 1994. Canopy modeling for estimating sensible heat flux from thermal infrared measurements. *Proc. Thermal Remote Sensing of the Energy and Water Balance over Vegetation in Conjunction with other Sensors.*, La Londe-Les Maures, France, 17-22.

Priestley, C.H.B. and Taylor, R. J. 1972. On the assessment of surface heat flux and evaporation using large-scale parameters. *Monthly Weath. Rev.*, **100,** 81-92.

Ross, J. 1981. The radiation regime and architecture of plants. In: *Tasks for Vegetation Sciences 3* (H. Lieth, Series Ed.). Dr. W. Junk, The Hague, Netherlands.

Raupach, M.R. and Finnigan, J.J. 1995. Scale issues in boundary-layer meteorology: Surface energy balances in heterogeneous terrain. *Hydrol. Processes*, **9,** 589-612.

Raupach, M.R. 1995. Vegetation-atmosphere interaction and surface conductance at leaf, canopy and regional scales. *Agric. Forest Meteorol.*, **73,** 151-179.

Sauer, T.J., Norman, J.M., Tanner, C.B. and Wilson, T.B. 1995. Measurement of heat and vapor transfer at the soil surface beneath a maize canopy using source plates. *Agric. For. Meteorol.* **75,** 161-189.

Shuttleworth, W.J. and Gurney, R.J., 1990. The theoretical relationship between foliage temperature and canopy resistance in sparse crops. *Q.J.Roy. Met. Soc.*, **116,** 497-519.

Stannard, D.I., Blanford, J.H., Kustas, W.P., Nichols, W.D., Amer, S.A., Schmugge, T.J. and Weltz, M.A. 1994. Interpretation of surface flux measurements in heterogeneous terrain during the Monsoon '90 experiment. *Water Resour. Res.*, **30,** 1227-1239.

Stull, R.B. 1988. *An introduction to boundary-layer meteorology*. Kluwer Academic Publishers, Dordrecht, The Netherlands.

Tillman, J.E. 1972. The indirect determination of stability, heat and momentum fluxes in the atmospheric boundary layer from simple scalar variables during dry conditions. *J. Appl. Meteorol.*, **11**, 783-792.

Weltz, M.A., Ritchie, J.C. and Fox, H.D. 1994. Comparison of laser and field measurements of vegetation height and canopy cover. *Water Resour. Res.*, **30,** 1311-1320.

Wieringa, J. 1986. Roughness-dependent geographical interpolation of surface wind speed averages. *Q.J. Roy. Met. Soc.*, **112,** 867-889.

# 9 Scintillometer measurements of sensible heat flux over heterogeneous surfaces

**J-P. LAGOUARDE[1], K.J. McANENEY[2] and A.E. GREEN[2]**
[1]INRA Bioclimatologie, Villenave d'Ornon, France
[2]Hort Research, Kerikeri Research Centre, Kerikeri, New Zealand

## INTRODUCTION

The idea of using optical scintillations to infer the surface heat flux should not surprise anyone who has watched the air shimmering above a hot surface in summer. The meteorological variable of interest here is $C_N^2$ which denotes the strength of refractive turbulence. In turbulence theory, it is common to describe the spatial variability of atmospheric variables by their relevant structure function, in this case the dimensionless $D_N(r)$. For separation distances r in the inertial sub-range of scales about a point x,

$$D_N(r) = \langle [N(x) - N(x+r)]^2 \rangle = C_N^2 r^{2/3} \qquad l_0 \ll r \ll L_0$$

where N is refractive index, the angle brackets indicate a time or ensemble average, and $l_0$ and $L_0$ are the inner and outer scales respectively. The inner scale marks the transition from the inertial to viscous range of eddy sizes, whereas $L_0$ describes the size of the dominant inhomogeneities. In the near surface atmosphere $l_0$ is typically around 1 cm and $L_0$ is about half the measurement height (Weseley and Derzko, 1975). At visible and near infrared wavelengths, fluctuations in N are due mainly to turbulent variations in air temperature and thus are intimately related to the sensible heat flux. It is this liaison that we seek to exploit in order to quantify sensible heat through measurement of optical scintillations.

Early efforts to derive turbulence information from scintillation measurements of $C_N^2$ have been reviewed by Green *et al.* (1994) and McAneney *et al.* (1995). These studies also show how, by complementing $C_N^2$ with a measure of mean windspeed U, both sensible heat H and the momentum fluxes $u_*$ may be derived by a simple iterative procedure. This approach has worked remarkably well above uniform surfaces whose aerodynamic characteristics (surface roughness and the zero plane displacement) are well-described as simple rule-of-thumb fractions of the vegetation height.

*Scaling up in hydrology using remote sensing*. Edited by J.B. Stewart, E.T. Engman, R.A. Feddes and Y. Kerr

Scintillometers using a point source of electromagnetic radiation saturate under high turbulence, a feature which limits the practical pathlength (L) of visible or near infrared laser scintillometers to around 100 m for near surface measurements. This study used a large aperture instrument (LAS) designed at the NOAA Wave Propagation Laboratory, Boulder, Colorado (Ochs and Cartwright, 1980) with the aim of maintaining linearity between $C_N^2$ and signal variance over a much larger range of refractive turbulence. McAneney *et al.* (1995) note its operational advantages of short averaging times to reduce uncertainty in turbulence statistics, insensitivity to the effects of flow distortion by obstacles near the transmitter or receiver, and the possibility of housing the instruments behind glass windows in order to achieve long-term monitoring of surface fluxes. However it is the promise of spatially-averaged fluxes at scales of the order of several kilometres that is our primary consideration here as we evaluate possibilities for spatial integration of surface fluxes. In particular the method offers a way of directly validating remote sensing estimates of surface fluxes at comparable spatial scales of resolution.

## PHYSICAL PRINCIPLES

The LAS transmitter consists of an incoherent source of near-infrared radiation positioned at the focus of a concave mirror (diameter D=15 cm). Inhomogeneities in refractive index caused by eddy motions along the path distort the wavefront of the transmitted beam and the resulting variations in signal intensity are recorded at a receiver comprising an identical mirror and a photodiode detector. Further details can be found in McAneney *et al.* (1995) and the theoretical basis is given by Clifford *et al.* (1974) and Wang *et al.* (1978).

Turbulent distortions to a wavefront by eddy motions in the optical path may be regarded as a collection of diverging and converging lenses. For a point source of light under conditions of weakly-refracting turbulence, it follows from geometrical optics that the smallest disturbances (of the scale of $l_0$) are most effective in causing intensity fluctuations at the receiver (Tatarskii, 1993). The incoherent source used in the LAS can be considered as the sum of a large number of point sources; the diverging wavefronts from these will be distorted by eddies of scale $l_0$. However since $D >> F > l_0$, the effects of fine-scale turbulence are smoothed out over the receiver aperture: F is the first Fresnel zone ($F = (\lambda L)^{1/2}$, $\lambda$ being the optical wavelength) and gives the scale of diffractional spreading. Thus for the LAS, the most effective eddies are those with a length similar to the aperture diameter and which by focussing or defocussing the entire beam can create a large change in the irradiance at the receiver.

Clifford *et al.* (1974) extended the weak turbulence theory of Tatarskii (1961) to show that the variance of the natural logarithm of the intensity incident at the receiver is

$$\sigma^2_{LnI} = 16\pi^2 k^2 L \int du \int dKK\phi_N(K)\sin^2(K^2Lu(1\text{-}u)/2k)[2J_1(x_1)2J_1(x_2)/(x_1x_2)]^2 \qquad (2)$$

where the integration over u goes between 0 and 1, and the second over K, between 0 and infinity; u is the normalised path distance from the transmitter; $k = 2\pi/\lambda$ is the optical wavenumber; $x_1 = KD_R u/2$ and $x_2 = KD_T(1-u)/2$, where $D_R$ and $D_T$ are the receiver and transmitter apertures; K is the 3-dimensional spatial wavenumber; $J_1$ is a Bessel function of the first kind of order one; and $\Phi_N$ describes the turbulent medium in terms of its Fourier components K

$$\Phi_N(K) = 0.033 C_N^2 K^{-11/3} \qquad (L_o)^{-1} << K << (l_o)^{-1}$$

Assuming equal transmitter and receiver apertures and on the condition that

$$D/F > (0.124 k^{7/6} L^{11/6} <C_N^2>)^{3/5}$$

Wang *et al.* (1978) evaluate Equation 2 to obtain:

$$\sigma^2_{LnI} = 0.892 <C_N^2> D^{-7/3} L^3$$

where

$$<C_N^2> = \int C_N^2(u) W(u) du \tag{3}$$

Equation 3 is the usual calibration equation for LAS, with the weighting function W(u) calculated by numerical integration of Equation 2.

The experimental configurations of most interest in this study were chosen expressly to have $C_N^2$ varying along the optical path. For a two component surface, we have

$$< C_N^2> = (C_N^2)_1 W_1 + (C_N^2)_2 W_2 \tag{4}$$

where weighting functions $W_1$ and $W_2$ follow from Equations 2 and 3 and the subscripts refer to two surfaces. The main object of this paper was to test Equations 2, 3 and 4.

Figure 1 shows the normalised path-weighting function for the case of equal receiver and transmitting apertures. In the case of different sized apertures, Wang *et al.* (1978) give the following expression:

$$<C_N^2> = C\sigma^2_{LnI} (D_R/D_T)^{7/3} L^{-3} \tag{5}$$

where C is a calibration constant that is a function of the ratio $D_R/D_T$. Figure 1 shows how unequal receiver and transmitter diameters allow some coarse spatial filtering by shifting the position of maximum weighting towards the smaller aperture end.

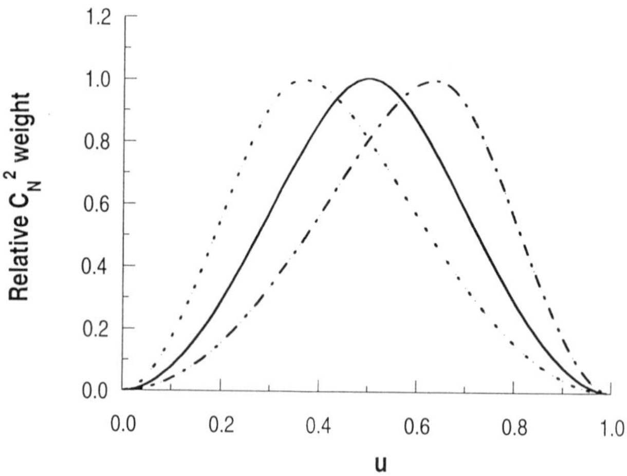


**Figure 1** Path-weighting functions for equal apertures (0.15 m) of transmitter and receiver (solid line), $D_T/D_R = 2$ (-·-·-) and $D_T/D_R = 0.5$ (····). All calculations were done for L = 450 m.

## FETCH CONSIDERATIONS

In addition to the spatial weighting along the optical path, it is also necessary to consider the source area contributing to the refractive turbulence responsible for scintillation. These turbulent eddies are representative of upwind sources and are advected down or across the optical path. Above a horizontally homogeneous surface this creates no problem but, in the case of changing terrain the situation is more complicated, with Brunet *et al.* (1994) finding that a transition between dry and irrigated vegetation introduced a step change in surface fluxes. Downwind of the transition the air flow gradually adjusts to the new boundary conditions. Calculations of the upwind source area affecting eddy correlation measurements have been proposed by Gash (1986) and Schuepp *et al.* (1990), amongst others, but the question remains as to whether the 'footprint' of a scintillometer is any different. In this respect we recall that a scintillometer is sensitive only to eddies with length scales comparable to the beam diameter. For these smaller eddies, the adjustment to changing surface conditions occurs rapidly but the greater response time or 'memory' of larger eddies means that the low frequency parts of the spectrum may never be completely in equilibrium with the underlying surface (Panofsky and Dutton, 1984). Given the restricted bandwidth of the scintillometer, and the iterative approach to calculation assuming Monin-Obukhov similarity, the flux estimations are based upon the presumption that the turbulence is fully adjusted to the underlying surface. Recalling the Brunet *et al.* (1994) result, we contend that under conditions of local advection, the scintillometer should be relatively insensitive to flux divergence between the surface and the measurement height and thus it provides a more locally representative measurement than eddy correlation measurements at the same height.

## EXPERIMENTAL

The experimental site was in the middle of a large (4 km by 5 km) clearing in the forest of *Les Landes*, roughly 25 km south of Bordeaux. We chose large circular fields of maize of radius 350 and 600 m irrigated by central pivot and bordered by fallow fields of tall fescue. The experiment took place in three phases: the first was conducted over a single uniform field, the second dealt with averaging fluxes from two adjacent surfaces of roughly similar aerodynamic characteristics, while the third explored the influence of several factors pertaining to the theoretical weighting curves (Equations 2 - 5). The different configurations are shown in Figure 2 and the pertinent experimental details are summarised in Table 1.

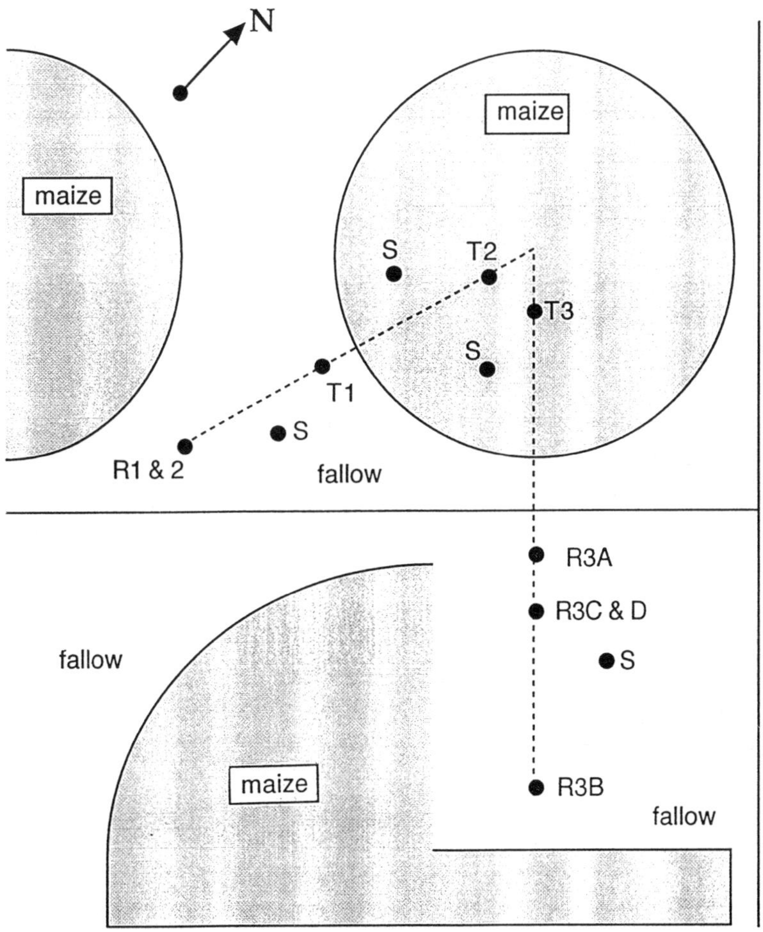


**Figure 2** Schematic plan showing experimental layout of the various configurations. T1, 2 and 3 and R1, 2 and 3A, B and C refer to the transmitter and receiver positions summarised in Table 1. S indicates positions of the eddy correlation and micro-meteorological stations.

**Table 1** Experimental configuration

| Configuration | Surface | Pathlength (m) | $D_{Trans}$ (m) | $D_{Rec}$ (m) | $W_1(W_2)$ |
|---|---|---|---|---|---|
| 1 | Fallow | 260 | 0.15 | 0.15 | |
| 2 | 50% Fallow 50% Maize | 600 | 0.15 | 0.15 | 0.50 (0.50) |
| 3A | 33% Fallow 67% Maize | 450 | 0.15 | 0.15 | 0.84 (0.16) |
| 3B | 67% Fallow 33% Maize | 870 | 0.15 | 0.15 | 0.21 (0.79) |
| 3C | 50% Fallow 50% Maize | 600 | 0.15 | 0.075 | 0.69 (0.31) |
| 3D | 50% Fallow 50% Maize | 600 | 0.075 | 0.15 | 0.35 (0.66) |

*Phase 1*

On 31 May, 1994, a single day of measurements were made over a fallow field with a transmitter-receiver distance of 260 m and light easterly winds. The fescue sward was in its second year of fallow and comprised reasonably dense clumps of foliage 30 cm high occupying some 70% of the soil surface. Above 30 cm, stalks constituted a lower density canopy up to a height of 0.8 to 1.1 m. The scintillometer was mounted at a height z =2.5 m at the top of a stayed micrometeorological mast.

*Phase 2*

Six days of measurement were undertaken between the 5th and 13th of June over a path crossing 300 m of fallow and 300 m of irrigated maize. Rows of maize were orientated NE-SW and spaced at 80 cm. During this measurement period, the maize grew rapidly from 45 to 80 cm high and by the final date of measurement had obscured almost all the inter-row soil surface. The scintillometer was again mounted at 2.5 m. Irrigation began on June 12 providing a good contrast in soil water availability for the last day of measurement (day of year 164). Prior to this date, occasional heavy rains mitigated expected contrasts in plant water status.

*Phase 3*

From July 27 to August 3, the trajectory crossed two very different surfaces. By this time the maize had attained a total height of 3.3 m and the scintillation measurement height z was raised to 5.6 m. The fescue field was in its first year of fallow with foliage to a maximum height of 30 cm and bare soil estimated to occupy some 30 to 40% of the surface. Several different experimental configurations were tested over trajectories across 273 m of maize, variable portions of dryland fallow ranging between 150 m and 570 m, and a small (27 m) interconnecting region comprising a partially water-filled ditch and pasture maintained in a well-watered condition by spray drift. In the analysis that follows, the well-watered pasture and ditch were not distinguished from the maize which was assumed to extend 300 m.

In addition to the scintillation measurements, sensible heat flux was measured independently over each surface by eddy correlation using a one-dimensional sonic anemometer and fine thermocouple (Campbell Scientific Inc., Utah, USA). Eddy correlation instruments were sited after considerations of fetch as calculated by Schuepp *et al.* (1990). Net radiation, soil heat flux and other supporting measurements, such as wet and dry-bulb air temperatures and mean windspeed, were also measured at these reference flux positions. During phases 1 and 2, air temperatures and windspeed were measured at two heights to enable backup flux estimations to be made by a simplified aerodynamic method (Panofsky and Dutton 1984). In phase 3, these variables were only monitored at a single height: windspeed at the height of the scintillometer path and estimates of surface roughness ($z_0$) and zero plane displacement ($d_0$) are necessary prerequisites to estimating the sensible heat and momentum fluxes by the procedure in Appendix A.

All data were recorded and processed on-line using portable data loggers (Campbell Scientific Inc., Utah) as described by Green *et al.* (1994) and McAneney *et al.* (1995). On occasions rapid (20 Hz) sampling of the vertical windspeed signal from the sonic anemometer was undertaken to allow estimates of surface shear stress ($u_*$) by the spectral dissipation method (see Appendix B).

## RESULTS AND DISCUSSION

*Phase 1*

This first phase of measurements accorded results in good agreement with the experience already obtained over uniform surfaces in New Zealand (data not shown). Comparison of the eddy correlation measurements of H with estimates from the simplified aerodynamic method established $d_0$ as 0.42 m for the tall fescue. Using this figure in conjunction with spectral measurements and the scintillometer-derived estimate of the Richardson number ($\xi$), we determined $z_0$ to be 0.10 m. This procedure was repeated at the completion of Phase 2 with identical results.

*Phase 2*

During these measurements, the maize grew rapidly, with attendant changes in its aerodynamic characteristics. The zero plane displacement was considered to be 0.65 of the measured crop heights giving values that varied from 35 cm on day 156 to 50 cm on day 164. These figures were then used to calculate $z_0$ from spectral measurements carried out on days 156 ($z_0$ = 5 cm), 161 ($z_0$ = 7 cm) and 164 ($z_0$ = 4 cm), as described in Appendix B. Values for other days were obtained by interpolating from these figures: the decrease observed on day 164 is consistent with the eventual closure of the inter-row space and the attainment of a relatively smooth aerodynamic surface. Wind directions were variable, but mostly between 30° and 90° to the LAS axis. On day 158 the wind blew directly down the rows of maize.

With the optical path crossing equal portions of maize and grass, weighting functions $W_1$ and $W_2$ should each have been 0.5. Figure 3 plots the scintillometer measure of $\langle C_N^2 \rangle$ against the arithmetic average of the $C_N^2$ for each field calculated from the measured values of H and U and using Equations A1 to A5. The accord is satisfactory with a slope of 1.11 (± 0.029) and a residual standard deviation of 0.09

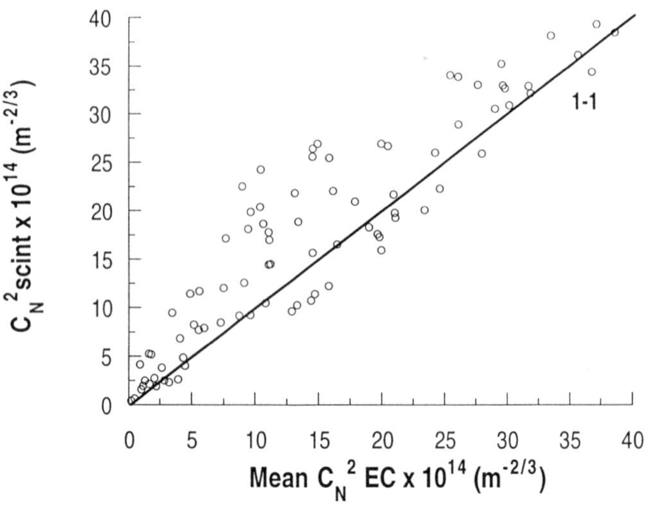


**Figure 3** Comparison of scintillometer measurements of $<C_N^2>$ against the mean of the micro meteorological estimates for each field

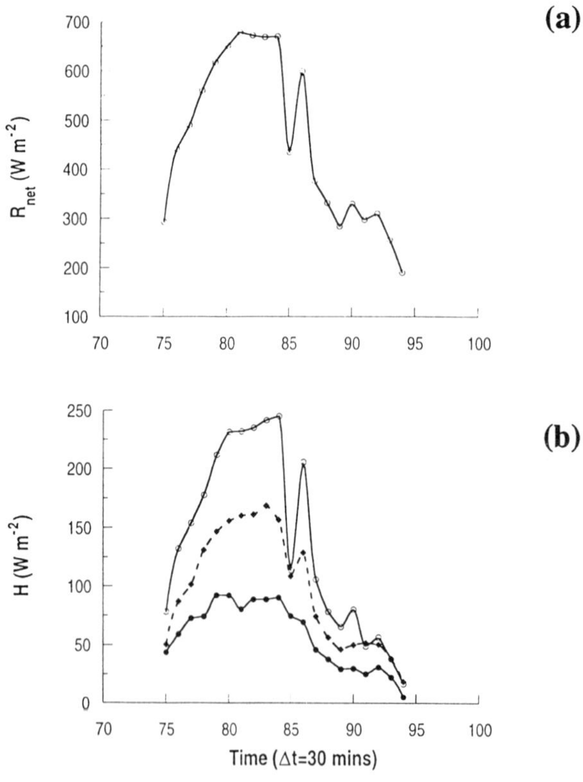


**Figure 4** Time series for day 164 of (a) net radiation, and (b) sensible heat derived from the scintillometer (solid diamonds) and direct measurement by eddy correlation in the maize (solid circles) and fallow (open circles) fields. All data points represent 30-minute flux measurements.

x $10^{-14}$ ($m^{-2/3}$). Eliminating day 158, a day when the wind blew parallel to the rows of maize and when a lower $z_o$ may well have been more appropriate than the interpolated value, improves the slope to 1.06 (±0.020). This suggests that much of the scatter between days is the result of an imperfect knowledge of $z_o$ and $d_o$ rather than because of inaccuracies in the directly measured value of $<C_N^2>$ by scintillation.

The only day with a significant between-field contrast in sensible heat flux was day 164 following overnight irrigation of the maize. On this day, despite having very different canopy architectures, the net aerodynamic characteristics — $\ln((z-d_o)/z_o)$ and hence U — were very similar for the two fields. Thus by means of irrigation, we had effectively imposed a large contrast in water availability upon a more-or-less uniformly rough surface. As can be seen in Figure 4, the scintillometer provides a plausible average of the eddy correlation measurements of sensible heat and all measurements faithfully follow the contemporaneous behaviour of net radiation.

Collecting all available data, Figure 5 illustrates very satisfactory agreement with regression analysis giving

$$H_{scint} = 1.12\ (\pm\ 0.020)H(\text{weighted EC})$$

with a residual standard deviation of 26.7 W $m^{-2}$, a figure not too different from the value we might expect if comparing two eddy correlation systems at different locations in the same field or at different heights in a constant flux layer. Excluding day 158 improves the slope to 1.03. More importantly, data from day 164 are all scattered about the 1 : 1 line with little bias.

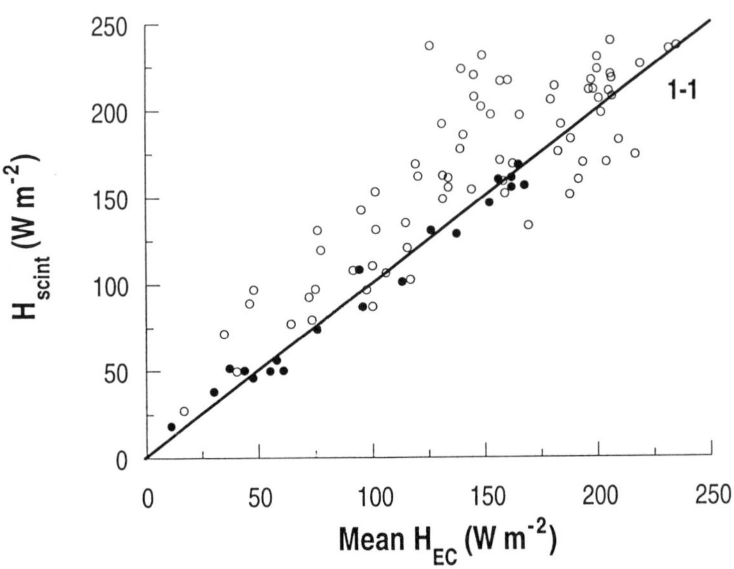


**Figure 5** Comparison of sensible heat flux derived from scintillometery against the mean of the eddy correlation values. Day 164 is represented by the solid circles.

*Phase 3* Our purpose here was to test Equations 2 - 5 by varying proportions of the trajectory traversing surfaces which differed significantly in both surface roughness and surface energy budget. Two days were also devoted to testing the influence of varying the ratio of the receiver and transmitter apertures. Winds were from the north-east, perpendicular to the path, so that local advection down the optical path was not of itself a concern. Nevertheless sensible heat fluxes were sometimes downwards over the maize. This was most common in the late afternoon on the second day after irrigation, when a combination of large surface roughness, high saturation deficits, bountiful soil water and a falling radiation load predisposed the crop towards such stable conditions (McAneney & Itier 1996). Over the fallow field, daytime conditions were always unstable. The changing sign of the sensible heat along the path poses problems when attempting to calculate a path-averaged heat flux from the scintillometer; we have stopped short of attempting this analysis. The previous success of the scintillometer in Phases 1 and 2 is taken to indicate that the instrument itself responds faithfully to $<C_N^2>$, with a symmetric weighting function when transmitter and receiver apertures are equal. The relevant theoretical weighting functions for the different experimental configurations are listed in Table 1.

The results are summarised in Figure 6. Regression using all the data gives

$$C_N^2 \text{ (weighted EC)} = 1.02(\pm 0.031) C_N^2 \text{ (scint)}$$

with a residual standard deviation of $1.263 \times 10^{-14}$ ($m^{-2/3}$) and a statistically non-significant intercept. While this global result is gratifying, there is a tendency for the weighted $C_N^2$, as determined by 'back-calculation' from the eddy correlation

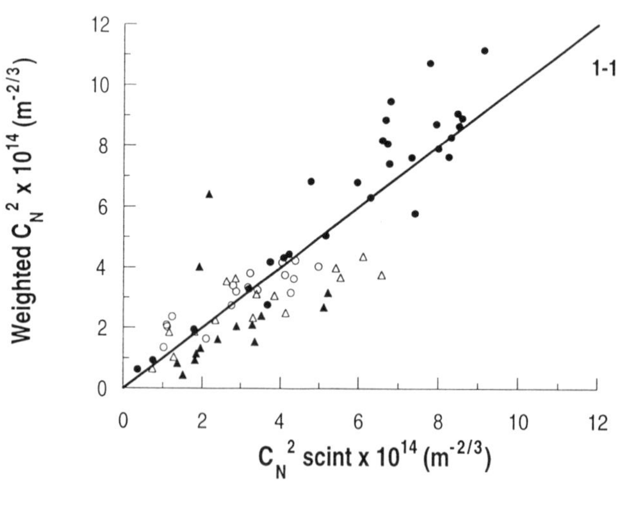


**Figure 6** Weighted average of the micro-meteorological estimates of $<C_N^2>$ versus from the scintillometer. Open circles represent configuration 3A (450 m), solid circles 3B (870 = m), and open and closed triangles 3C and 3D.

measurements, to underestimate the scintillometer response when aperture diameters were unequal. The reverse seems true for the longest pathlength configuration. Two factors may explain those apparent deviations from theory. In the former case the apertures were stopped down with a plastic cap having an off-centred circular hole of diameter 7.5 cm. The hole was off-centred in order to minimise blockage by either the central source or the detector mounting assemblies (~2.5 cm diameter) but obviously did not eliminate the effect entirely. The effective D (Equation 3) is somewhat smaller than 7.5 cm because of this, and/or because of an unequal radiance pattern of the light emitting diode (Kohseik, 1987). For the long pathlength case, there is some small uncertainty as to the true path length — 850 m or 900 m — and the authors have negotiated a middle path: the shorter distance would bring the two independent sets of measurements almost perfectly into line. There is an important lesson here and we have since adopted the precaution of a surveyor's range-finder to determine the pathlength accurately.

## CONCLUDING REMARKS

The collective results of tests conducted using different path lengths and aperture diameters above surfaces of varying aerodynamic properties and water budgets leads us to conclude that the scintillometer indeed provides a sensitive measure of the path-weighted $C_N^2$ as suggested by theory. We stress the importance of knowing path-length accurately, as the signal variance varies as the cube of this distance. When the underlying surfaces have similar roughness, a path-averaged sensible heat flux can be extracted without difficulty. This is more problematic should the direction of the sensible heat flux change sign along the path because the scintillometer cannot distinguish between refractive turbulence created under stable or unstable conditions. These problems, plus those created by large contrasts in surface roughness, will be scrutinised in subsequent studies.

## ACKNOWLEDGMENTS

The authors gratefully acknowledge the help of Alistair Hall (Hort Research, Palmerston North) in matters theoretical and heretical.

## REFERENCES

Brunet, Y., Itier, B., McAneney, K.J. and Lagouarde, J.P. 1994. Downward evolution of scalar fluxes and surface resistance under conditions of local advection. Part II: measurements over barley. *Agric. For. Meteorol.*, **71**,227-245.

Clifford, S.F., Ochs, G.R. and Lawrence, R.S. 1974. Saturation of optical scintillation by strong turbulence. *J. Opt. Soc. Am.*, **64**, 148-154.

Fairall, C.W., Edson, J.B., Larsen, S.E. and Mestayer, P.G. 1990. Inertial dissipation air-sea flux measurements: a prototype system using realtime spectral computations. *J. Atmos. and Ocean Techn.*, **7**, 425-453.

Gash, J.H.C. 1986. A note on estimating the effect of limited fetch on micrometeorological evaporation measurements. *Boundary-layer Meteorol.*, **35**, 409-413.

Green, A.E., McAneney, K.J., Astill, M.S. 1994. Surface layer scintillation measurements of daytime heat and momentum fluxes. *Boundary-layer Meteorol.*, **68**, 357-373.

Kaimal, J.H.C., Wyngaard, J.C., Izumi, Y. and Coté, O.R. 1972. Spectral characteristics of surface layer turbulence. *Quart. J. Roy. Met. Soc.*, **98.** 563-589.

Kohseik, W. 1987. A 15 cm aperture LED scintillomer for $C_N^2$ and crosswind measurements. KNMI Scientific Report WR 87-3, KNMI, de Bilt, The Netherlands.

McAneney, K.J., Green, A.E. and Astill, M.S. 1995. Large-aperture scintillometry: the homogeneous case. *Agric. For. Met.*, **76**,149-162.

McAneney, K.J. and Itier, B. 1996. Operational limits to the Priestley and Taylor formula. *Irrig. Sci.* In press.

Ochs, G.R. and Cartwright, W.D. 1980. *Optical system model IV for space-averaged wind and $C_N^2$ measurements.* NOAA Tech. Memo. ERL WPL-52, NOAA Environmental Research Laboratories, Bounder CO.

Panofsky, H.A and Dutton, J.A. 1984.*Atmospheric Turbulence: Models and Methods for Engineering Applications.* John Wiley & Sons, Inc. NY, 397pp.

Schuepp, P.H., Leclerc, M.Y., Macpherson, J.I. and Desjardins, R.L. 1990. Footprint predictions of scalar fluxes from analytical solutions of the diffusion equation. *Boundary-layer Meteorol.*, **50**, 355-373.

Tatarskii, V.I. 1961. *Wave propagation in turbulent medium.* McGraw-Hill, New York.

Tatarskii, V.I. 1993. Review of scintillation phenomena, pp. 2-15. In: Wave Propagation in Random Media (Scintillation) V.I. Tatarskii, A. Ishimaru and V.U. Zavorotny (eds.) Co published by SPIE - The International Soc. for Optical Engineering and Institute of Physics Publishing.

Wang, Ting-i, Ochs, G.R. and Clifford, S.F. 1978. A saturation-resistant optical scintillometer to measure $C_N^2$. *J. Opt. Soc. Am.*, **68**, 334-338.

Weseley, M.L. and Derrzko, Z.I. 1975. Atmospheric turbulence parameters from visual resolution. *Appl. Optics,* 14 847-853.

Weseley, M.L. 1976. The combined effect of temperature and humidity fluctuations on refractive index. *J. Appl. Meteorol.* 15 43-49.

## APPENDIX A

In what follows we describe the estimation of H and u* from optical measurements of $C_N^2$ and U. Exactly the same equations can be used to calculate $C_N^2$ from U and H. Both forms of calculations were used for analyses discussed in the text.

For scintillation at visible or near-infrared wavelengths, the temperature structure parameter $C_T^2$ is related to $C_N^2$ by:

$$C_T^2 = C_N^2(T^2/\gamma P)^2(1+0.03/\beta)^2 \qquad \text{(A1)}$$

where T is the average air temperature (K), P the atmospheric pressure (Pa), the refractive index parameter $\gamma$ is $7.9 \times 10^{-7}$ K Pa$^{-1}$, and $\beta$ is the Bowen ratio (Weseley, 1976).

Since the sensible heat and momentum fluxes together determine atmospheric stability, which in turn influences turbulent transport, an iterative procedure is necessary to calculate fluxes from $C_T^2$ and U. Following Panofsky and Dutton (1984), we begin by defining a temperature scale:

$$T_* = -H/\rho c_p u_* \tag{A2}$$

where $\rho$ is the density of air and $c_p$ is the heat capacity at constant pressure. $C_T^2$ is written in dimensionless form and assumed to be a universal function of $\xi$ (= $z/L_{mo}$, where $L_{mo}$ is the Monin-Obukhov length)

$$C_T^2/T_*^2 z^{-2/3} = f(\xi)$$

where

$$f(\xi) = 4.9(1 + 7|\xi|)^{-2/3} \qquad \xi \leq 0$$

or

$$f(\xi) = 4.9(1+2.5\xi) \qquad \xi > 0 \tag{A3}$$

and

$$\xi = z/L_{mo} = (z-d_o)g\kappa T_*(1+0.07/\beta)/u_*^2 T \tag{A4}$$

where von Karman's constant $\kappa$ is taken as 0.4 and g is the acceleration due to gravity.

The vertical profile of mean wind velocity is given by

$$u_* = \kappa U/[\ln((z - d_o)/z_o) - ¥_m(\xi)] \tag{A5}$$

where $¥_m$ the appropriate integrated stability correction. For uniform vegetative surfaces, $z_o$ and $d_o$ can be taken as simple fractions of the vegetation height ($h_c$):

$$d_o = 0.65h_c \qquad \text{and} \qquad z_o \approx 0.13h_c \tag{A6}$$

A5 then allows $u_*$ to be determined given a measurement of U, estimates of $z_o$ and $d_o$ and an initial guess for $\xi$. $C_T^2$ is calculated from the measured optical signal (A1) and an assumed initial value for $\beta$. T* follows from (A3) and H from (A2). This leads to a second approximation of $\xi$ from (A4). The procedure is repeated until the estimated and calculated values for $\xi$ converge. For stable conditions, the Bowen ratio correction in A4 was ignored to ensure convergence. The calculations adapt easily to a computer-based spreadsheet.

## APPENDIX B

Here we describe the spectral dissipation method for determining $u_*$. After Panofsky and Dutton (1984), spectral densities of the vertical windspeed $S_w(f)$ in the inertial range of frequencies ($f$) are written as

$$fS_w(f)/u_*^2 = 0.2\phi_\varepsilon(\xi)^{2/3}\kappa^{-2/3}(f(z-d_o)/U)^{-2/3} \tag{B1}$$

where for unstable air

$$\phi_\varepsilon(\xi) = 1 - \xi$$

and $S_w(f)$ has units of velocity squared per unit frequency; all other units have been defined previously.

The $S_w(f)$ data were then plotted against frequency on a log-log graph, a –5/3 line fitted by eye through the data and $S_w(f)$ determined at the point where $f$ = 2U. (Kaimal *et al.* (1972) and Fairhall *et al.* (1990) expect inertial subrange behaviour provided that $f$ exceeds 2U/z.) Shear stress follows after substituting for $S_w(f)$ in B1 and assuming a value for $d_o$ based on plant height (see Equation A6). Finally, Equation A5 was solved to give a value of $z_o$ to consistent with the above choice of $d_o$. Clearly $z_o$ and $d_o$ are not independent but this is not a drawback for this particular application. The iterative calculations described in Appendix A provided the estimate of $\beta$.

# 10 Canopy to region scale translation of surface fluxes

**J.-P. LHOMME, A. CHEHBOUNI and B. MONTENY**
*ORSTOM, Montpellier, France*

## INTRODUCTION

Since most of the relevant variables in hydrology, such as precipitation, evaporation, infiltration, runoff are both measured and modelled at a local scale, the canopy-region (or patch-region) scale translation has become an important issue, essential to comprehend, describe and model hydrological phenomena at regional scale. This issue has become very acute in relation to the parameterisation of land surface processes in large-scale atmospheric models. The topic discussed in this chapter is the areal averaging problem in land-atmosphere interaction, applicable to canopy-region scale translation. More precisely, it deals with convective fluxes of sensible and latent heat estimated by the Penman-Monteith formulae. The strategy used is based upon two basic principles: (1) matching of the model between scales, and (2) the scalar conservation, which implies that the spatially averaged flux of a given scalar (heat or water vapour) is simply the area-weighted mean of the flux contributions from each patch. The strategy used also involves the concept of «blending height», which has proved to be successful in calculating regionally averaged surface fluxes (Mason, 1988; Claussen, 1991; Blyth *et al.*, 1993). This concept stipulates that there exists a height at which the characteristics of the air flow (wind velocity, temperature, humidity) become approximately independent of horizontal position and can be extrapolated from one patch to another. Several aggregation schemes based upon this strategy have been recently published (Raupach, 1991; McNaughton, 1994; Lhomme *et al.*, 1994; Raupach, 1995; Braden, 1995). They apparently differ for reasons that have not been clearly explained. Here we examine these differences to show in what way they differ and why.

## BASIC EQUATIONS AND GENERIC CONSTRAINTS

The convective flux of latent heat is expressed in the form of the Penman-Monteith equation

$$\lambda E = \frac{sA + \rho c_p D_a / r_a}{s + \gamma (r_v / r_a)} \tag{1}$$

*Scaling up in hydrology using remote sensing.* Edited by J.B. Stewart, E.T. Engman, R.A. Feddes and Y. Kerr

where $A$ is the available energy, $D_a$ is the water vapour pressure deficit of the air in the well-mixed layer, $\rho$ is the mean air density, $c_p$ is the specific heat of air at constant pressure, $\gamma$ is the psychrometric constant, *and* $s$ is the rate of change of the saturated vapour pressure with temperature. $r_a$ is the aerodynamic resistance to heat and water vapour transfer from the surface to the well-mixed layer. $r_v$ is the total resistance to water vapour transfer expressed as $r_v = r_a + r_s$, where $r_s$ is the surface resistance. The sensible heat flux is given by a similar equation

$$H = \frac{\gamma A - \rho c_p D_a / r_v}{\gamma + s(r_a / r_v)} \tag{2}$$

Both equations are based on a linearisation of the saturated vapour pressure curve between the surface temperature $T_s$ and the air temperature in the well-mixed layer $T_a$. This linearisation is justified as a first approximation (Paw U and Gao, 1988). The available energy $A$ is detailed as

$$A = R_n - G = (1 - \alpha)R_s + \epsilon(R_l - \sigma T_s^4) - G \tag{3}$$

with $\alpha$ and $\varepsilon$ the albedo and the emissivity of the surface, $R_s$ and $R_l$ the incoming short-wave and long-wave radiations, $T_s$ the surface temperature, $\sigma$ the Stefan-Boltzmann constant and $G$ the soil heat flux.

We now aim to derive the aggregation procedures to be used to calculate the effective parameters of the energy balance at regional scale. At this scale they will be denoted by angle brackets. The air characteristics ($T_a$, $D_a$) in the well-mixed layer and the incoming radiations ($R_s$, $R_l$) are assumed to be constant over the whole area. Since surface emissivity ($\varepsilon$) does not vary markedly among the natural surfaces and much less than the other parameters, it is also assumed to be constant over the whole area. The strategy used consists in determining the effective parameters in such a way that the flux equations, valid at local scale, give the correct areal fluxes at grid-square scale. From the definition of a flux and the principle of scalar conservation, the spatially-averaged flux over a region or a grid-square is simply the area-weighted mean of the flux contributions from each patch. So, it is possible to write the following conservation equation

$$\langle\phi\rangle = \sum_i a_i \phi_i \qquad \textit{with} \qquad \phi = H, \lambda E, A \tag{4}$$

where $a_i$ is the fractional area of patch $i$. It is worthwhile noting that the energy balance equation ($H + \lambda E = A$) is always closed at regional scale when each individual flux is preserved according to Equation 4, since the following equations hold

$$\langle\lambda E\rangle + \langle H\rangle = \sum_i a_i \lambda E_i + \sum_i a_i H_i = \sum_i a_i(\lambda E_i + H_i) = \sum_i a_i A_i = \langle A\rangle \tag{5}$$

Available energy can be preserved according to Equation 4 as an independent magnitude, irrespective of flux equations ($H$ and $\lambda E$). This preservation leads to

$$<A>=(1-<\alpha>)R_s+\epsilon(R_l-\sigma<T_s>^4)-<G>=\sum_i a_i[(1-\alpha_i)R_s+\epsilon(R_l-\sigma T_{s,i}^4)-G_i] \tag{6}$$

from which the following averaging procedures can be derived by matching term by term

$$<X>=\sum_i a_i X_i \qquad for \qquad X=\alpha, G, T_s^4 \tag{7}$$

Replacing $T_s^4$ by $T_s$ in the averaging procedure above does not lead to a significant error. We can then say that all the parameters average linearly in the preservation of available energy.

# FLUX-MATCHING AVERAGING PROCEDURES

## LATENT HEAT FLUX

The Penman-Monteith equation is written with effective parameters as

$$<\lambda E>=\frac{s<r_a><A>+\rho c_p D_a}{s<r_a>+\gamma<r_v>} \tag{8}$$

since $D_a$ is supposed not to change over the whole area. Using Equation 4, with $\phi=\lambda E$, and putting

$$\omega_i=\frac{1}{sr_{a,i}+\gamma r_{v,i}} \tag{9}$$

the areal latent heat flux also reads as

$$<\lambda E>=s\sum_i a_i\omega_i\, r_{a,i}\, A_i+\rho c_p D_a\sum_i a_i\omega_i \tag{10}$$

Matching Equations 8 and 10, term by term, gives two equations:

$$\frac{<r_a><A>}{s<r_a>+\gamma<r_v>}=\sum_i a_i\omega_i\, r_{a,i}\, A_i \tag{11}$$

and

$$\frac{1}{s<r_a>+\gamma<r_v>}=\sum_i a_i\omega_i \tag{12}$$

Assuming $<A>$ is known and given by Equation 6, the following averaging procedure is obtained for the effective resistances ($<r_a>$, $<r_v>$ and $<r_s>$)

$$<r_x>=\frac{\sum_i a_i\omega_i(A_i/<A>)r_{x,i}}{\sum_i a_i\omega_i} \qquad with \qquad x=a,v \text{ or } s \tag{13}$$

It is the solution given by Raupach (1991, 1995) and McNaughton (1994). In fact, Equation 12 is not quite the equation given by Raupach (1995) because he uses the «equilibrium departure» form of the combination equation and the isothermal net radiation, which leads to a slightly different algebraic expression, but the result is fundamentally the same.

Another solution can be obtained by replacing $<A>$ in Equation 8 and $A_i$ in Equation 10 by the detailed expression of available energy (Equation 3), before matching term by term (Lhomme *et al.*, 1994). This matching leads to the following set of equations

$$\frac{<r_a>}{s<r_a>+\gamma<r_v>}=\sum_i a_i\omega_i\, r_{a,i} \tag{14}$$

$$\frac{<r_a><\alpha>}{s<r_a>+\gamma<r_v>}=\sum_i a_i\omega_i\, r_{a,i}\alpha_i \tag{15}$$

$$\frac{<r_a><T_s>^4}{s<r_a>+\gamma<r_v>}=\sum_i a_i\omega_i\, r_{a,i}T_{s,i}^4 \tag{16}$$

$$\frac{<r_a><G>}{s<r_a>+\gamma<r_v>}=\sum_i a_i\omega_i\, r_{a,i}G_i \tag{17}$$

$$\frac{1}{s<r_a>+\gamma<r_v>}=\sum_i a_i\omega_i \tag{18}$$

This set of five equations with five unknowns (in angle brackets) has the following solution. For $<\alpha>$, $<G>$ and $<T_s>^4$, the averaging scheme is given by

$$<X>=\frac{\sum_i a_i\omega_i r_{a,i}\, X_i}{\sum_i a_i\omega_i r_{a,i}} \qquad with \qquad X=\alpha, G \text{ or } T_s^4 \tag{19}$$

and for $<r_a>$, $<r_v>$ and $<r_s>$, by

$$<r_x>=\frac{\sum_i a_i\omega_i\, r_{x,i}}{\sum_i a_i\omega_i} \qquad with \qquad x=a, v \text{ or } s \tag{20}$$

Calculating $<\alpha>$, $<G>$ and $<T_s>^4$ according to Equation 19 is strictly equivalent to calculating $<A>$ in Equation 8 as

$$<A>_E=\frac{\sum_i a_i\omega_i r_{a,i}\, A_i}{\sum_i a_i\omega_i r_{a,i}} \tag{21}$$

The method developed by Braden (1995) is very similar to the one proposed by Lhomme *et al.* (1994), with the difference that the surface temperature (expressed as net radiation) is eliminated by linearisation. An additional term thus appears which is the radiative resistance (Monteith and Unsworth, 1990; Lhomme, 1992) in parallel with the convective resistance to heat transfer. It should also be pointed out that Braden (1995) does not develop the basic expressions to derive the effective values of the physical parameters.

## SENSIBLE HEAT FLUX

The same procedure is used as for latent heat flux. It consists in matching term by term the equation written with effective parameters,

$$\langle H\rangle = \frac{\gamma\langle r_v\rangle\langle A\rangle - \rho c_p D_a}{s\langle r_a\rangle + \gamma\langle r_v\rangle} \tag{22}$$

and the equation obtained from the conservation principle,

$$\langle H\rangle = \gamma\sum_i a_i \omega_i\, r_{v,i}\, A_i - \rho c_p D_a \sum_i a_i \omega_i \tag{23}$$

with $\omega_i$ given by Equation 9. When $\langle A\rangle$ is not replaced by its expression (Equation 3), and assumed to be known, the solution for the effective resistances is the same as for latent heat, and is given by Equation 13. It is the solution proposed by Raupach (1991, 1995). When $\langle A\rangle$ is developed according to the procedure described for latent heat flux, the solution for the effective resistances is also the same as for latent heat, and is given by Equation 20. It is the solution proposed by Lhomme *et al.* (1994). The other parameters ($a$, $G$ and $T_s$) are calculated as

$$\langle X\rangle = \frac{\sum_i a_i \omega_i r_{v,i}\, X_i}{\sum_i a_i \omega_i r_{v,i}} \qquad \textit{with} \qquad X = \alpha, G, T_s^4 \tag{24}$$

which means, in this case, that $\langle A\rangle$ in Equation 22 must be replaced by $\langle A\rangle_H$ given by

$$\langle A\rangle_H = \frac{\sum_i a_i \omega_i r_{v,i}\, A_i}{\sum_i a_i \omega_i r_{v,i}} \tag{25}$$

Braden (1995) finds a similar result for sensible heat flux with the restriction stated concerning latent heat flux. The two methods developed above (Raupach's and Lhomme *et al.*'s) are summed up and compared in Table 1.

**Table 1** Comparison of the two methods of calculating the effective parameters of the Penman-Monteith equations for sensible and latent heat flux.

| | $\langle X\rangle = \langle\alpha\rangle$, $\langle G\rangle$ or $\langle T_s^4\rangle$ | $\langle r_x\rangle$ $(x=a, v, s)$ | $\langle A\rangle$ in $\langle H\rangle$ equation | $\langle A\rangle$in$\langle\lambda E\rangle$ equation |
|---|---|---|---|---|
| Raupach's method | $a_iX_i$ | $a_i\omega_iA_ir_{xi}$ / $\langle A\rangle\, a_i\omega_i$ | $a_iA_i$ | $a_iA_i$ |
| Lhomme *et al.*'s method | $a_iX_i$ | $a_i\omega_i r_{xi}$ / $a_i\omega_i$ | $a_i\omega_i r_{vi} A_i$ / $a_i\omega_i r_{vi}$ | $a_i\omega_i r_{ai} A_i$/ $a_i\omega_i r_{ai}$ |

with $\omega_i = 1/(s\, r_{ai} + \gamma r_{vi})$

# DISCUSSION

## COMPARISON OF THE TWO APPROACHES

A numerical simulation has been carried out to assess the performance of the two averaging procedures. A heterogeneous surface consisting of just two patches with the same area has been considered. Three cases, the same as in Lhomme *et al.* (1994), corresponding to three different combinations of two contrasted patches, have been analysed: crop-desert, forest-water, desert-water. The patch characteristics are given in Table 2. The aerodynamic resistance $r_a$ is calculated between the surface and a reference height ($z_m$) within the well mixed layer by the following equation which assumes conditions of neutral atmospheric stability

$$r_a = \frac{\ln^2(z_m/z_0)}{k^2 u_m} \tag{26}$$

$z_0$ is the roughness length, $k$ is the von Karman constant (0.4), and $u_m$ is the wind velocity at the reference height $z_m$ set to 50 m. Soil heat flux is calculated as a given fraction of the isothermal net radiation $G = \zeta R_n^*$ ($R_n^*$ being defined by Equation 3 in which $T_s$ is replaced by $T_a$). This parameterisation is substantiated by the results from Clothier *et al.* (1986). For each component of the whole area, the albedo ($\alpha$), the surface resistance ($r_s$), the roughness length ($z_0$) and the coefficient $\zeta$ are specified as input. The surface temperature of each patch is calculated by solving the energy balance equation.

The results of the simulations are presented in Table 3 which gives the three fluxes ($H$, $\lambda E$ and $A$) for each patch and for the whole area. For the whole area the

**Table 2** Characteristics of the surfaces considered in the numerical simulations

| Parameters | forest | crop | desert | water |
|---|---|---|---|---|
| $\alpha$ | 0.1 | 0.2 | 0.3 | 0.05 |
| $\zeta$ | 0.01 | 0.05 | 0.3 | 0.6 |
| $z_0$(m) | 1.0 | 0.1 | 0.01 | 0.001 |
| $r_s$(s m$^{-1}$) | 100 | 100 | 10000 | 0 |

$\alpha$: albedo; $\zeta=G/R_n^*$; $z_0$: roughness length; $r_s$: surface resistance.

flux is obtained by areally averaging the component fluxes according to Equation 4. This value is considered as the true areal flux and is used as a reference, with which to compare the estimates. Three areal estimates are given. The first (a) is based upon a simple areal averaging of the surface parameters (albedo, soil heat flux, surface temperature and resistances). The second (b) is based on Raupach's averaging procedure. The third (c) uses Lhomme *et al.*'s averaging scheme. In each

**Table 3** Comparison between the areal fluxes calculated by different means. The areal value is obtained from Equation 4. The areal estimates use three types of aggregation procedure: (a) simple areal averaging of the parameters, (b) averaging procedures according to Raupach's method, and (c) averaging procedures according to Lhomme *et al.*'s method. The climatic conditions used are : $R_s$=800 W m$^{-2}$, $R_l$=350 W m$^{-2}$, and at $z_m$=50 m, $T_a$=25 C, $e_a$=1500 Pa, $u_m$=5 m s$^{-1}$.

| | Patch 1 | Patch 2 | Areal value | Areal estimates (a) | (b) | (c) |
|---|---|---|---|---|---|---|
| *Case 1:* | *crop* | *desert* | | | | |
| $\lambda E$ (Wm$^{-2}$) | 334 | 8 | 171 | 18 | 171 | 171 |
| $H$ (Wm$^{-2}$) | 149 | 216 | 183 | 337 | 183 | 183 |
| $A$ (Wm$^{-2}$) | 483 | 224 | 354 | 355 | 354 | 354 |
| $r^*$ (%) | 0 | 0 | 0 | 0 | 0 | 0 |
| *Case 2:* | *forest* | *water* | | | | |
| $\lambda E$ (Wm$^{-2}$) | 353 | 244 | 299 | 354 | 299 | 299 |
| $H$ (Wm$^{-2}$) | 242 | 12 | 127 | 72 | 127 | 27 |
| $A$ (Wm$^{-2}$) | 595 | 257 | 426 | 426 | 426 | 426 |
| $r$ (%) | 0 | 0 | 0 | 0 | 0 | 0 |
| *Case 3:* | *desert* | *water* | | | | |
| $\lambda E$ (Wm$^{-2}$) | 8 | 244 | 126 | 19 | 126 | 126 |
| $H$ (Wm$^{-2}$) | 216 | 12 | 115 | 222 | 114 | 114 |
| $A$ (Wm$^{-2}$) | 224 | 257 | 241 | 242 | 241 | 241 |
| $r$ (%) | 0 | 0 | 0 | 0 | 0 | 0 |

* $r=1-(H+\lambda E)/A$

case, the closure of the energy balance equation is tested by calculating the ratio $r=[A-(H+\lambda E)]/A$ expressed in percentage. The numerical simulations show that the simple areal averaging of the parameters (a) leads to a rather poor estimate of the areal fluxes of sensible and latent heat whereas methods (b) and (c) give perfect and equivalent results.

If the two methods are numerically equivalent for preserving the fluxes, they do not have the same characteristics and do not yield the same values for the effective resistances. In Raupach's scheme, the resistances are weighted not only by surface characteristics, as in Lhomme's scheme, but also by climatic conditions through available energy (which complicates the averaging procedure since the weighting coefficients depend upon the climatic inputs). Moreover, according to Raupach's averaging scheme, when all the patches have the same resistance $r_x$ ($x$ stands for $a$, $v$ or $s$), the corresponding effective resistance $<r_x>$ is not necessarily this common value $r_x$, since $\Sigma a_i\, \omega_i\, A_i$ is not necessarily equal to $<A>\Sigma a_i \omega_i$. This result, a little surprisingly, is not produced by Lhomme *et al.*'s scheme.

In the averaging procedures obtained and summarized in Table 1, the effective surface temperature and the effective albedo are simple area-averages of component values, which intuitively conform to the physics of the measurements. Over a heterogeneous surface, the radiometer receives a radiative flux proportional to the relative area of each patch and to the fourth power of its absolute temperature and, considering constant emissivity, the measured temperature can be approximated by

$$<T_s>_{meas.} = \sum_i a_i T_{s,i} \qquad (27)$$

As to the albedo, it is measured by the radiometer as an area-average of the component albedos, like temperature. So, for both parameters there is equivalence between satellite-measured and effective parameters.

## SCHEMES NOT BASED UPON THE FLUX CONSERVATION PRINCIPLE

Lhomme (1992) proposed an averaging scheme based on the preservation of surface temperature expressed by developing the energy balance equation. $<T_s>$ is defined as the simple area-weighted temperature, such as would be measured by a remote infrared thermometer (Equation 27). Developing the energy balance and matching term by term leads to the following averaging scheme for the aerodynamic resistance

$$<r_a> = \frac{\sum_i a_i \omega_i^l\, r_{v,i}\, r_{a,i}}{\sum_i a_i \omega_i^l\, r_{v,i}} \qquad (28)$$

and for the global resistance to water vapour transfer

$$<r_v> = \frac{\sum_i a_i \omega_i' \, r_{a,i} \, r_{v,i}}{\sum_i a_i \omega_i' \, r_{a,i}} \tag{29}$$

where $\omega'_i$ is defined as

$$\omega_i' = \frac{1}{s r_{a,i} + \gamma r_{v,i}(1 + r_{a,i}/r_0)} \tag{30}$$

$r_0$ being the radiative resistance obtained by linearisation of $T_s^4$ (Monteith and Unsworth, 1990; Lhomme, 1992). When there is no radiative coupling ($T_s = T_a$ in the available energy formulation), $r_0$=infinity and $\omega'_i = \omega_i$. In this averaging procedure, available energy is no more the area-weighted mean of patch contributions. The weighting coefficients are more complex and given by $a_i \omega'_i r_{a,i} r_{v,i}$. Braden (1995) has shown that this method can lead to strong deviations with respect to the correct areal fluxes. Another procedure, based upon similar principles, was proposed by Chehbouni *et al.* (1995). The averaging scheme is based solely on the preservation of the energy balance equation and leads to a set of effective parameters, where $<\alpha>$, $<G>$, $1/<r_a>$, $1/<r_v>$ are direct area-weighted averages of the individual values, and the effective surface temperature is weighted by the component resistances in the following way :

$$<T_s> = \frac{\sum_i a_i T_{s,i} \, /(\omega_i' \, r_{a,i} \, r_{v,i})}{\sum_i a_i \, /(\omega_i' \, r_{a,i} \, r_{v,i})} \tag{31}$$

This aggregation scheme seems to perform better than the one proposed by Lhomme (1992), when fluxes are calculated by the basic diffusion equations (Chehbouni *et al.*, 1995). Nevertheless, this procedure does not preserve the fluxes when calculated by the Penman-Monteith equations.

## CONCLUSION

When the Penman-Monteith equations are used to express sensible and latent heat fluxes, there are two different ways of calculating the effective parameters. The two methods, which are summarized in Table 1, are strictly equivalent from a flux conservation standpoint, yielding the same numerical results. In both schemes, the effective parameters for available energy (surface temperature, albedo and soil heat flux) are direct area averages of the component values, but the effective resistances are not calculated in the same way. In Lhomme *et al.*'s scheme they are weighted only by surface characteristics (relative areas and component resistances), whereas in Raupach's procedure, the weighting factors are also a function of climatic conditions through available energy. This difference is compensated for by the fact that effective available energy in the Penman-Monteith equations is not expressed in the same way in the two schemes. With Raupach's scheme, $<A>$ is a simple area weighted mean of the component available energies, whereas in Lhomme *et al.*'s scheme the weighting factors also involve surface resistances as detailed in Table 1. It seems that the latter averaging scheme is sounder and more straightforward, firstly since it does not involve climatic conditions and secondly, because when the different patches of the regional mosaic have the same (surface or aerodynamic) resistances, it restores this common value as the effective resistance, while Raupach's scheme does not.

## REFERENCES

Blyth, E.M., Dolman, A.J. and Wood, N. 1993. Effective resistance to sensible and latent heat flux in heterogeneous terrain. *Q. J. R. Meteorol. Soc.*, **119**, 423-442.

Braden, H. 1995. Energy fluxes from heterogeneous terrain: averaging input parameters of the Penman-Monteith formula. *Agric. For. Meteorol.*, **75**, 121-133.

Chehbouni, A., Njoku, E.G., Lhomme, J.P. and Kerr, Y.H. 1995. An approach for averaging surface temperature and surface fluxes over heterogeneous terrains. *J. Climate*, **8**, 1386-1393.

Claussen, M. 1991. Estimation of areally-averaged surface fluxes. *Boundary-Layer Meteorol.*, **54**, 387-410.

Clothier, B.E., Clawson, K.L., Pinter, P.J., Moran, M.S., Reginato, R.J. and Jackson, R.D. 1986. Estimation of soil heat flux from net radiation during the growth of alfalfa. *Agric. For. Meteorol.*, **37**, 319-329.

Lhomme, J.P. 1992. Energy balance of heterogeneous terrain: averaging the controlling parameters. *Agric. For. Meteorol.*, **61**, 11-21.

Lhomme, J.P., Chehbouni, A. and Monteny, B. 1994. Effective parameters of surface energy balance in heterogeneous landscape. *Boundary-Layer Meteorol.*, **71**, 297-309.

McNaughton, K.G. 1994. Effective stomatal and boundary-layer resistances of heterogeneous surfaces. *Plant Cell Environ.*, **17**, 1061-1068.

Mason, P.J. 1988. The formation of areally-averaged roughness lengths. *Q. J. R. Meteorol. Soc.*, **114**, 399-420.

Monteith, J.L. and Unsworth, M.H. 1990. Principles of Environmental Physics. Edward Arnold, London.

Paw U, K.T. and Gao, W. 1988. Applications of solutions to non-linear energy budget equations. *Agric. For. Meteorol.*, **43**, 121-145.

Raupach, M.R. 1993. The averaging of surface flux densities in heterogeneous landscapes. In: H.J. Bolle, R.A. Feddes and J.D. Kalma (Eds), Exchange Processes at the Land Surface for a Range of Space and Time Scales (Proceedings of the Yokohama Symposium, July 1993). *IAHS Publ.* no. 212.

Raupach, M.R. 1995. Vegetation-atmosphere interaction and surface conductance at leaf, canopy and regional scales. *Agric. For. Meteorol.*, **73**, 151-179.

# 11 Responses of the convective boundary layer and the surface energy balance to large-scale heterogeneity

**K.G. McNAUGHTON[1] and M.R. RAUPACH[2]**
[1]*Horticulture and Food Research Institute of New Zealand, Kerikeri, New Zealand*
[2]*CSIRO Centre for Environmental Mechanics, Canberra, Australia*

## INTRODUCTION

The modern phase of research into land-atmosphere interactions can be characterised by a general recognition of the pivotal role of the atmospheric boundary layer. Its importance has long been recognised in air-pollution meteorology, where the capping inversion at the top of the daytime convective boundary layer (CBL) so clearly limits observed haze layers. The importance of the CBL to the regional surface energy balance was recognised later (Deardorff, 1978; Troen and Mahrt, 1986; McNaughton and Spriggs, 1986). Attempts to infer the surface fluxes of trace gases by measuring their accumulation rates in the CBL are now under way (e.g. Raupach, *et al.*, 1992).

In the context of land-atmosphere interactions, the CBL has two key general properties. First, it acts as a natural integrator by accumulating entities exchanged at the surface, such as heat, water vapour, $CO_2$, trace gases or pollutants. This means that the concentrations of these entities in the CBL change quite slowly even when their fluxes at the surface change rapidly, so that CBL concentrations are relatively insensitive to short-term details of the surface fluxes, and are mainly determined by the long-term history of the surface fluxes over the lifetime of the CBL (the time since dawn).

Second, for the key entities involved in the surface energy balance (SEB) - heat and water vapour - the CBL concentrations influence the fluxes themselves, forming a negative-feedback loop. The coupled concentrations and fluxes therefore tend to evolve toward equilibrium values, at rates determined by characteristic time scales. The nature of SEB-CBL feedback also depends fundamentally on the entrainment processes at the top of the CBL, because these are largely driven by the surface heat flux (McNaughton and Jarvis, 1983; 1991; McNaughton and Spriggs, 1986; Raupach, 1991; Raupach and Finnigan, 1995).

Much of our current knowledge of the coupled SEB-CBL system is based on the

*Scaling up in hydrology using remote sensing.* Edited by J.B. Stewart, E.T. Engman, R.A. Feddes and Y. Kerr

study of land surfaces that are statistically homogeneous on the scale of the CBL turbulence. A pressing question is the extension of these ideas to more general situations in which the landscape is heterogeneous at large scales. In the present paper, we approach this question by adopting a "semi-Lagrangian column" view: the CBL is envisaged as a well-mixed column of air moving across the landscape, accumulating and integrating scalar entities along its path as it passes over a succession of different surface types. As the underlying surface changes, the coupled SEB-CBL system is perturbed away from development toward an equilibrium state determined by the "old" upwind surface, and commences development toward a different state determined by the "new" downwind surface. The nature of both the SEB and the CBL in these general circumstances clearly depends critically on the adjustment processes following surface changes.

We address here two basic questions about the adjustment process: what is the nature of the equilibrium state of the coupled SEB-CBL system in the presence of entrainment, and what are the time scales controlling evolution towards this equilibrium state?

## CBL TIME SCALES AND THE SEMI-LAGRANGIAN COLUMN VIEW

For our purposes the CBL has two significant time (and therefore length) scales. The most widely known is the dynamical time scale $h/w_*$ a scale for the overturning time of convective motions in the CBL, and also for the time needed for these motions to adjust dynamically after a sudden change in surface heat flux (e.g. Briggs, 1988; Raupach, 1991; Garratt, 1992). Here $h$ is the height of the CBL and $w_*$ is the convective velocity scale $(hw\theta_v g/T_0)^{1/3}$, where $g$ is acceleration due to gravity, $T_0$ is a reference absolute temperature and $w\theta_v$ is the surface-layer vertical flux of virtual potential temperature $\theta_v$. Full adjustment occurs in a time around $4h/w_*$ which is also the time required for complete vertical mixing of scalars through the depth of the CBL. At midday, taking $h$ = 1000 m and $w_*$ = 2 m s$^{-1}$ as typical values, full adjustment takes about half an hour, though it is quicker in the early morning when $h$ is lower. During this time the CBL air mass moves with the mean wind in the CBL, $U$, through a distance $4hU/w_*$. If $U$ = 5 m s$^{-1}$, then the adjustment distance is about 10 km at midday. To resolve processes at scales less than these, we must use a model which describes the mixing in detail; if our scale of interest is larger, then the assumption of a well-mixed CBL is satisfactory.

The second important time scale, and the main focus of this paper, is the time $T_q$ for thermodynamic adjustment of the CBL to changes in the surface energy balance. We define $T_q$ as the time scale for the saturation deficit $D(t)$ to become independent of the initial saturation deficit $D_0$, following a transition in surface properties. The saturation deficit is the significant quantity because of its link with evaporation and thence the SEB. If $T_q$ is short, then the local CBL state is controlled by local conditions only and is influenced little by upwind surfaces; if it is long, then the state of upwind surfaces is significant and the past history of the CBL air mass has a significant influence on the local CBL state, and thence, via $D(t)$, on the local SEB.

Our approach is through a slab or mixed-layer model of the CBL, using the semi-Lagrangian view in which we track the properties of columns of air as they move across the landscape. Two constraints upon this approach must be noted. First, a mixed-layer model only has meaning over averaging time (and therefore length) scales large enough for the CBL to be treated as fully mixed, that is, over times larger than about $4h/w_*$ and lengths larger than $4hU/w_*$. Mixed-layer models cannot describe CBL evolution on scales shorter than this. Therefore, in considering the evolution of a moving column of air with a mixed-layer model, we are implicitly speaking of a column at least $4hU/w_*$ in linear extent.

Second, we must ask to what extent a column of air remains intact during its motion, given that the column is continually smeared or mixed with neighbouring columns by the horizontal dispersion induced mainly by wind shear in combination with vigorous vertical mixing. This smearing is the reason for calling the approach 'semi-Lagrangian'. Within a CBL, the wind vector turns by perhaps 10° from top to bottom, and its speed is slowed by friction near the ground. Estimating the resulting horizontal dispersion velocity to be about $0.1U$, the time required for the column to double its linear extent ($4hU/w_*$) is around $40h/w_*$, typically 6 to 10 hours. We conclude that the semi-Lagrangian column view is viable for examining resolving changes in CBL properties at intermediate time scales of several hours.

## ANALYTICAL EVALUATIONS OF THE CBL THERMODYNAMIC TIME SCALE

Even the simplest models of the CBL are complex enough that properties like $T_q$ are not immediately obvious. Response times can be explored by repeated simulations with a range of parameter values, but there are so many parameters that the patterns found are not easy to characterise. An alternative approach is by analysis of the model equations, though here again, complexity limits the analysis to simple situations. Fortunately, the two approaches are complementary. Here we develop the analytical approach, using the mixed-layer model of the coupled SEB-CBL system (McNaughton and Spriggs, 1986) in its linearised form (Raupach, 1991). We first write the model in a suitable form, and then analyse $T_q$ in three simplified cases with steady forcing: (a) no entrainment; (b) prescribed entrainment, and (c) interactive entrainment but constant height.

The model represents the CBL as a well-mixed slab of air, capped by an inversion layer, with the transition marked by sharp jumps in potential temperature and specific humidity. The CBL grows by entraining air through the inversion base. Near the ground is a surface layer which is thin compared with $h$ and is characterised by an aerodynamic resistance to heat and vapour transport, $r_a$. Water availability at the surface is described by an effective surface resistance, so evaporation from the ground can be described by the Penman-Monteith equation.

The budget for the saturation deficit in the CBL is

$$h\frac{dD}{dt} = \frac{D_q - D}{r_D} + (D_+ - D)(\frac{dh}{dt} - W_+) \tag{1}$$

where $D(t)$ is the linearised potential saturation deficit

$$D = Q_{sat}(\Theta_0) - Q + (\varepsilon c_p/\lambda)(\Theta - \Theta_0) \tag{2}$$

This is an approximation to the true saturation deficit, linearised about a mean temperature at the ground, $\Theta_0$. The 'equilibrium' saturation deficit, $D_q$, is defined as

$$D_q = \frac{\varepsilon F_A r_s}{(\varepsilon + 1)\rho\lambda} \tag{3}$$

and $r_D$ is given by

$$r_D = r_a + r_s / (\varepsilon + 1) \tag{4}$$

where $r_D$, $r_a$ and $r_s$ are effective averages over small scale ($hU/w_*$) variations in the landscape. Here $h$ is the height of the mixed layer; $D_+$ and $W_+$ respectively, are the values of $D$ and the subsidence velocity immediately above the mixed layer,; Q is specific humidity and $\Theta$ potential temperature; $Q_{sat}$ is the saturation specific humidity); $r$ is air density; $l$ is the latent heat of vaporisation of water; $c_p$ the isobaric specific heat of air; $\varepsilon = (\lambda/c_p)dQ_{sat}/d\Theta_0$, evaluated at the mean surface temperature $\Theta_0$; $F_A = F_H + F_E$ is the total energy flux from the surface; $F_H$ is the sensible heat flux; and $F_E$ the latent heat flux. These equations follow Raupach (1995), but ignore radiative coupling (the factor $p$ in that paper) for simplicity.

Entrainment and the growth of the CBL in Equation 1 is described by specifying the entrainment velocity $W_e = dh/dt - W_+$. For this, we use the encroachment model

$$W_e = \frac{dh}{dt} - W_+ = \frac{F_v}{\rho c_p \gamma_v h} \tag{5}$$

where $F_v = F_H + 0.07F_E$ is the virtual heat (buoyancy) flux and $\gamma_v = d\Theta_v/dz$ is the strength of the inversion in virtual potential temperature above the CBL. This very simple entrainment hypothesis facilitates analytical work and usually gives good descriptions of the thermodynamic coupling between the SEB and the CBL (McNaughton and Spriggs, 1986), though more sophisticated expressions are preferable for accurate prediction of $h(t)$ itself. The subsidence velocity $W_+$ is often neglected in Equation 5 but it is formally necessary because entrainment proceeds in relation to the subsiding air mass.

Many quantities vary with time in these equations, either as independent driving variables such as $F_A(t)$, $r_s(t)$ and $r_a(t)$, or as part of the solution such as $D(t)$, $h(t)$,

$D_+(h(t))$ and $W_+(h(t))$. Consequently general analytic solutions are not available. Our present limited aim is to use this model to investigate the adjustment of a CBL to a new surface, after a sudden change in surface properties. To make the analysis workable, we hold the driving variables $F_A(t)$, $r_s(t)$ and $r_a(t)$ constant except at the transition point. Even after this major simplification, the pair of Equations 1 and 5 have two dependent variables, $D(t)$ and $h(t)$. Our three simplified solution schemes all reduce the system to just one variable, $D$.

*Scheme 1: CBL without entrainment*

A solution of Equation 1 is easily found when $W_+ = 0$ and $h$ = constant, so that $W_e = 0$ and the entrainment term vanishes (e.g. McNaughton and Jarvis, 1983; Raupach, 1991). Equation 1 then simplifies to

$$dD / dt = (D_q - D) / (hr_D) \quad (6)$$

with the solution

$$D(t) = D_q + (D_0 - D_q)exp(-t / hr_D) \quad (7)$$

where $D_0$ is the initial saturation deficit, at the time of the step change. Hence, $D(t)$ exponentially approaches the equilibrium value $D_q$, with a time constant $T_{q(a)} = hr_D$ Å $r_a+r_s/3$ (since $\varepsilon \approx 2$ at 20°C). Taking a typical midday value of $h$ to be 1000 m, and $r_D$ to be 100 s m$^{-1}$, this time constant is about 28 hours. Therefore, equilibration without entrainment is usually slow compared to the diurnal time scale, 1/2p days or about 3.8 hours (the time scale for changes at the frequency of the diurnal cycle). Faster equilibration occurs early in the morning when the CBL is shallow, but with $r_D = 100$ s m$^{-1}$ we need $h$ to be less than 140 m for $T_{q(a)}$ to be less than the diurnal time scale. This situation would be approached only beneath a "lid" such as a very intense inversion. More commonly the CBL grows rapidly during the main part of the day, so a more realistic solution must include growth of the CBL.

*Scheme 2: CBL with prescribed entrainment*

We have found two ways to reduce Equation 1 to a soluble equation while retaining entrainment. In the first of these we prescribe entrainment realistically by matching to the known entrainment at the time ($t = 0$) when the CBL air mass passes across a step in surface properties. Following Raupach (1991), we take

$$W_+ = 0 \text{ , } h(t) = [K(t + t_0)]^{1/2} \quad (8)$$

where $t_0$ is an offset to the time origin. This is a good description of the growth patterns of real CBLs when the surface virtual heat flux is constant (Tennekes, 1973). The values of $K$ and $t_0$ are found by matching both $h$ and $dh/dt$ in Equation 8 to the encroachment model (Equation 5) at $t = 0$, when the entrainment velocity is $W_{e0} = F_{v0}/(\rho c_p \gamma_v h_0)$, with the subscript 0 denoting conditions just after transition.

This gives

$$K = 2h_0W_{e0}\,, \quad t_0 = h_0 / (2W_{e0}) \tag{9}$$

Next, we transform the time variable in Equation 1 to the dimensionless quantity $\tau = (2/r_D)[(t+t_0)/K]^{1/2}$, obtaining

$$\frac{dD}{d\tau} = \left(D_q + \frac{K\gamma_D r_D}{2} + \frac{D_e}{\tau}\right) - D\left(1 + \frac{1}{\tau}\right) \tag{10}$$

where the profile of saturation deficit above the CBL is given by $D_+(h) = D_e + \gamma_D h$ with $\gamma_D = (ec_p/\lambda)\gamma_Q\Theta - \gamma_Q$. Here $\gamma$ denotes the slope of a profile in the troposphere just above the CBL: $\gamma_\Theta = d\Theta/dz$, $\gamma_Q = dQ/dz$.

Equation (10) can be solved analytically, yielding a solution with three terms:

$$D(\tau) = D_0M_0(\tau) + D_eM_e(\tau) + D_pM_p(\tau) \tag{11}$$

where $D_p = D_q + K\gamma_D r_D/2$, and

$$\begin{aligned} M_0(\tau) &= \frac{\tau_0}{\tau}\exp(\tau_0 - \tau) \\ M_e(\tau) &= \frac{1}{\tau}(1 - \exp(\tau_0 - \tau)) \\ M_p(\tau) &= 1 - M_0(\tau) - M_e(\tau) \end{aligned} \tag{12}$$

with $\tau_0 = 1/(r_D W_{e0})$, the value of $\tau$ at $\tau = 0$ from Equations 9 and 10. The terms in Equation 11 describe, in order, the influence of the initial saturation deficit $D_0$, the effect of entrainment of air with saturation deficit $D_0$, and progress towards a final equilibrium state, $D_p$. At any time $t$, the actual deficit $D(t)$ in the CBL is given by a weighted average of these three fixed, given deficits, with time-dependent weights $M_0$, $M_e$ and $M_p$. An earlier analysis on these lines by Raupach (1991) assumed $D_0 = 0$ and $D_e = 0$ for simplicity, and therefore gave only the last term of Equation 11.

The nature of the solution can be shown by an example in which $M_0$, $M_e$ and $M_p$ are plotted against time (Figure 1) for $\tau_0 = 0.50$, a value for typical conditions specified in the caption. For these conditions, memory of the initial saturation deficit $D_0$ decays within a few hours after the step in surface properties, while the influence of the entrainment increases and then declines more slowly, yielding eventually to a long-term equilibrium deficit $D_p$ in a time well over 50 hours. Of course, this time far exceeds the time over which it is reasonable to assume constant $D_q$ and $r_D$ in Equation 10: thus, no long-term equilibrium is possible within a diurnal cycle. Nevertheless, this solution is instructive in showing how the CBL deficit $D(t)$ moves through three stages following a transition: an initial stage of typically a few hours when the deficit $D_0$ at transition is influential, a stage when the deficit is

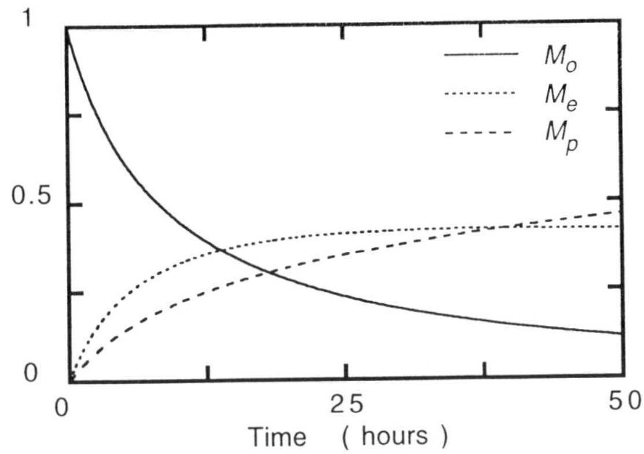


**Figure 1** Plot of the three functions $M_0$, $M_e$ and $M_p$ for $\tau_0 = 0.50$. Parameter values are: $h_0 = 1000$ m; $F_A = 350$ W m$^{-2}$; $\Theta_0= 20$ C; $Q_0 = 0.003$ kg kg$^{-1}$ $\Theta_{+0}= 23$°C; $Q_{+0}= 0.002$ kg kg$^{-1}$; $\gamma_\Theta= 0.005$ K m$^{-1}$; $g_q = 0$ m$^{-1}$, $r_a = 40$ s m$^{-1}$ $r_s = 150$ s m$^{-1}$.

dominated by that of the entrained air, and a final stage when the deficit approaches the equilibrium value $D_q$, slightly modified by entrainment to the value $D_p$.

We can estimate the thermodynamic time scale $T_{q(b)}$ for this solution, the time scale for $D(t)$ to become independent of $D_0$, by finding the value of $t$ at which $M_0(\tau) = 1/2$. By methods to be described in a forthcoming paper by Raupach and McNaughton, it can be shown that

$$T_{q(b)} \approx \frac{3h_0}{W_{e0} + 2.3\,/\,r_D} \tag{13}$$

to a good approximation. Thus, when entrainment is vigorous ($W_{e0} >> 1/r_D$), entrainment dominates the time scale and $T_{q(b)}$ $3h_0/W_{e0}$. When entrainment is low ($W_{e0} << 1/r_D$), $T_{q(b)}$ relaxes to a value close to $hr_D = T_{q(a)}$, the closed-box time scale.

*Scheme 3: CBL with interactive entrainment but constant height*

Scheme 2 entrainment is prescribed at a rate determined by conditions just after the step, by Equation 8. This remains a good approximation for some time after transition, but does not account for subsequent changes in entrainment induced by changes in the SEB. We can retain the dependence of entrainment on the SEB by substituting for $dh/dt$ in Equation 1 using Equation 5 before prescribing $h(t)$. The entrainment term in Equation 1 then remains nonlinear. An analytical solution is possible if both $h(t)$ and $D_+$ are held constant after the step, which in effect models the case where CBL growth by entrainment is exactly offset by subsidence of air with constant deficit.

Substituting (5) into (1), we obtain

$$h\frac{dD}{dt} = \frac{D_q - D}{r_D} + (D_+ - D)\frac{F_v}{\rho c_p \gamma_v h} \tag{14}$$

where we now take both $h$ and $D_+$ to be constant. The essence of the problem is that $F_v$ depends on $D$. From the energy balance, we have

$$F_V = F_A - 0.93F_E = \frac{\varepsilon F_A}{\varepsilon + 1}\left(\frac{1}{\varepsilon} + 0.07\right) - \frac{0.93\rho\lambda(D - D_q)}{(\varepsilon + 1)r_D} \tag{15}$$

Substitution into (14) leads to an equation of the form

$$a\frac{dx}{dt} = x^2 - bx + c \tag{16}$$

where

$$X(t) = D(t) - D_q, \quad X_+ = D_+ - D_q, \quad X_0 = D_0 - D_q \tag{17}$$

$$a = \frac{r_D(\varepsilon + 1)\rho c_p \gamma_v h^2}{0.93\rho\lambda} \tag{18}$$

$$b = \frac{a}{hr_D} + \frac{(1 + 0.07\,\varepsilon)F_A r_D}{0.93\rho\lambda} + X_+ \tag{19}$$

$$c = \frac{X_+(1 + 0.07\,\varepsilon)F_A r_D}{0.93\rho\lambda} \tag{20}$$

Equation (15) can be solved by separation of variables to give

$$X(t) = \frac{X_1 - X_2 k(t)}{1 - k(t)} \tag{21}$$

where

$$X_{1,2} = \frac{b}{2} \pm \frac{1}{2}\sqrt{b^2 - 4c} \tag{22}$$

$$k(t) = \frac{X_1 - X_0}{X_2 - X_0}\exp\left(\frac{(X_1 - X_2)t}{a}\right) \tag{23}$$

In this solution, time is scaled by $a/(x_1 - x_2) = a(b^2 - 4c)^{-1/2}$, which is our estimate for the time scale $T_{q(c)}$. To indicate the nature of this time scale, we note that the dependence of $T_{q(c)}$ upon $X_+$ is weak in practice (as confirmed by calculations), and that the main properties of $T_{q(c)}$ are revealed by considering the case $X_+ = 0$, or $D_+ = D_q$. Then

$$\frac{1}{T_{q(c)}} = \frac{b}{a} = \frac{1}{hr_D} + \frac{(1+0.07\,\varepsilon)F_A}{(\varepsilon+1)\rho c_p \gamma_v h^2} \tag{24}$$

Thus $T_{q(c)}$ is reduced below the closed-box time scale $T_{q(a)} = hr_D$, by an amount given by the second term on the right hand side. For the conditions specified in the caption to Figure 2, $T_{q(c)}$ is about 1/3 of $T_{q(a)}$, or around 8 hours. This is still significant compared with the diurnal time scale. However, the analysis shows two important features of CBL equilibration when interactive feedback is included. First, the feedback substantially accelerates equilibration. Second, the acceleration is relatively greater for smaller $h$ (lower CBL depth). This follows because the ratio of the second to the first terms in the expression for $1/T_{q(c)}$ is proportional to $1/h$, so the second term dominates and makes $T_{q(c)}$ proportional to $h^2$, rather than $h$, at small $h$.

This solution should provide realistic results for short times, before the limitation of a constant depth becomes too severe. At longer times the effect of too small a depth will cause the solution to approach an equilibrium too rapidly, because it does not account for the increasing capacitance of the CBL with increasing $h$.

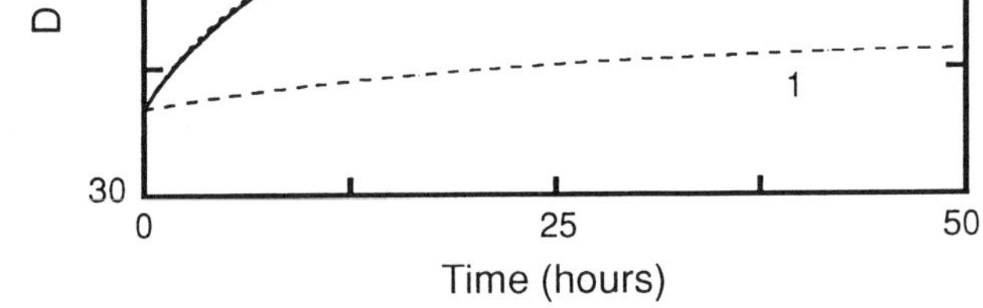


**Figure 2** Comparison of four solutions to the mixed-layer model of the CBL for a wet-to-dry transition. From the bottom the solutions are: (1) CBL without entrainment; (2) CBL with prescribed entrainment; (3) CBL with interactive entrainment but fixed height; (4) full model. Parameter values are the same as those in the caption to Figure 1.

*Comparison of solutions*

We now compare sample results from the three analytical solutions with results from numerical simulations using the full model, Equations 1 to 5. We take as examples two cases: one where the CBL moves onto a fairly dry surface with $r_s$ = 150 s m$^{-1}$; the other where the new surface is wetter with $r_s$ = 50 s m$^{-1}$. The four solutions are shown in Figures 2 and 3.

These results illustrate several points. Firstly, the solution without entrainment (i) is always a poor approximation to the full solution by simulation; that is, it is important that entrainment be represented in any model of the CBL. Secondly, both analytical solutions with entrainment (ii and iii) agree well with the full model at short times: this is not surprising since entrainment is initially matched to the full solution in both cases. Thirdly, the solution for prescribed-entrainment (ii) matches the full solution better over the dryer landscape where the CBL grows more vigorously and it is more important to account for the saturation deficit of the entrained air; the solution with interactive entrainment, (iii) matches better where the CBL grows less vigorously and entrainment is more intimately controlled by the surface energy balance. Fourthly, the final values of $D$ are very different in the three solutions, though which is more faithful to the full solution depends on the circumstances. Finally, all solutions show half-response times longer than the diurnal time scale 12/π hours.

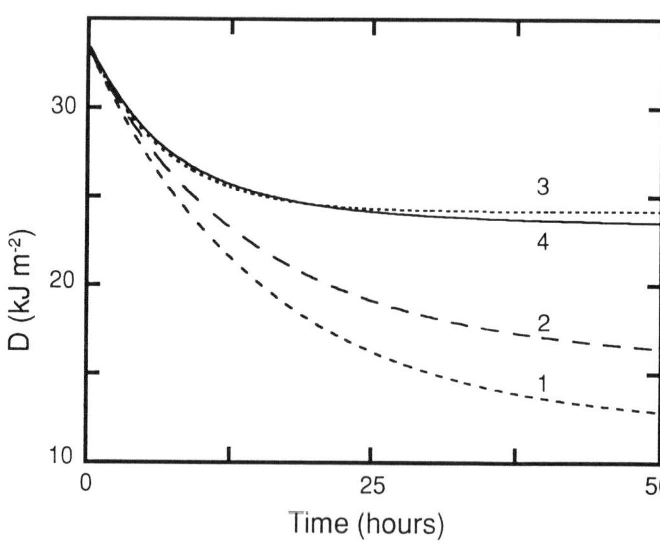


**Figure 3** As for Figure 2 but for a dry-to-wet transition. Parameter valuesare the same as in Figures 1 and 2, except that $r_s$ = 50 s m$^{-1}$.

## CLOSING COMMENTS

The results can be interpreted in the context of the diurnal behaviour of the CBL. In the early morning, when the CBL is shallow, it responds rapidly to surface changes and comes close to being in continuous equilibrium with the local surface. A one-dimensional Eulerian model of the CBL should then give good results. However, as the CBL grows during the morning, the time scale for adjustment increases dramatically (roughly as $h^2$), and with it increases the area of uniform terrain needed for successful use of an Eulerian model. By mid morning the response time is frequently similar to the time since dawn; thereafter the state of the CBL reflects the whole history of the CBL air mass since dawn. By mid afternoon it will have travelled, typically, more than 100 km. This limits the use of one-dimensional Eulerian CBL models to only the largest of uniform regions. In most cases a successful one-dimensional model of SEB-CBL interaction must be cast in a Lagrangian framework which moves with the CBL air mass.

The results obtained here are relevant to problems of measuring and modelling energy fluxes from heterogeneous landscapes. Previous discussions have considered landscapes that are 'statistically homogeneous' at various scales (e.g. Raupach & Finnigan, 1995). These studies support the view that the CBL-SEB system can be modelled, to good accuracy, by assuming that the landscape consists of a set of homegeneous sub-areas, each interacting vertically with a mixed layer that is uniform across the whole landscape. The present work is relevant to surfaces which are not statistically homogeneous: ones with secular trends in surface properties. Heterogeneity which displays secular trends over distances comparable to the thermodynamic distance scale will certainly affect the development of the CBL as its air masses traverse these gradients, and the averaging question in these circumstances may require a different treatment from that for a statistically uniform landscape. This paper finds that the thermodynamic distance scale is typically of the order 100 km - a scale comparable to that over which rainfall gradients can be quite large.

## REFERENCES

Briggs, G.A. 1988. Surface inhomogeneity effects on convective diffusion. *Bound.-Layer Meteor.*, **45**, 117-135.

Deardorff, J.W. 1978. Efficient prediction of ground surface temperature and moisture, with inclusion of a layer of vegetation. *J. Geophys. Res.* **83**, 1889-1903.

Garratt, J.R. 1992. *The Atmospheric Boundary Layer.* Cambridge University Press, Cambridge.

McNaughton, K.G. and Jarvis, P.G. 1983. Predicting effets of vegetation changes on transpiration and evaporation. In: *Water Deficits and Plant Growth*, Vol. VII (ed.) T.T. Kozlowski. Academic Press, New York. 1-47.

McNaughton, K.G., Spriggs, T.W. 1986 A mixed-layer model for regional evaporation. *Bound.-Layer Meteor.*, **34**, 243-262.

McNaughton, K.G. and Jarvis, P.G. 1991 Effects of spatial scale on stomatal control of transpiration. *Agric. For. Meteorol.*, **54**, 279-301.

Raupach, M.R. 1991. Vegetation-atmosphere interaction in homogeneous and heterogeneous terrain. *Vegetatio,* **91**, 105-120.

Raupach, M. R.: 1995 Vegetation-atmosphere interaction and surface conductance at leaf, canopy and regional scales. *Agric. For. Meteorol.,* **73**, 151-179.

Raupach, M.R. and Finnigan, J.J. 1995. Scale issues in boundary-layer meteorology: surface energy balances in heterogeneous terrain. *Hydrol. Processes,* **9**, 589-612.

Raupach, M.R., Denmead, O.T. and Dunin, F.X. 1992. Challenges in linking atmospheric CO2 concentrations to fluxes at local and regional scale. *Aust. J. Bot.,* **40**, 697-716.

Tennekes, H. 1973. A model for the dynamics of the inversion above a convective boundary layer. *J. Atmos. Sci.,* **53**, 149-160.

Troen, I.B. and Mahrt, L. 1986. A simple model of the atmospheric boundary layer; sensitivity to surface evaporation. *Bound.-Layer Meteor.,* **37**, 129-148.

# 12 Soil moisture variability

**T.J. SCHMUGGE and T.J. JACKSON**
*USDA/ARS Hydrology Lab, Beltsville, USA*

## INTRODUCTION

There are many methods for the *in situ* measurement of soil moisture: Time Domain Reflectometry (TDR), capacitance probes, resistance probes, tensiometers etc. but the most reliable and the one most often used as the standard is the gravimetric sample (Schmugge *et al.* 1980). We present results from a number of experiments where large numbers of gravimetric samples were collected to study the spatial variability of soil moisture. The implications of this variability for remote sensing are discussed concentrating on the zero to 5-cm layer of the soil since this is the layer that is accessible by remote sensing techniques. The data from various field experiments ranging over some 30 years are presented first, followed by some remotely sensed results. Soil moisture values are presented as % water of the dry soil weight unless otherwise noted.

Consider Figure 1, which presents contour plots of the surface soil moisture for a bare soil field with a loam texture in central South Dakota in the Northern Great Plains of the United States. The contour plots were generated from a 6 x 6 m sampling pattern over a 350 by 350 m square. The field was sampled three times over an 18 hour period and in addition the height contours were surveyed. The first sampling took place at 5 p.m. on May 2, 1978, the second at 8 a.m. on May 3 and the third at 10 - 11 a.m. on the same day. The moisture variation ranged from 16 to 30% gravimetric soil moisture. Note that the basic pattern of the variation did not change over the three samplings and that there was little correlation of the moisture with the height contours, $r < 0.4$ for the upper layers. Thus, the moisture variations are presumably mostly due to small scale topographic variations which were not measured in the field survey on the same 6 x 6 m grid. The limited topographic relief — there was only about a 2-metre variation across the field — contributes to the lack of correlation with height. There were two samples taken at each location and the average of the two is plotted in Figure 1. The Root Mean Square (RMS) difference between the two samples at the 36 points was generally found to be about 70% of the overall spatial standard deviation, indicating that there is almost as much variation within a metre or so as there was over the entire field. The variogram for these data showed no spatial correlation.

Similar results are obtained for the other two depths which were sampled, i.e 0-

*Scaling up in hydrology using remote sensing*. Edited by J.B. Stewart, E.T. Engman, R.A. Feddes and Y. Kerr

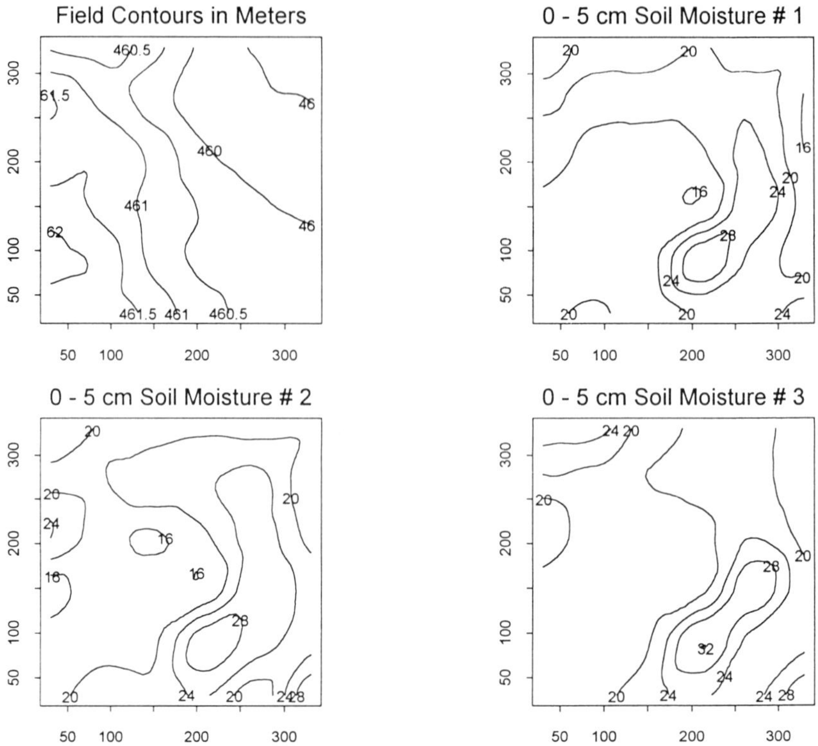


**Figure 1** Contour plots of 0 to 5 cm soil moisture for a 350 x 350 m bare field in Hand county South Dakota. The values are in gravimetric percent. The height contours in metres are also shown. Sampling #1 occurred at 5 PM on 2 May 1978, #2 at 8 AM on 3 May 1978 and #3 at 10 - 11AM on 3 May 1978.

2.5 cm and 0-10 cm. The results are summarized in Figure 2 where boxplots of the results for the three depths in the three samplings are given. These plots show that the medians were consistent for the three samplings and for the three layers, with the 0 to 10 cm layer having the highest moisture and least variability. Note that the early morning sampling (#2) had the greatest variability for each depth. Also the spatial patterns for the three depths were similar.

An early comprehensive study of soil moisture variability was presented by Hills and Reynolds (1969) who performed an intensive sampling program at Chew Stoke catchment south of Bristol, England. They collected 60 randomly located samples for the 0 to 5-cm layer over four different spatial scales: 1x1 m, 30x30 m, three drainage classes (estimated to be 300x300 m) and the entire catchment. For the 1x1 m patch, they effectively took the entire patch. They stratified the sampling by drainage classes: A – poorly drained, B – imperfectly drained and C – freely drained. The results summarized in Table 1 show that the Standard Deviation (SD) increases dramatically, by about a factor of two, when area increases from the patch scale to a drainage area. The soil moisture values for this experiment were high, 30 to 70% by weight. For drainage class A, the moisture was high (> 60%) for all plot sizes

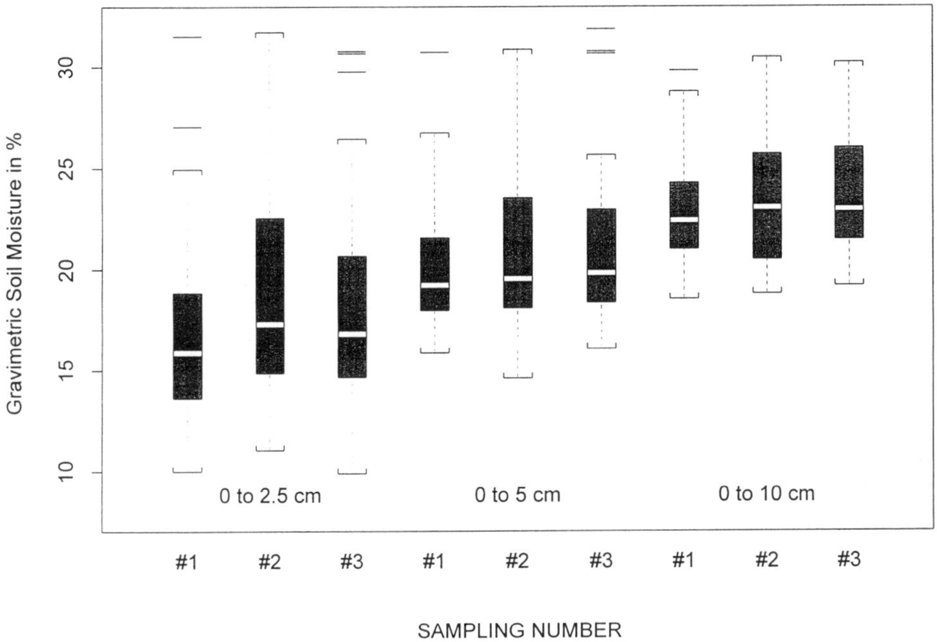


**Figure 2** Boxplots of the soil moisture values for the three depths, 0-2.5, 0-5, and 0-10 cm, from the three samplings conducted in May 1978 in Hand County South Dakota. The box shows the range for half of the data in each case with the median value indicated by the open bar. The range of the data is indicated by the closed brackets with outliers indicated by the lines outside the brackets.

while for the other classes the smaller plots had lower moistures < 40%. In another experiment they took ten samples from plots ranging from 15x15 cm to 31x31 m, all having average moisture contents of about 70% by weight. The SD ranged from 3 to 7%, yielding a Coefficient of Variation (CV = SD/mean) between 4 and 10% with no correlation to plot size.

In the late 1970s a number of fields from several sites in the Midwestern part of the USA were intensively sampled (> 20 samples for a 40-acre field). The results are summarized in Figure 3 where we have plotted CV versus the mean soil moisture for about 60 fields (Bell *et al.*, 1980; Owe *et al.*, 1982). The pattern that appears is that at high moisture levels the CV is low because of the large mean and at low moistures the CV can be very large (~50%). The SD for these fields ranged between 1 and 5%, indicating the level of what might be expected for the magnitude of the variability for ground sampling. In this figure about 90% of the points lie below the curve, indicating that it could be used to estimate the maximum CV to be expected for a given sampling scheme.

Another large area ground sampling occurred during the First ISLSCP Field Experiment (FIFE) (Sellers *et al.* 1992). Soil moisture measurements were made at

**Table 1** Gravimetric Soil Moisture Values for the Chew Stoke Catchment (from Hills and Reynolds, 1969) Results for six random samples: 0 to 5-cm layer.

| Category | Mean | Std. Dev. | CV |
|---|---|---|---|
| *Whole Catchment* | | | |
| All Classes | 62.7 | 22.0 | 35.1 |
| *Drainage classes 300x300 m* | | | |
| Class A | 67.4 | 20.8 | 30.9 |
| Class B | 65.5 | 16.7 | 25.5 |
| Class C | 63.9 | 16.2 | 25.4 |
| *30x30 m plot* | | | |
| Class A | 64.6 | 7.7 | 11.9 |
| Class B | 38.3 | 3.0 | 7.8 |
| Class C | 40.8 | 6.2 | 15.2 |
| *1x1 m plot* | | | |
| Class A | 73.0 | 6.4 | 8.7 |
| Class B | 39.8 | 2.5 | 6.2 |
| Class C | 30.7 | 4.9 | 15.9 |

A: Poorly drained soils
B: Imperfectly drained soils
C: Freely drained soils

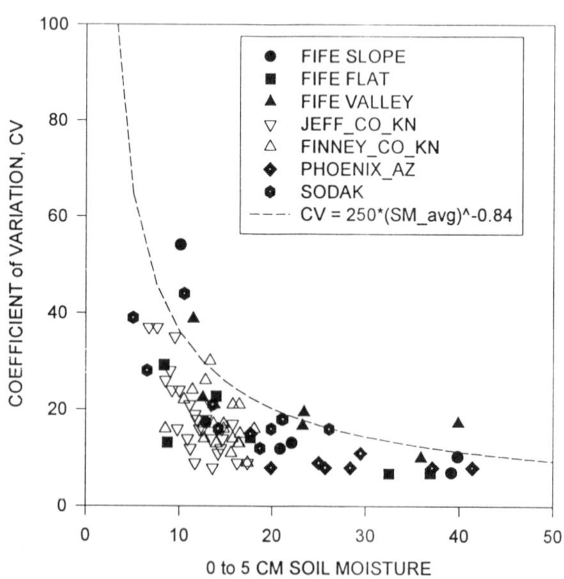


**Figure 3** Comparison of the Coefficient of Variation (CV) and the mean 0-5 cm soil moisture in % by weight for a large number of intensively sampled fields in Great Plains of the USA

20 locations, chosen to represent the range of slope and aspect conditions present over the site. Five samples were taken at each location, one at the centre and four at the points of the compass: a distance of 30 metres from the centre. Thus, we have local scale variation within the patch at each location and the variation over the entire 15x15 km area. The average soil moistures for the entire region ranged from 14% to more than 40% with CVs ranging from 10 to 30% for sample sizes between 50 and 100. As is usually the case, the larger CVs were for the drier situations. Davis *et al.* (1992) performed a stratified analysis of these data and found that the stratification of the data did not reduce the variability. The stratification was done on the basis of whether the location was burned or unburned and its topography. They hypothesize that the lack of an effect was due to high soil moisture conditions present during most of the 1987 IFCs and they observed that most of the variation occurred among the five soil samples at a single site.

A portion of the data are presented in Figure 4 where boxplots of both the ground measurements of soil moisture and microwave brightness temperature, $T_B$, are given

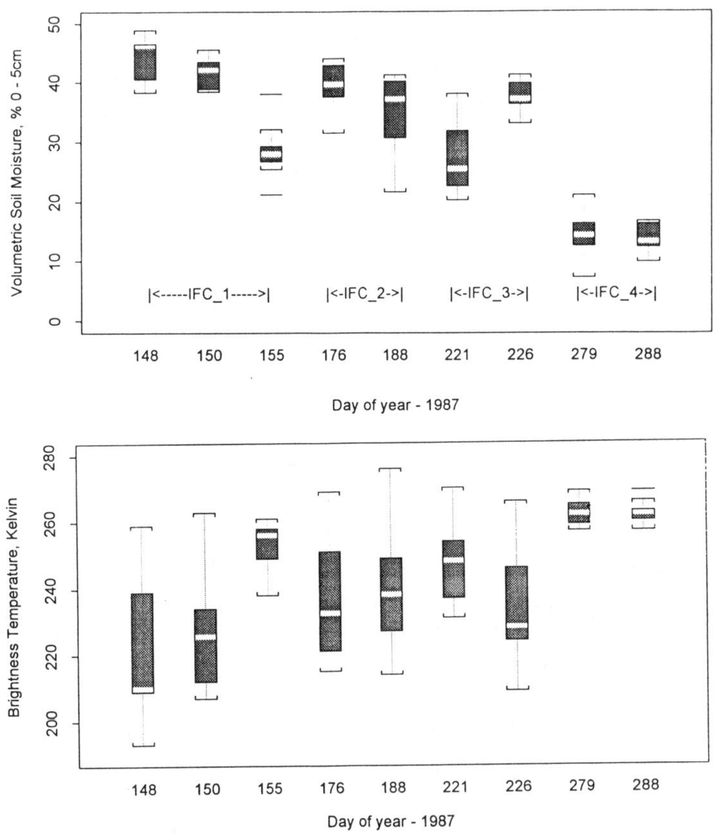


**Figure 4** Boxplots showing the range of soil moisture values (upper panel) and aircraft observations of microwave ($\lambda = 21$ cm) brightness temperature (lower panel) for nine days during the three intensive field campaigns (IFC) of FIFE during 1987

for a number of sites, 9 to 13. The upper plot shows the range of moisture conditions during the 1987 experiments which were mostly wet. The values presented are averages of five points, hence the variability is reduced. The values of $T_B$ were obtained from the PushBroom Microwave Radiometer (PBMR) operating at the 21 cm wavelength (1.412 GHz) (Wang *et al.* 1989; 1990). The median values of $T_B$ track the variation of soil moisture during the course of the experiment, i.e. high soil moisture values produce low values of $T_B$. However $T_B$ is biased by the inclusion of points where the PBMR did not respond to moisture variations due to a heavy thatch layer in the soil, which caused high $T_B$ values even for high moisture conditions. For example, on day 148 there were three sites from the unburned Kings Creek watershed in the Konza preserve which had $T_B > 245$ K while the remaining ten sites were between 240 and 190 K. All had roughly the same moisture content, > 40%. The same behaviour for these three sites was observed on the other days also.

In addition to this large area sampling there was a small intensive programme by Charpentier and Goffman (1992) who studied three 66 x 66m areas with a 11 x 11 m grid of 49 points in a burned watershed in the Konza. They looked at a flat site, a slope site and a valley bottom. Their objective was to understand the variation which can occur within a PBMR pixel. They found that the slope and valley pixels had the highest variations. The CVs ranged from about 7 to more than 50%, with the drier situation in October 1987 having the greatest variability. Approximately the same range of variability was seen in this plot sized sampling as was seen over the entire site. They also found a lack of any spatial correlation as represented by a variogram at the different sites.

## IMPLICATIONS FOR REMOTE SENSING

These results indicate that there is a limit to the accuracy which can be expected when ground samples are used to verify spatially integrated remotely-sensed observations of soil moisture. The SD of the ground measurements can be reduced by larger numbers of samples, but that it will be typically in the range of 1 to 5% and this will produce uncertainty in the resulting calibrations of remote sensors. This result was demonstrated by Mo and Schmugge (1983) in a Monte Carlo simulation of the expected variation of $T_B$ and soil moisture. The values of $T_B$ were calculated for known profiles and were plotted against the appropriate soil moisture values, however, the latter were allowed to have a relative variation of ±25%. When this was done the scatter in the calculated values approximated the scatter observed in the ground measurements. This uncertainty was also shown to have an effect on the slope of the derived regression line for the data.

An example of this behaviour is shown in Figure 5, which is a comparison of the $T_B$ obtained with the PBMR instrument during FIFE with the ground measurements made over the entire site, as described above. The points for the non-responding unburned watersheds in the Konza nature preserve have been eliminated. The

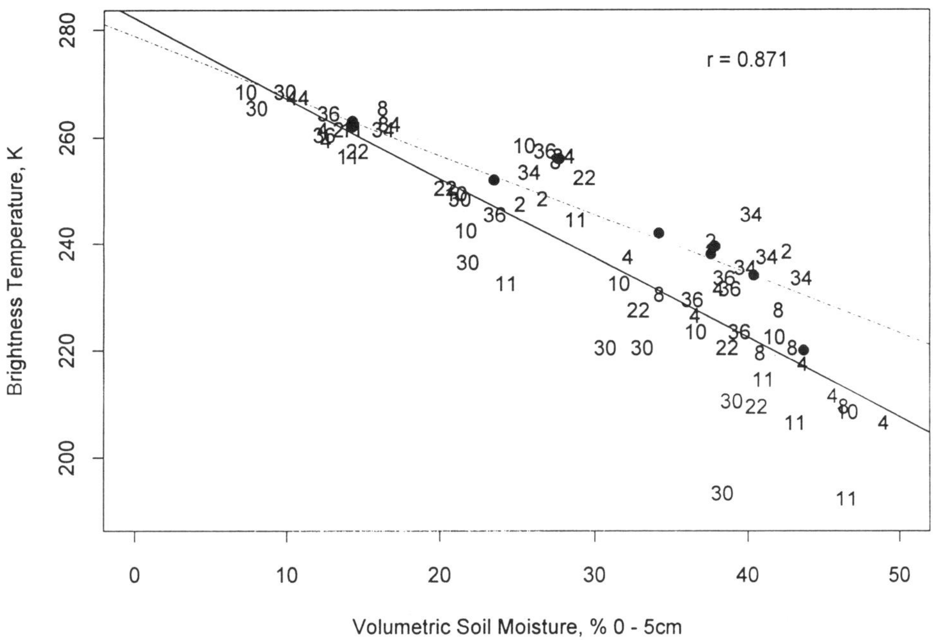


**Figure 5** Comparison of the microwave brightness temperature, $T_B$ as observed by the PBMR during FIFE with ground measurements of soil moisture for seven to ten sites during the nine days represented in Figure 4. The numbers identify the sites; not all the sites were covered each day. The solid symbols are for the global averages of $T_B$ and soil moisture for each of the nine days.

general linear decrease of $T_B$ with soil moisture is seen with a correlation coefficient of 0.87. Much of the scatter seen in the plot is due to the variations of ground measurement which were made over a 30 m circle as described above. The microwave footprint at the altitude flown in FIFE (600 m) is about 180 m so that it averages over a much larger area than the ground sampling which is from five points over the 30 m circle. The numbers on the graph indicate the sample locations and there does appear to be consistent differences among the sites, mostly due to vegetation differences. The solid symbols in Figure 5 represent the average $T_B$ over the entire area for each day compared with average soil moisture from the 50 to 100 samples taken on the ground for that day. A much smoother variation of $T_B$ with soil moisture is seen. This behaviour has been observed in the Washita experiments and the Monsoon 90 experiment and indicates the type of behaviour which would be seen with a spacecraft sensor having a resolution comparable to the area covered in FIFE, approximately 10 by 20 km (Jackson *et al.* 1995). These results also indicate that the ground measurements over a large area can be simply averaged to get a representative value of the soil moisture for the area.

The larger areas observed by a remote sensor enables one to test for spatial correlations in the remotely-sensed data. An example is seen in Figure 6 where a

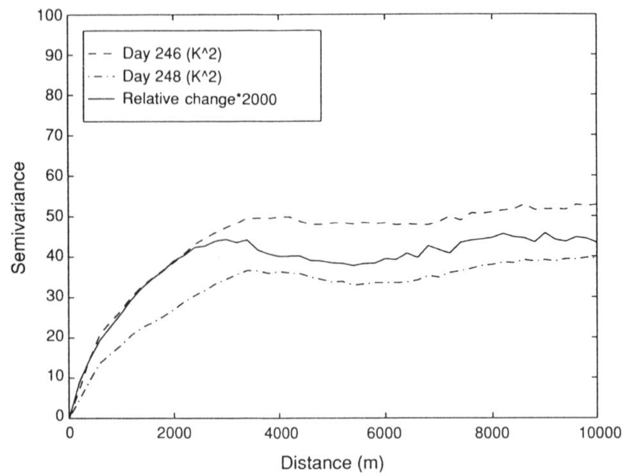


**Figure 6** Semivariograms of microwave brightness temperature observed by the PBMR during the HAPEX-Sahel experiment in 1992 (Goutorbe *et al.*, 1994). The dashed lines are for $T_B$ observed on 2 and 4 September 1992 over the East and West Central sites. The solid line is the relative change of $T_B$ between the two days and is related to soil hydraulic properties from Hollenbeck *et al.*, 1996.

spatial correlation is seen in PBMR data acquired during the HAPEX-Sahel experiment in Niger during the summer of 1992 (Goutorbe *et al.* 1994). The results show a relatively large range (~3500 m) in the variogram of the brightness temperature for the two days and in the relative change of $T_B$. The latter is a qualitative indicator of the soils' hydraulic properties. The sill value of the $T_B$ variograms decreases as the soil dries from day 246 to 248. The stability of the correlation length in the $T_B$ variograms during the drydown is indicative that it is related to a physical feature in the scene. The observed large range (about 3500 m) indicates that the scale of the feature is much larger than the pixel scale (Hollenbeck *et al.*, 1996). Similar variograms were observed for the FIFE 1987 PBMR data with a somewhat larger correlation length (about 4500 m). In this case the magnitude of the semivariance decreased dramatically as the soil dried during IFC-1.

The microwave data from the Washita experiment were used to make soil moisture maps for an area of 46 by 19 km with 200 m resolution in central Oklahoma. There were data for eight days following a series of saturating rains. These data have been used by Rodriguez-Iturbe *et al.* (1995) to study the statistical structure of soil moisture fields. They did this by aggregating the pixels up to a 1 x 1 km size and looking at the variance in the data as a function of pixel area. They found a power law relationship between the variance and pixel area with the variance decreasing as the area increased. The slope of the relationship changed, generally increasing as the soils dried. Working with the same data, Dubayah *et al.*(1996) found that multiscaling, i.e. the log-log linearity of statistical moments as functions of scale, can be used to infer spatial variability at scales other than the measurement. For example they showed how a coarse, remotely-sensed field of soil moisture could be used to estimate variability at finer grid scales.

## CONCLUSIONS

The results presented here at the field scale show that there is usually no spatial correlation in point gravimetric soil moisture samples but that when the data are aggregated, as with a remote sensor, there does appear to be some structure to the soil moisture field. At the field scale the variability appears to be random, due to micro-topography and variations in soil properties at the metre scale. As one aggregates to larger scales these differences are averaged and the variation in soils, e.g. in the Washita basin, or rainfall e.g. Monsoon 90 & 91, cause the variations in soil moisture (Jackson *et al.* 1993). The recent studies with brightness temperature /soil moisture derived from the PBMR and ESTAR airborne instruments indicate that there may be some large scale spatial structures in the data which warrants further study. The remote sensing results presented here were for a range of lightly vegetated conditions and demonstrate that under these conditions microwave radiometers are very robust indicators of the surface (0 to 5 cm) soil moisture conditions. From the results comparing a large number of ground samples with a global average of the microwave response it would appear that the ground measurements over a large area can be simply averaged to get a representative value of the soil moisture for the area. The question arises about how well can this surface layer value be related to deeper profile moisture state. For fully vegetated conditions the moisture profile is more uniform and reasonable estimates can be made. For bare soils the problem is more difficult but can be done, a review of possible methods is presented in Kostov and Jackson, (1993). Ragab (1995) explores the possibility of using these methods on an operational basis using remotely-sensed observations.

## ACKNOWLEDGMENT

The gravimetric soil moisture data were collected for FIFE by the staff and students of the Evapotranspiration Laboratory at Kansas State University under the direction of Dr. Edward Kanemasu. The dedicated effort of A. Nelson and G. Harbers in the collection and preparation of these data is particularly appreciated.

## REFERENCES

Bell, K.R., Blanchard, B.J., Witczak, M.W. and Schmugge, T. J. 1980. Analysis of surface moisture variations within large field sites, *Wat. Resour. Res.*, **16**, 796-810.

Charpentier, M.A. and Groffman, P.M. 1992. Soil moisture variability within remote sensing pixels. *J. Geophys. Res.*, **97**, 18987-18995.

Davis, F.W., Schimel, D.D., Friedl, M.A., Michaelsen, J.C., Kittel, T.G.F., Dubayah, R., and Dozier, J. 1992. Covariance of biophysical data with digital topographic and land use maps over the FIFE site. *J. Geophys. Res.*, **97** 19,009-19,021.

Dubayah, R., Wood, E.F. and Lavallée, D. 1996. The Spatial scaling properties of remotely sensed and modelled near-surface soil moisture state for the Little Washita Basin. In: Scaling of Remote Sensing Data for GIS, (ed.) D. Quattrochi and M. Goodchild, Lewis Publishers. In press.

Goutorbe, J.P., Lebel, T., Tinga, A., Bessemoulin, P., Brouwer, J., Dolman, A.J., Engman, E.T., Gash, J.H.C., Hoepffner, M., Kabat, P. Kerr, Y.H., Monteny, B., Prince,S., Said, F., Sellers, P.J. and Wallace, J.S. 1994. HAPEX-SAHEL: a large scale study of land-atmosphere interactions in the semi-arid tropics. *Annales Geophysicae*, **12,** 53-64.

Hills, R.C. and Reynolds, S.G. 1969. Illustrations of soil moisture variability in selected areas and plots of different sizes. *J. Hydrol.*, **8,** 27-47.

Hollenbeck, K.J., Schmugge, T.J., Hornberger, G.M. and Wang, J.R. 1996. Identifying soil hydraulic heterogeneity by detection of relative change in passive microwave remote sensing observations. *Wat. Resour. Res.*, **32,** 139-148.

Jackson, T.J., Le Vine, D.M., Griffis, A., Goodrich, D.C., Schmugge, T.J., Swift, C. and O'Neill, P.E. 1993. Soil moisture and rainfall estimation over a semiarid environment with the ESTAR microwave radiometer. *IEEE Tran. Geosci. Remote Sensing*, **31**, 836-841.

Jackson, T.J., LeVine, D.M., Swift, C.T., Schmugge, T. and Schiebe, F.R. 1995. Large area mapping of soil moisture using the ESTAR passive microwave radiometer. *Remote Sens. Environ.*, **54,** 27-37.

Kostov, K. and Jackson, T.J. 1993. Estimating profile soil moisture from surface layer measurements: a review. *Proc SPIE Symp on Ground Sensing*, Orlando, Florida, April 1993, SPIE **1941,** 125-136.

Mo, T. and Schmugge, T.J. 1983. Monte Carlo simulation of the effect of soil moisture and surface roughness variations on the microwave emission from soils. *IEEE Trans. on Geoscience and Remote Sensing*, ***GE-21***, 473-479.

Owe, M., Jones, E.B. and Schmugge, T.J. 1982. Soil moisture variation patterns observed in Hand County, South Dakota. *Wat. Resour. Bull.*, **18**, 949-954.

Ragab, R. 1995. Towards a continuous operational system to estimate the root-zone soil moisture from intermittent remotely sensed surface moisture. *J. Hydrol.*, **173,** 1-25.

Rodriguez-Iturbe, I., Vogel, G.K., Rigon, R., Entekhabi, D., Castelli, F. and Rinaldo, A. 1995. On the spatial organization of soil moisture fields. *Geophysical Research Letters*, **22,** 2757-2760.

Schmugge, T.J., Jackson, T.J. and McKim, H.L. 1980. Survey of methods for soil moisture determination. *Wat. Resour. Res.*, **16**, 961-979.

Sellers, P.J., Hall, F.G., Asrar, G. Strebel, D.E. and Murphy, R.E. 1992. An overview of the First International Satellite Land Surface Climatology Project (ISLSCP) Field Experiment (FIFE), *J. Geophys. Res.* **97**,18345-18371.

Wang, J.R., Shiue, J. C., Schmugge, T.J. and Engman, E.T. 1989. Mapping surface soil moisture with L-band radiometric measurements. *Remote. Sens Environ.*, **27**, 305-312.

Wang, J.R., Schmugge, T.J., Shiue, J.C. and Engman, E.T. 1990. The L-band PBMR measurements of surface soil moisture in FIFE. *IEEE Trans. Geoscience and Electronics*, **GE-28**, 906-914.

# 13 Spatial variability of saturated soil hydraulic conductivity derived from remotely-sensed data

**N.M. MATTIKALLI[1], E.T. ENGMAN[1], L.R. AHUJA[2], K. JOHNSEN[2] and T.J. JACKSON[3]**
[1]*NASA Goddard Space Flight Center, Greenbelt, USA*
[2]*USDA-Agricultural Research Service-GPSR, Fort Collins, USA*
[3]*USDA-Agricultural Research Service, Beltsville, USA*

## INTRODUCTION

Studies on spatial heterogeneity of soil and its properties date back to the 1950s (Miller and Miller, 1956). However, spatial variability of soil and other hydrological parameters of the water cycle have emerged as one of the primary research topics in hydrology (e.g. Beven *et al.*, 1988; Wood *et al.*, 1988; Famiglietti and Wood, 1995; Seyfried and Wilcox, 1995). Spatial variability imparts significant consequences in hydrologic modelling, especially with the availability of distributed hydrologic data derived from remotely-sensed information, and physically based distributed models to handle such data sets. Several components of the water budget, including infiltration, soil moisture storage, deep drainage etc. are functions of soil hydraulic properties, which are typically measured in the laboratory and less often in the field. Since laboratory and field measurement of soil hydraulic properties is both tedious and expensive, establishing methods of relating soil properties at various sites to each other within a catchment – and eventually to the scale of a catchment, region and continent – would be an important source of input information for hydrologic models.

Scaling is a means to relate saturated soil hydraulic conductivity of a single soil type at different spatial locations by use of simple conversion factors, called scaling factors. It is a simple but powerful tool for describing (approximately) field spatial variability of soil hydraulic properties (Warrick *et al.*, 1977; Russo and Bresler, 1981) and for establishing characteristics derived from these, such as infiltration, drainage, and available moisture content (Sharma *et al.*, 1980). Previous studies (e.g. Peck *et al.*, 1977; Sharma and Luxmoore, 1979; Luxmoore and Sharma, 1980) have derived scaling factors for water retention and hydraulic conductivity, and described variations in the soil water balance components. Although relationships between soil properties, such as saturated hydraulic conductivity, and some water

*Scaling up in hydrology using remote sensing*. Edited by J.B. Stewart, E.T. Engman, R.A. Feddes and Y. Kerr

balance terms (e.g. soil evaporation) do not have a physically explicit description, soil properties play a significant role in the redistribution of soil moisture and hence availability of moisture content for evaporation from the surface soil layer. The frequency distribution and spatial correlation of scaling factors describe the variability of saturated hydraulic conductivity in the field, and therefore result in both considerable simplicity and enhanced understanding as well as convenience in modelling a heterogeneous catchment for its hydrologic response (Peck *et al.*, 1977; Sharma and Luxmoore, 1979; Ahuja *et al.*, 1984). Inversely, scaling can also be used to estimate soil hydraulic properties at different locations in a catchment from measurement of these properties at one representative location having mean properties and limited data at other locations (Ahuja *et al.*, 1985; Williams and Ahuja, 1991). Although strictly similar media conditions do not hold under typical field conditions of varying soil types within a catchment, some earlier studies (e.g. Sharma and Luxmoore, 1979; Luxmoore and Sharma, 1980) suggest that approximate scaling factors can be used to scale a large number of water retention and unsaturated hydraulic conductivity data for different soil types, and that the modified scaling approach based on the concept of similar media has been useful for describing spatial variability of hydraulic properties of different soil types in catchment modelling. Therefore, the modified scaling approach has a significant potential for application of remotely-sensed data to study the effect of spatial variability of soil hydraulic properties across a catchment on various components of the soil water balance.

The effects of spatial variability in saturated soil hydraulic conductivity determined by remote sensing on some components of the soil water balance such as available moisture content, deep drainage and evaporation are examined by using a state-of-the-art soil hydrology model (the Root Zone Water Quality Model or RZWQM), and a raster-based Geographical Information System (GIS).

## BACKGROUND

### ESTIMATION OF HYDRAULIC CONDUCTIVITY FROM REMOTELY-SENSED SOIL MOISTURE CONTENT

Recent studies using the data collected from an airborne passive microwave sensor for the Little Washita River catchment, Oklahoma, revealed that spatial and temporal soil moisture contents closely follow the pattern of soil texture in the catchment (Jackson *et al.*, 1995; Mattikalli *et al.*, 1995a). This suggests that temporal changes in surface soil moisture, obtainable from microwave remote sensing, can be used as an indicator of soil properties (Mattikalli *et al.*, 1995b). Specifically, initial changes in surface soil moisture hold a significant potential to determine the saturated hydraulic conductivity ($K_s$). Ahuja *et al.* (1993) investigated a relationship, based on the generalized Kozeny-Carman equation, between an effective average $K_s$ of a soil profile and drainable porosity of the surface soil as a result of drainage (or redistribution of water) for two days following a thorough wetting of the profile. The generalized form of the relationship is as follows:

$$K_s = B\,\Phi_e^{\,n} \tag{1}$$

where B and $n$ are empirical constants, and $\Phi_e$ is the effective porosity. Based on both theoretical and experimental data, Ahuja *et al.* (1993) have established statistically significant relationships between $K_s$ and changes in initial surface moisture content. Such approaches based on remote sensing that have capabilities of deriving spatial distribution of $K_s$ would be an extremely important data source for hydrological applications.

## SCALING OF SOIL HYDRAULIC CONDUCTIVITY

Scaling theory, based on the similar media concept (Miller and Miller, 1956), is used in this study to derive scaling factors. These factors are based on the assumption of a geometric similarity existing within a soil type located at different spatial locations (termed as similar media). The essence of scaling theory is a microscopic dimension which could be a particle, pore, aggregate size or some average of one of these parameters for a particular soil sample. Let $\lambda$ be the microscopic dimension and $\lambda'$ be some reference value of $\lambda$ (usually $\lambda'$ is the microscopic dimension characterizing a reference soil). Then a dimensionless scaling factor, $\alpha$, can be defined as

$$\alpha = \frac{\lambda}{\lambda'} \tag{2}$$

The scaling factor $\alpha$ allows us to approximate $K_s$ of a soil from that of a reference soil. If $K_s$ of the reference soil is 1 unit, a scaled soil at the same moisture content will have a conductivity of $\alpha^2$ units. Therefore, if $K_{si}$ is the $K_s$ of a certain soil profile at site $i$, $K_{sm}$ is a scaled mean hydraulic conductivity, the spatial variability of $K_s$ can be characterized in terms of scaling factors (Naney *et al.*, 1992) as:

$$K_{si} = K_{sm}\,\alpha_i^{\,2} \tag{3}$$

where $\alpha_i$ is a scaling factor for site $i$. Setting the average $\alpha_i$ values for N sites within the soils of a catchment to 1.0, $K_{sm}$ can be obtained as:

$$K_{sm} = \frac{1}{N^2}\left[\sum_{i=1}^{N}\sqrt{K_{si}}\right]^2 \tag{4}$$

From Equation 3 and Equation 4, $\alpha_i$ can be obtained as

$$\alpha_i = \frac{N \sqrt{K_{si}}}{\sum_{i=1}^{N} \sqrt{K_{si}}} \tag{5}$$

Therefore, the scaling factors can be obtained from the two-day initial drainage data measurable from microwave remote sensing. The application of scaling theory allows us to investigate, in a systematic manner, the effect of varying $K_s$ on components of the water balance.

## MATERIALS AND METHODS

### STUDY AREA AND DATA AVAILABILITY

The study area is the Little Washita River catchment located in the southern part of the Great Plains in southwest Oklahoma. Drainage area of the catchment is 610 km$^2$, and its climate is classified as moist and subhumid with an average annual rainfall of about 750 mm (Allen and Naney, 1991). The Washita'92 research experiment was carried out in the catchment during June 10-18, 1992, with joint cooperation between National Aeronautics and Space Administration (NASA), United States Department of Agriculture (USDA) and various universities. The study site is of prime importance to the USDA Agricultural Research Service and the USDA Global Climate Change initiative, and has a long term (more than 24 years) hydrological monitoring facility. The site is adjacent to or part of many other major science activities. This is also a super site for NASA's SIR (Shuttle Imaging Radar) C/X Synthetic Aperture Radar (SAR) experiments.

During the Washita'92 experiment, land cover in the Little Washita catchment was dominated by pasture and senescent or harvested winter wheat (Jackson and Schiebe, 1993). The forest cover within the catchment is very sparse and constitutes a small proportion of the catchment. As part of Washita'92, multi-temporal airborne passive microwave data were collected with a ground spatial resolution of 200 m x 200 m using the Electronically Steered Thinned Array Radiometer (ESTAR). The microwave data were employed to derive daily surface soil moisture data for the catchment (Jackson and Schiebe, 1993). In addition, a large number of ground soil moisture measurements (more than 700 gravimetric samples per day) were carried out. Profile soil moisture has been measured at 13 sites using a Resonant Frequency Capacitance (RFC) probe (Jackson and Schiebe, 1993). Soil moisture measurements were made at 15 cm intervals to a depth of 115 cm beginning at 7.5 cm below the soil surface. Soil temperatures were collected at each soil moisture sampling site at depths of 5 cm and 15 cm using metal dial type thermometers as well as digital probes. The study area experienced heavy rainfall (more than 30 mm) on 5 June, 1992, while moderate rainfall continued till 9 June, 1992. However, no rainfall

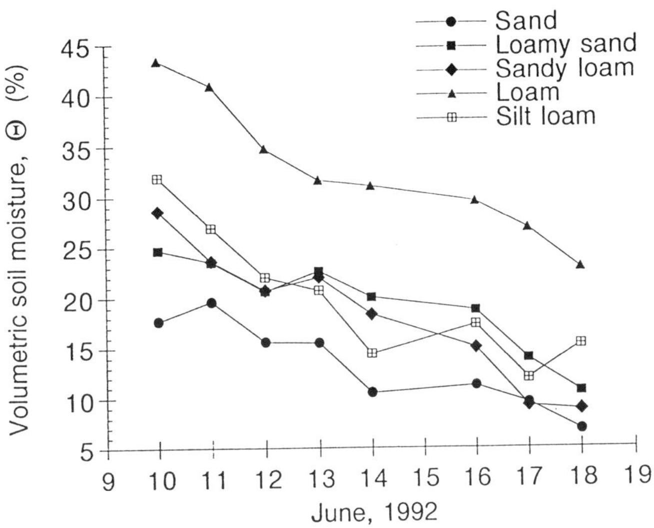


**Figure 1** Dry-down pattern of field measured volumetric soil moisture content for various soil types during the Washita'92 experiment.

occurred in the catchment for the entire duration of the Washita'92 experiment (Jackson and Schiebe, 1993). Therefore, the hydrological conditions in the catchment were ideal because it was possible to follow a drying period from very wet (about 30% volumetric) to dry (about 10% volumetric) over a period of nine days (Figure 1).

## ESTIMATION OF SOIL HYDRAULIC PROPERTIES

Our recent research has established methods of determining $K_s$ from remotely-sensed surface soil moisture data (Mattikalli *et al.*, 1995b). We exploited the close correspondence between characteristic temporal changes of surface soil moisture and soil texture encountered in the study area (i.e. sand, loamy sand, sandy loam, silt loam and loam — see Figure 1) and developed relationship between $K_s$ of the soil profile and two-day initial drainage. The main difficulty encountered in this effort was the absence of data concerning properties of the soil profile. Therefore, a state-of-the-art soil hydrology model, the RZWQM, and a GIS were employed to carry out soil moisture simulation studies to estimate $K_s$ for the soil profile. RZWQM is a physically-based hydrology model developed to simulate movement of water through the unsaturated zone (Decoursey and Rojas, 1990).

The RZWQM simulations were performed to optimize the approximate profile soil properties of texture classes obtained from soil survey maps. Details of the modelling procedure are reported in Mattikalli *et al.* (1995b). In the model simulations, soil survey data were used to obtain typical descriptions (horizon and texture) of all thirteen sites, and soil properties from Rawls and Brakensiek (1989)

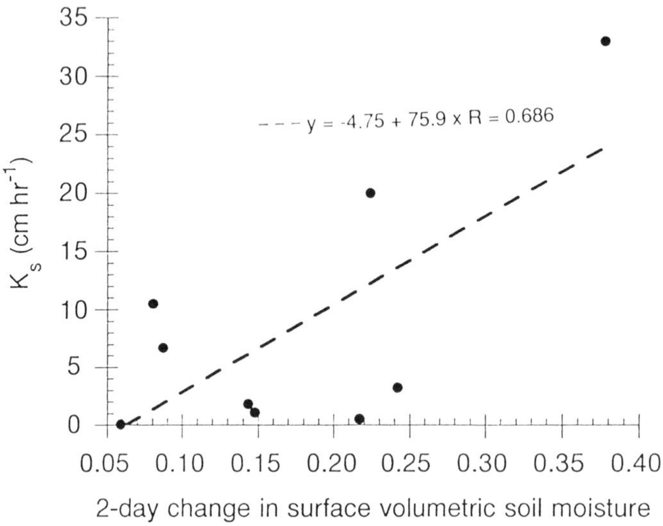


**Figure 2** Relationship between the harmonic-mean profile saturated hydraulic conductivity $K_s$ and two-day change in surface moisture content because of initial drainage as measured from microwave remote sensing (after Mattikalli *et al.*, 1995b).

were utilized as first approximations. Using data for various meteorological parameters, initial soil moisture and temperature profiles, model simulations were carried out to estimate the profile soil moisture distributions for the nine-day period on a daily time step. Four important hydraulic properties — bulk density, water content at 1/3 bar, $K_s$ and pore size distribution index — were optimised by repetitive model simulations, which resulted in a unique set of values for these properties. Validation was carried out by comparing the model derived soil moisture distributions for various depths with those measured in the field. This approach yielded a very good match between the model predicted and field-measured soil moisture variations for all depths, which suggested that the unique set of soil hydraulic properties obtained were fair estimates of field soil properties. Analysis of the results yielded a positive relationship between the model derived $K_s$ and two-day change in surface moisture content because of initial drainage and evaporation (Figure 2). Such a relationship suggests the potential application of microwave remotely-sensed soil moisture for estimating $K_s$. The model-estimated soil properties have been used to derive scaling factors of $K_s$ and the soil water balance components in the RZWQM simulations.

## DERIVATION OF SCALING FACTORS AND COMPONENTS OF THE WATER BALANCE

The scaling factors for all 13 sites have been derived using the harmonic-mean $K_s$. Since the depth of the soil profile varied between 60 cm to 110 cm, a constant profile depth of 60 cm was chosen in the current study, and the harmonic-mean values were computed to represent average conditions of the profile. Calculation of harmonic-mean $K_s$ assumes that the microscopic dimensions of each soil horizon at any site increase or decrease together, i.e. that similar media conditions hold constant. Therefore, the same scaling factor is applied to the entire soil profile at any time. The mean $K_s$ values were employed to generate the scaling factors using Equation 5.

The water balance components were derived by performing the RZWQM simulations. It was assumed that there are no roots in the layers below 60 cm and the lowest layer ending at 60 cm is always assumed to drain under a unit potential gradient. In the simulation studies, the field saturation factor was set to 0.95. It was also assumed that all pan evaporation data represented soil evaporation only, and that transpiration was spatially and temporally negligible during the experiment. We consider that this assumption does not bias our results since all sites were characterized by either senescent/ harvested wheat or pasture. This condition does not always hold well if land cover types are different, and NDVI information could be used in the model simulations to incorporate the differences in vegetation conditions. The dry mass of the residue on the surface was set to zero so that the residue would not interfere with the evaporation conditions. Further, the initial volumetric soil moisture content was set to 0.5 for all horizons. Although this is higher than the saturated moisture contents ($\Theta_s$), the model will re-initialise the initial soil moisture content down to 0.95 * $\Theta_s$. This will enable us to start the simulations at 95% saturation and follow the drainage pattern on a daily time step for the nine days of the Washita'92 experiment. The model simulates redistribution of soil moisture between horizons by a finite difference numerical solution to the Richards' equation:

$$\frac{\partial \Theta}{\partial t} = \frac{\partial}{\partial z}\left[K(h, z)\frac{\partial h}{\partial z} - K(h, z)\right] - S(z, t) \tag{6}$$

where $\Theta$ = volumetric soil moisture content, t = time, z = soil depth, h = soil-water pressure head, K = unsaturated hydraulic conductivity as a function of h and z, and S(z, t) = root water uptake rate considered to be zero in this experiment. The modelled moisture content at each horizon was used to calculate the weighted-average moisture content for the entire profile for each day of simulation.

Evapotranspiration (ET) is calculated using the Penman-Monteith equation. The approach of Shuttleworth and Wallace (1985) which provides both canopy transpiration and soil evaporation is used. It provides an estimate of actual transpiration and evaporation by incorporating stomatal resistance and a soil surface evaporation rate that is calculated using soil resistance as a function of the soil moisture condition. Although the basic ET expressions developed by Shuttleworth and Wallace have been used without any change, the input parameters were corrected to reflect a substrate that is made up of both surface residue (that was set to zero) and bare soil. Determination of actual soil evaporation ($E_{SA}$) was carried out as given below:

$$E_{SA} = E_{SP}$$

if the conductivity of the soil is sufficient to meet the demand at the surface, or:

$$E_{SA} = E_{SS} \qquad \text{if } E_{SS} < E_{SP}$$

where

$E_{SP}$ = potential soil evaporation, and
$E_{SS}$ = flux produced by the soil which is calculated in the soil moisture redistribution component by placing a flux at the surface equal to $E_{SP}$.

If tensiometric gradients limit $E_{SP}$, then $E_{SA}$ is set equal to that produced by the soil flux calculations.

## RESULTS AND DISCUSSION

Relationships between scaling factors of $K_s$ and three components of the water balance (i.e. daily average moisture content for the 60 cm profile, deep drainage below the profile, and evaporation from soil surface) generated by the model simulations have been examined. Figure 3 shows the relationships between the scaling factor and the three components of the water balance for each day between 10-18 June, 1992. Since there was no rainfall during the experiment, surface runoff did not occur, and the three components of the water balance have gradually decreasing magnitudes over the nine-day period. Note that increasing values of scaling factors (i.e., increasing microscopic characteristic length) represent soils characterized by larger mean pore size distribution and therefore higher $K_s$ values.

It is clear from Figure 3a that the scaling factor, $\alpha$, has a negative linear relationship with the average moisture content of the soil profile, although there exist one outlying set of points for $\alpha = 0.95$. This supports the fact that fine grained clayey soils characterized by lower $K_s$ drain slowly and retain much of the initial moisture content, while the coarse grained sandy soils drain very quickly because of their high $K_s$ and retain less moisture in the profile.

The scaling factor $\alpha$ shows high deep drainage values for 10 June, 1992 (Figure 3b), and deep drainage maintains low values for the remainder of the experiment. However, deep drainage was found to be independent of the $K_s$ scaling factors. One can only speculate as to why this is the case. One possibility is that after initial drainage the soil profile is dominated by unsaturated conditions and field saturated conditions no longer hold. In this case, some other parameter that characterizes the unsaturated hydraulic conductivity might be used to generate scaling factors. Since the unsaturated hydraulic conductivity changes with soil water content, the slope of the water retention curve might be an alternative parameter to derive scaling factors

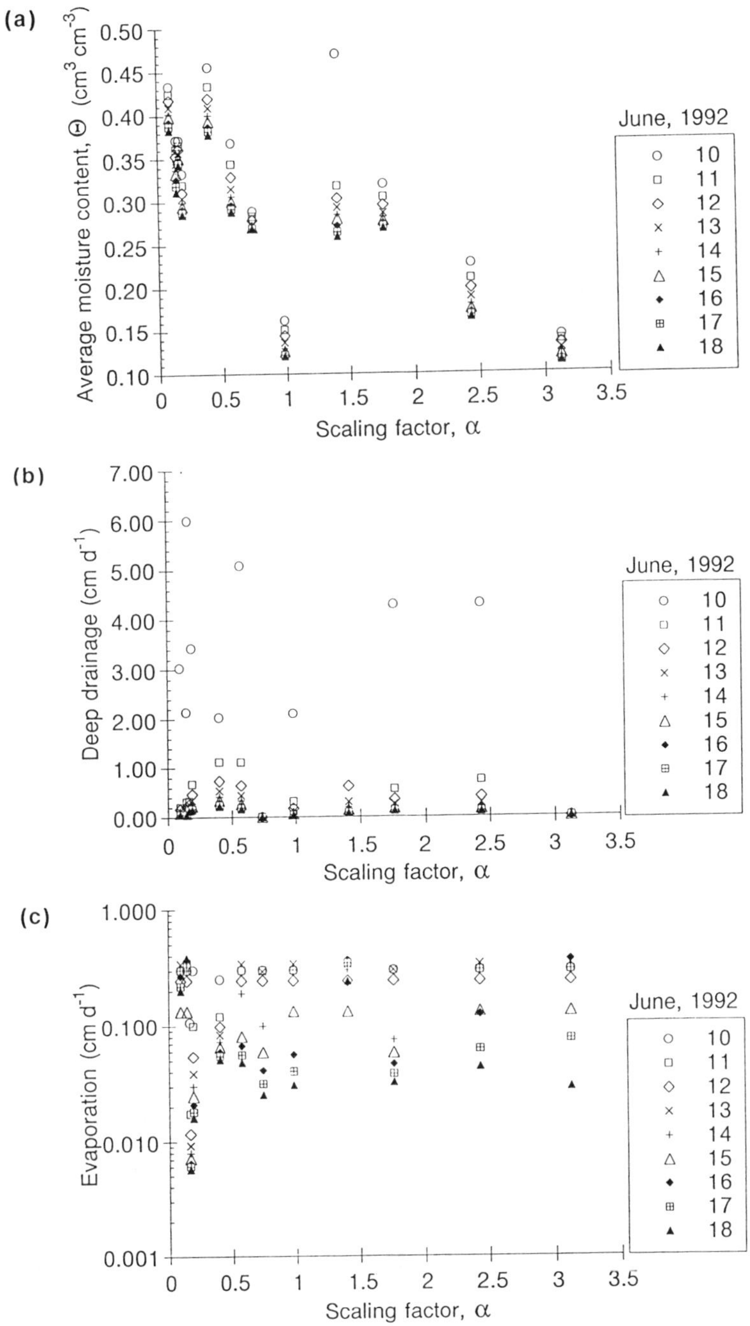


**Figure 3** Relationship between the scaling factors of $K_s$ and some components of the water balance for period between June 10-18, 1992: (a) average soil moisture content of the 60 cm profile; (b) deep drainage below the 60 cm deep profile; and (c) evaporation from the soil surface. The scaling factors correspond to various textural classes (i.e. sand, loamy sand, sandy loam, silt loam and loam) encountered in the study area.

to better understand the effect of spatial variability on deep drainage and evaporation.

Figure 3c suggests that evaporation from the surface soil remains invariant and is generally independent of scaling factors. The change in evaporation for the nine-day period tends to reduce for $\alpha < 1.0$ in most cases. Based on our studies, the effects of soil variability on evaporation are small. Again, one can only speculate on the lack of any scaling effect on evaporation. One possibility is that in no case during the experimental period was soil moisture limiting, i.e. the soil was controlling the evaporation rate. In this case potential soil evaporation always satisfied. Results for deep drainage and evaporation are similar to those obtained by Peck *et al.* (1977) who concluded that soil spatial variability had a rather small effect on drainage and evapotranspiration for two average months. Our results are based on simulations in which vegetation parameters are assumed to be the same for all scaling factors, while we recognize that in real world situations, vegetation characteristics are known to vary in response to varying soil properties. Although this suggests that our results should be taken as approximations only, the results provide some insights into the processes of soil spatial variability and issues on derivation of scaling factors for soil properties.

The assumption of the existence of similar media throughout the soil profile will not be valid in most field cases (Ahuja *et al.*, 1984). This might be another reason why deep drainage and evaporation remained independent of scaling factors derived on the similar media concept. Although similar media scaling has been used in earlier studies of the water balance modelling (e.g. Peck *et al.*, 1977; Sharma and Luxmore, 1979), recent research (e.g. Ahuja and Williams, 1991; Williams and Ahuja, 1992) has shown that the one-parameter model of Gregson *et al.* (1987) might yield better estimation of soil water content. Therefore, scaling based on the one-parameter model might provide improved explanation of the effects of spatial variability of soil properties on the components of the water balance considered in this study.

Relationships between the scaling factor $\alpha$ and various components of the water balance (e.g. Figure 3a) indicate that there is potential to employ remotely-sensed information to determine either scaled-up or scaled-down components. Figure 4 shows a schematic diagram of the scaling-up (i.e. aggregation) and scaling-down (i.e. disaggregation) of the soil water balance components which employs both the $K_s$ soil moisture change relationship (from Figure 2) and the scaling factor water balance component relationship (from Figure 3). If the original remotely-sensed soil moisture data are at a high spatial resolution (e.g. 200 m), the approach can be used to derive the spatial distribution of some water balance components at this scale, which can be then aggregated to larger scales (into 10 km pixels for example). On the other hand, the low resolution data (derived from either remote sensing or mesoscale/general circulation models) having larger foot prints (e.g. 10 km pixels) could be employed in the approach to determine the sub-pixel variability of the water balance components at smaller scales (e.g. 200 m pixels). For example, a major research issue facing land surface process modellers is how to parameterise the spatial variability observed in nature and how to interpret satellite data that may

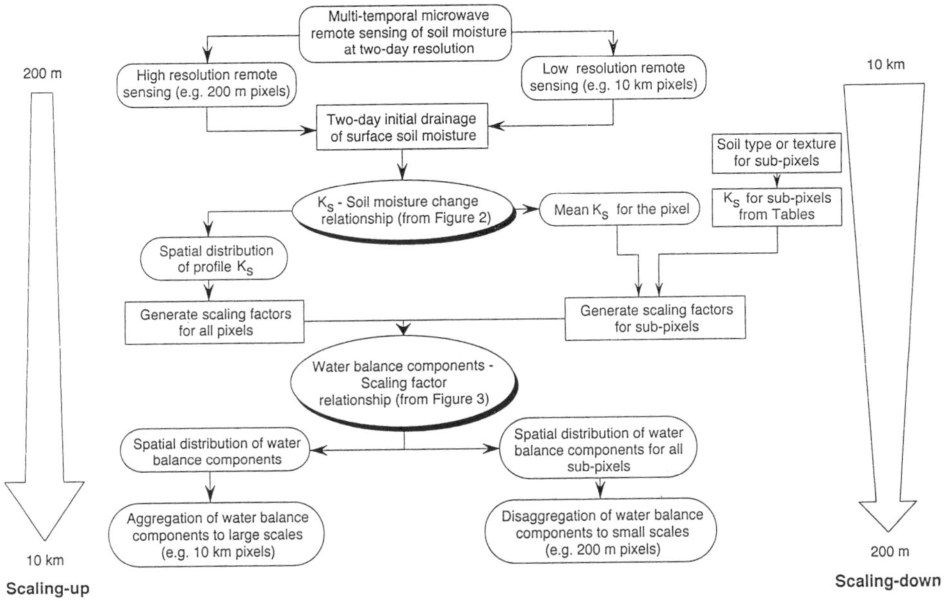


**Figure 4** Schematic diagram of the approach for scaling-up (aggregation) or scaling-down (disaggregation) of the water balance components from remotely-sensed soil moisture information.

have a spatial resolution of tens of kilometres. The issue is that a large pixel represents some sort of average surface condition, for example soil moisture. In reality we know that the surface probably varies from some parts of the pixel that are quite wet to other parts that are quite dry. The evaporation from the dry areas may be considerably less than that from the wet areas, and the relationship between evaporation and soil water content is non linear. The procedure proposed here suggests a way to use scaling factors to spatially distribute a pixel average soil moisture content to realistic values based on the soil texture class.

## CONCLUSIONS

It has been shown in our earlier work that temporal changes in surface soil moisture could be used as an indicator of soil properties. Spatial and temporal remotely-sensed soil moisture data were employed in the RZWQM and GIS to estimate profile $K_s$ for 13 sites in the Little Washita catchment. The modelling procedure for the optimization of $K_s$ could be performed using any other physically-based soil hydrology model which handles the redistribution of soil moisture using a numerical solution to the Richard's equation and which has the capability of discritizing the soil profile into several horizons. The modified similar media concept was used to derive $K_s$ scaling factors and to evaluate the effects of spatial variability of $K_s$ on average moisture content, deep drainage, and evaporation components of the water balance. The scaling factor $\alpha$ had a negative linear relationship with average moisture content of the soil profile, and maintained low values for deep drainage for the entire duration of simulation except the first day. Evaporation from the surface soil remained constant and was generally independent of scaling factors. Deep drainage and evaporation showed insignificant relationships with scaling factors possibly because of the assumptions of existence of similar media conditions throughout the soil profile and development of unsaturated conditions after initial drainage. Scaling factors for some other parameters (e.g. the slope of the water retention curve) that characterizes the unsaturated hydraulic conductivity and use of a one-parameter model of Gregson *et al.* (1987) might provide a better explanation of the effects of spatial variability of soil properties. The conclusions drawn here are highly dependent upon the nature of the distribution and the extent of soil variability, as well as on the water balance component of interest. Nevertheless, the relationships between scaling factors and the water balance components are potentially useful to derive scaled-up or scaled-down components from remotely-sensed soil moisture information. A schematic outline of a procedure for scaling-up (aggregation) and scaling-down (disaggregation) of the water balance components from remotely-sensed soil moisture measurements has been presented.

## ACKNOWLEDGEMENT

This work was performed while Nandish Mattikalli held a National Research Council - NASA GSFC Research Associateship. The support was provided by the Science Division of NASA's Office of Mission to Planet Earth.

## REFERENCES

Ahuja, L.R. and Williams, R.D. 1991. Scaling water characteristic and hydraulic conductivity based on Gregson-Hector-McGowan approach. *Soil Sci. Soc. Am. J.*, **55**, 308-319.

Ahuja, L.R., Naney, J.W. and Nielsen, D.R. 1984. Scaling soil water properties and infiltration modelling. *Soil Sci. Soc. Am. J.*, **48**, 970-973.

Ahuja, L.R., Naney, J.W. and Williams, R.D. 1985. Estimating soil water characteristics from simpler properties or limited data. *Soil Sci. Soc. Am. J.*, **49**, 1100-1105.

Ahuja, L.R., Wendroth, O. and Nielsen, D.R. 1993. Relationship between initial drainage of surface soil and average profile saturated conductivity. *Soil Sci. Soc. Am. J.*, **57**, 19-25.

Allen, P.B. and Naney, J.W. 1991. Hydrology of the Little Washita River watershed, Oklahoma: data and analysis. USDA/ARS-90, 74 pages.

Beven, K., Wood, E.F. and Sivapalan, M. 1988. On the hydrological heterogeneity - catchment morphology and catchment response. *J. Hydrol.*, **100**, 353-375.

Decoursey, D.G. and Rojas, K.W. 1990. A Root Zone Water Quality Model (RZWQM), Proc. of Symp. on Towards a Sustainable Agriculture for the Great Plains, Jan. 19-20, Fort Collins, CO.

Famiglietti, J.S. and Wood, E.F. 1995. Effects of spatial variability and scale on areally averaged evapotranspiration. *Wat. Resour. Res.*, **31**, 699-712.

Gregson, K., Hector, D.J. and McGowan, M. 1987. A one-parameter model for the soil water characteristic. *J. Soil Sci.*, **38**, 483-486.

Jackson, T.J. and Schiebe, F.R. (eds.). 1993. Hydrology data report: Washita'92. USDA-Agri. Res. Service, Nat. Agri. Water Quality Lab., NAWQL 93-1, Durant, OK.

Jackson, T.J., Le Vine, D.M., Swift, C.T., Schmugge, T.J. and Schiebe, F.R. 1995. Large area mapping of soil moisture using the ESTAR passive microwave radiometer in Washita'92. *Remote Sensing Environ.*, **53**, 27-37.

Luxmoore, R.J. and Sharma, M.L. 1980. Runoff responses to soil heterogeneity: experimental and simulation comparisons for two contrasting watersheds. *Water. Resour. Res.*, **16**, 675-684.

Mattikalli, N.M., Engman, E.T. and Jackson, T.J. 1995a. Application of a GIS for spatial and temporal monitoring of remotely-sensed soil moisture. Proc. ISPRS Commission IV/1 Workshop on Mapping and Environmental Applications of GIS data, Sept., 28-29, Madison, WI, Int. Archives of Photo. and Remote Sensing, vol. XXX, Part 4W2, Dept. of Geog., Univ. of Georgia, Athens, 44-53.

Mattikalli, N.M., Engman, E.T., Ahuja, L.R. and Jackson, T.J. 1995b. Estimating soil properties from microwave remote sensing of soil moisture. In: Engman, E.T., Guyot, G. and Marino, C.M. (Eds) *Remote Sensing for Agriculture, Forestry, and Natural Resources*, Proc. SPIE vol. 2585, 89-101.

Miller, E.E. and Miller R.D. 1956. Physical theory for capillary flow phenomena, *J. Appl. Phycs.*, **27**, 324-332.

Naney, J.W., Ahuja, L.R. and Williams R.D. 1992. Estimating spatial distribution of hydraulic conductivity in a field using effective porosity data. In: van Genuchten M. T. and Leij F. J. (Ed.). Indirect methods for estimating the hydraulic properties of unsaturated soils, Univ. Calif. Press, Riverside, CA, October 11-13, 515-527.

Peck, A.J., Luxmore, R.J. and Stolzy, J.L. 1977. Effects of spatial variability of soil hydraulic properties in water balance modelling. *Wat. Resour. Res.*, **13**, 348-354.

Rawls, W.J. and Brakensiek, D.L. 1989. Estimation of soil water retention and hydraulic properties. In: Morel-Seytoux, H.J. (Ed.) *Unsaturated flow in hydrological modelling - theory and practice.* Kluwer Academic Publications, 275-300.

Russo, D. and Bresler, E. 1981. Soil hydraulic properties as stochastic process. 1. An analysis of field spatial variability. *Soil Sci. Soc. Am. J.*, **45**, 682-687.

Seyfried, M.S. and Wilcox, B.P. 1995. Scale and the nature of spatial variability: field examples having implications for hydrological modelling. *Wat. Resour. Res.*, **31**, 173-184.

Sharma, M.L. and Luxmoore, R.J. 1979. Soil spatial variability and its consequences on simulated water balance. *Wat. Resour. Res.*, **15**, 1567-1573.

Sharma, M.L., Gander, G.A. and Hunt, C.G. 1980. Spatial variability of infiltration in a watershed. *J. Hydrol.*, **45**, 101-122.

Shuttleworth, W.J. and Wallace, J.S. 1985. Evaporation from sparse crops - an energy combination theory. *Q. J. Roy. Met. Soc.*, **3**, 839-855.

Warrick, A.W., Mullen, G.J. and Nielsen, D.R. 1977. Scaling field-measured soil hydraulic properties using a similar media concept. *Wat. Resour. Res.*, **13**, 355-362.

Williams, R.D. and Ahuja, L.R. 1992. Estimating soil water characteristics using easily measured physical properties and limited data. In: van Genuchten M.T. and Leij F.J. (Eds). Indirect methods for estimating the hydraulic properties of unsaturated soils, Univ. Calif. Press, Riverside, CA, October 11-13, 405-416.

Williams, R.D. and Ahuja, L.R. 1992. Evaluation of similar-media scaling and a one-parameter model for estimating the soil water characteristics. *J. Soil Sci.*, **43**, 237-248.

Wood, E.F., Sivapalan, M., Beven, K. and Band, L. 1988. Effects of spatial variability and scale with implications to hydrological modeling. *J. Hydrol.*, **102**, 29-47.

# 14 Errors in estimation of areal soil water content from SAR data

**P.J. VAN OEVELEN, D.H. HOEKMAN and R.A. FEDDES**
*Wageningen Agricultural University, The Netherlands*

## INTRODUCTION

In 1991 and 1992 in the framework of climate prediction models two large field experiments took place: EFEDA-Spain (Bolle and Streckenbach, 1992) and HAPEX-Sahel (Goutorbe *et al.*, 1994), respectively. The objectives of these experiments were (1) improvement in the parameterisation of surface hydrological processes in semi-arid areas and (2) development and application of methods for monitoring surface hydrology at a large scale using remote sensing.

This chapter explores the errors involved in determination of areal soil water content by means of active microwave remote sensing. Comparisons are made between the use of the JPL-AIRSAR (P-, L-, and C-band, polarimetric) and the ERS-1 SAR (C-band, single polarization VV). The simulations and comparisons are limited here to C-band (a frequency of about 5.3 GHz). The heterogeneity of the surface parameters and the radar speckle is considered to be the main source of errors in soil moisture estimation.

The influence of vegetation in the detection of soil moisture can be significant and may limit the applicability of the sensors. Sensitivity analysis of the applied soil moisture retrieval algorithm using the Monte Carlo approach can show where problems will arise and how they may be circumvented. This knowledge of the limitations of soil moisture retrieval by active microwave remote sensing is invaluable in hydrological and climatological modelling, since it gives insight into attainable resolutions and accompanying errors.

## THEORY

### *Microwave backscatter modelling*

In microwave remote sensing the earth's surface is usually described in terms of a random rough surface or dielectric half space. Where this roughness is described as the deviation from a reference plane it can be characterized statistically by measures such as the root mean square of the height differences (s), the surface correlation

*Scaling up in hydrology using remote sensing.* Edited by J.B. Stewart, E.T. Engman, R.A. Feddes and Y. Kerr

length (l) or the height probability density function. Objects that are on top of the rough surface such as vegetation, are usually modelled as statistical ensembles of simple shapes like ellipsoids and cylinders. In the case of vegetation this requires extensive knowledge of its dielectric properties, the density, shape, size and orientation angle distributions of leaves and branches. Based upon this information the contribution of each of these objects to total scattering can be described.

We have limited the description of microwave backscattering to that of bare soil surfaces. Various empirical and theoretical approaches have been developed to describe the backscatter from these random rough surfaces, e.g. Fung (1994), Ulaby *et al.* (1986), Oh *et al.* (1992). The integral equation method (IEM) model of Fung *et al.* (1992) is a theoretical physically based model that describes the backscatter coefficient ($\sigma^0$) as a function of surface roughness, soil dielectric properties (expressed as a complex dielectric constant, $\varepsilon = \varepsilon' - j\varepsilon''$) and incidence angle ($\theta_{inc}$). The IEM model can be simplified for certain ranges of roughness conditions to three other models, namely the Geometrical Optics (GO) model being valid for 'very rough surfaces', the Physical Optics (PO) model being valid for 'intermediate rough surfaces' (both models are based upon the Kirchhoff approximations), and the Small Perturbation Model (SPM) being valid for 'smooth surfaces' (Oh *et al.*, 1992). The surface roughness parameters can be adequately described for those surfaces whose autocorrelation functions are normally distributed. The wavenumber is defined as $k = 2\pi/\lambda$, with $\lambda$ as the wavelength. Consequently, the surface roughness can be expressed as the root mean square (rms) of the height differences relative to the wave length ks. The autocorrelation length relative to the wave length is kl and the rms slope is $m = (2\ s/l)^{0.5}$.

*Inverse modelling of radar backscatter*

In general, forward modeling of radar backscatter already poses a tremendous problem when dealing with an object of a complex nature such as a vegetated surface. The most descriptive theoretical models available require detailed information about the object and are of such a nature that an analytical inversion to retrieve parameters of interest such as soil moisture is virtually impossible. To circumvent these problems one can either simplify the models and their boundary conditions and apply numerical inversion techniques, or use empirical approaches linking field data directly to measured radar backscatter, or use semi-empirical approaches which combine both previous methods (Dubois *et al.*, 1995, Oh *et al.* 1992). The first approach has been chosen here and IEM inverted to estimate soil moisture for bare soil surfaces. This model has been chosen because it can be applied to data from single as well as multiple frequency sensors.

The inverse IEM model (hereafter referred to as INVIEM) is based upon the forward solution of the IEM model and uses *a priori* knowledge of the surface roughness characteristics (ks and kl) and the incidence angle. The first step is to simulate the radar backscatter for the required frequencies and polarization as a function of soil moisture and incidence angle for a number of distinct roughness states. These simulated data are then tabulated in the form of a look-up table (LUT)

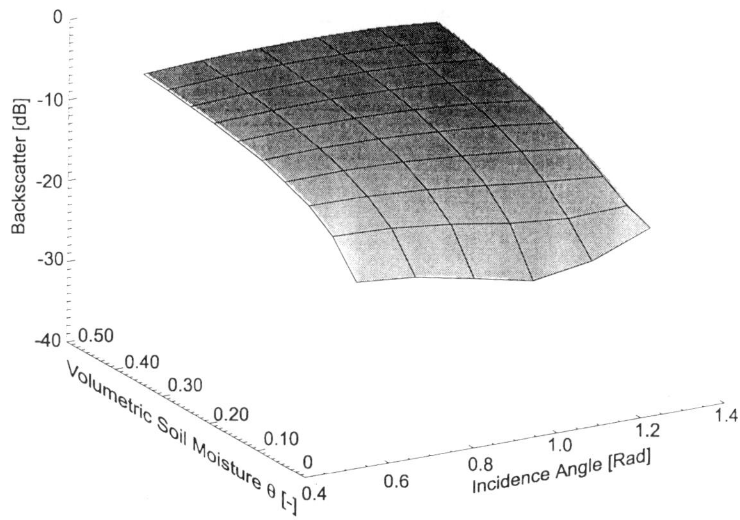


**Figure 1** The microwave backscatter coefficient $\sigma^0$ of C-band VV-polarization as a function of volumetric soil moisture content $\theta$ for various incidence angles for one surface roughness case (ks = 2.52 and kl= 9.71).

which shows per frequency and polarization the backscatter coefficient $\sigma^0$ as a function of volumetric soil moisture content $\theta$ for the different roughness states and the various incidence angles. In Figure 1 $\sigma^0(\theta)$ is shown for different incidence angles and one roughness case. Figure 1 shows that for higher soil water contents the relationship between $\sigma^0(\theta)$ and incidence angle is less sensitive than at lower $\theta$-values. The relationship of the backscatter coefficient with the incidence angle depends strongly upon the surface roughness.

Soil mixing models have been used to convert the fraction soil moisture and soil type into a dielectric constant as is required by the IEM model (Dobson *et al.*, 1985, Hallikainen *et al.*, 1985, Wang and Schmugge, 1980). Naturally the INVIEM model is valid only where the IEM model is valid (see Chapter 6 of Fung, 1994). The IEM model has been validated for certain roughness states where the physical optics, geometrical optics and small perturbation approaches are valid (Ulaby *et al.*, 1986). Periodic roughness characteristics of surfaces such as ploughed fields can be outside the validity range of IEM models, depending upon the view angle of the sensor.

The INVIEM in its current form is only valid for bare soil surfaces, hence vegetated areas have to be eliminated by using either groundtruth and/or other remotely sensed data. The preferred approach is to use polarimetric microwave data which makes it possible to show when vegetation has no significant influence on the signal. When these polarimetric data are available, the scattering behaviour that dominates a pixel can be deduced from the amplitude and phase information by comparison with simple classes of scattering such as even number of reflections, odd number of reflections, and diffuse scattering (Van Zyl, 1989). Vegetation is usually characterised by diffuse scattering. Using multiple frequencies this scattering classification can be used to develop a mask which selects pixels which do not show diffuse scattering in any of the bands. These pixels are then considered not to be influenced significantly by vegetation. On these pixels the inverse model can be applied in any of the bands. When polarimetric data are not available the use of remotely sensed optical data, e.g. in the form of vegetation indices, can be employed to discriminate between vegetated and non-vegetated pixels.

*Estimation of systematic and random errors*
Consider a bare soil field with a field averaged soil moisture content $<\theta>$ and field averaged roughness parameters $<ks>$ and $<kl>$. The field consists of n pixels, each with a soil moisture content $\theta_i$ (i=1,...,n) and roughness parameters, $ks_i$ and $kl_i$ that deviate from the field average values. The distribution of the variables is considered to have a Gaussian distribution with the mean equal to the field averages and a standard deviation determined from field observations. Furthermore, each pixel has an incidence angle $(\theta_{inc})_i$ which is not exactly known. The above mentioned variation in surface parameters can be considered as random errors. Another source of errors in radar observations is speckle, the statistics of which are well known (Hoekman, 1991). For logarithmically scaled data, i.e. $\sigma^o$ in dB, the standard deviation is a constant depending only on the number of looks and the distribution can be approximated by a Gaussian distribution. The following method is employed as a first attempt at simulating the influence of errors and error propagation on the model performance (Bastiaanssen *et al.*, 1993).

Given a random variable X, e.g. the radar backscatter, which is a function of the input parameters $x_i$, e.g. $\theta$, ks etc., with i=1,2,...n:

$$X = f(x_1,...x_i,...x_n) \quad (1)$$

and a set of reference input parameters of $x_i$, called $r_i$, e.g. field averages of $\theta$, ks etc., determined from field observations which give as a result the reference value of X hereafter called R:

$$R = f(r_1,...r_i,...r_n) \quad (2)$$

Thus R is the backscatter coefficient of the field using the field averaged values of $\theta$, ks, etc.. The input parameters are assumed to have a random error that is normally distributed with the reference value as a mean and the standard deviation set at a percentage of the mean. Using sets of randomly distributed input parameters to simulate errors in model input parameters, model output X can be generated from which the probability density function or distribution of X can be determined. A large number of calculations of X is necessary to retrieve meaningful statistics: one important condition is that the number of calculations must be large enough for the statistics to become constant. The simulation method described above is generally known as the Monte-Carlo method.

However, one is usually not interested in the distribution of X but in the distribution of the deviation $S_e$ from the reference value. The error $S_e$ is defined as:

$$S_e = [X - R] \tag{3}$$

The cumulative probability density function or distribution of $S_e$ can thus be calculated.

It is assumed that the reference value is the true value for X. However, the reference value for the backscattering coefficient $\sigma^o$ is determined using the reference input parameters, which were considered to be the true values. For the soil moisture retrieval algorithm the areal or field averaged value used to generate the backscattering coefficients is used as the reference value. This implies that only the true value of X is retrieved when a perfect model is used.

Two basic different situations are discussed. The first situation is the estimation of the soil moisture on a pixel by pixel basis both for pixels with or without speckle (referred to as pixel-based). The second situation is the estimation of soil moisture on a field or object basis using field averaged backscatter values also for fields (number of pixels per field is varied up to a maximum of $10^4$) with or without speckle (referred to as field-based).

## RESULTS AND DISCUSSION

### ERROR ANALYSIS

As was done in the previous study by Bastiaanssen *et al.* (1995) the statistics needed to perform the simulation were estimated from field data (Van Oevelen *et al.*, 1993; Bastiaanssen *et al.*, 1993). The standard deviations of the parameters that are considered as sources of random errors in this simulation are listed in Table 1; the field average or mean values of these parameters can be found in Table 2. For both

**Table 1** The parameters considered to have a random error with the standard deviation of the mean used in the simulation.

| Parameter | Standard Deviation (as % of mean) | |
|---|---|---|
| | Spain JPL-AIRSAR | Niger, ERS-1 |
| 1. Autocorrelation length kl | 32 | 70 |
| 2. RMS of height differences ks | 39 | 60 |
| 3. Volumetric moisture content $\theta$ (converted to a complex dielectric constant) | 20 | 20 |
| 4. Incidence angle $\theta_{inc}$ | 1 | 1 |

**Table 2** The mean parameter values used in the simulation of the backscatter coefficient $\sigma^o$.

| | Spain | | | | Niger | | | |
|---|---|---|---|---|---|---|---|---|
| Case | ks [-] | kl [-] | $\theta$ [-] | $\theta_{inc}$ [Rad] | ks [-] | kl [-] | $\theta$ [-] | $\theta_{inc}$ [Rad] |
| 1 | 1.6 | 4.4 | 0.1 | 0.78 | 0.3 | 4.5 | 0.1 | 0.40 |
| 2 | 1.6 | 11.0 | 0.1 | 0.78 | 0.8 | 11.2 | 20.1 | 0.40 |
| 3 | 1.1 | 16.9 | 0.1 | 0.78 | 1.2 | 17.9 | 50.1 | 0.40 |

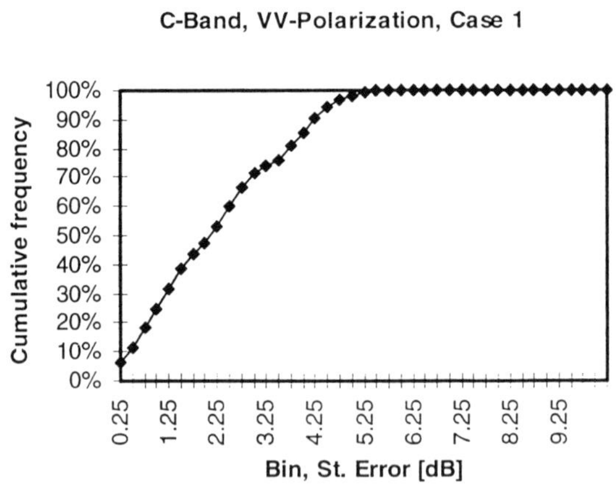


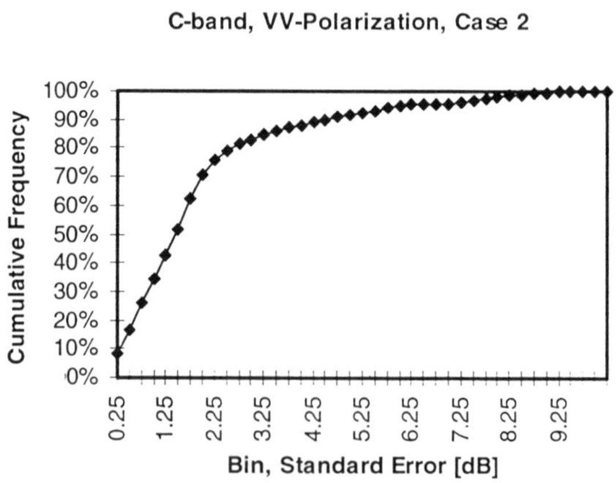


**Figure 2** Probability density functions for C-band VV polarization for two surface roughness cases 1 and 2 for Niger (reference values are given in Table 2).

ERS-1 and JPL-AIRSAR, three fields with different roughnesses have been chosen: each field can thus be considered to be representative of a certain roughness case. The simulation for 10,000 pixels for these fields or roughness cases will give a distribution of the backscatter coefficient. INVIEM is applied to these backscatter coefficients to yield a distribution of soil moisture estimates. The speckle is determined from six bare soil fields both in Spain for JPL-AIRSAR data and in Niger for ERS-1 data and compared with theoretical values from Hoekman (1991). For ERS-1 the standard deviation is about 2.5 dB (3 looks) and for the JPL-AIRSAR it is about 1.33 dB (16 looks). Hoekman (1991) gives a standard deviation of 2.73 dB and 1.10 dB for 3-and 16-look data, respectively and thus the experimental data are in good agreement with the theoretical values.

Two examples of the pdf of error of estimates for VV-polarized C-band for two surface roughness cases are shown in Figure 2. The results for three different roughness cases are listed in Table 3 which shows that the error in backscatter coefficient so for an individual pixel can become very large. On average, ERS-1 has a standard error $S_e$ varying from 1.91 dB to 2.94 dB. For the JPL-AIRSAR it is considerably less, namely from 0.82 dB to 1.56 dB for VV-polarization and from 1.03 dB to 1.72 dB for HH-polarization. The contribution of speckle to the standard error varies from approximately 0.4 dB to 0.9 dB.

Soil moisture content estimates $\theta$ are listed in Table 4 and two examples of the pdf are given in Figure 3. Clearly, the average standard errors are in most cases much larger for the individual pixel soil moisture estimates from Niger compared with those from Spain. However, if we look at the field estimates, the error decreases significantly to become even lower than those of the field estimates from Spain. The most likely explanation is the higher standard deviations for the surface parameters for Niger which give large deviations from the true mean backscatter coefficient for a low number of pixels and thus introduce an error when inverted. The standard error in estimating the soil moisture content for an individual pixel ranged from 10-15%

**Table 3** The standard deviation (sd), the mean standard error ($S_e$) and mean of the backscatter coefficient simulation (n=$10^4$) for C-band and three different roughness cases with and without speckle (see Table 2). For the JPL-AIRSAR (Spain) both results for both HH an VV polarization are given, for ERS-1 (Niger) only for VV-polarization.

| Case | Spain VV Mean | sd | $S_e$ | HH Mean | sd | $S_e$ | Niger VV Mean | sd | $S_e$ |
|---|---|---|---|---|---|---|---|---|---|
| 1 | -7.67 | 2.26 | 0.82 | -10.77 | 2.43 | 1.28 | -10.42 | 2.48 | 3.37 |
| speckle | -7.69 | 2.36 | 1.25 | -10.77 | 2.53 | 1.46 | -10.34 | 2.86 | 3.61 |
| 2 | -8.74 | 1.22 | 0.36 | -11.49 | 1.75 | 1.22 | -7.95 | 2.62 | 1.92 |
| speckle | -8.78 | 1.27 | 0.89 | -11.54 | 1.48 | 1.40 | -8.05 | 2.68 | 2.36 |
| 3 | -12.27 | 1.70 | 0.52 | -17.00 | 2.12 | 1.03 | -7.26 | 3.04 | 1.91 |
| speckle | -12.29 | 1.75 | 0.91 | -17.04 | 1.90 | 1.21 | -7.34 | 2.99 | 2.44 |

**Table 4** The standard error $S_e$ for the results of the soil moisture content θ estimation using C-band and three different roughness cases (see Table 2). Given is a comparison between pixel based estimates and field based estimates, i.e. using the average backscatter over all the pixels ($n=10^4$). For case 3 in Spain, the field average backscatter coefficients did not give a solution. For this case the pixel based estimation yielded only a result in 60% of the pixels. In all the other inversions the outfall was less then 15%.

| Case | Spain pixel | field | Niger pixel | field |
|---|---|---|---|---|
| 1 | 0.09 | 0.08 | 0.13 | 0.03 |
| speckle | 0.10 | 0.08 | 0.13 | 0.03 |
| 2 | 0.04 | 0.03 | 0.15 | 0.01 |
| speckle | 0.05 | 0.03 | 0.10 | 0.01 |
| 3 | 0.05 | - | 0.10 | 0.03 |
| speckle | 0.05 | - | 0.12 | 0.03 |

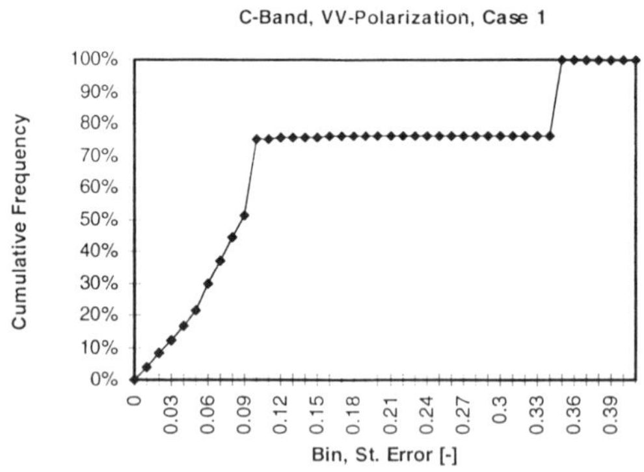


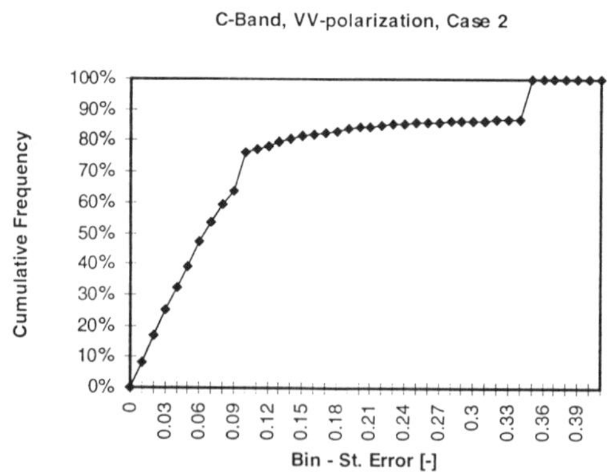


**Figure 3** Probability density functions for C-band soil moisture estimates for the surface roughness cases of Figure 2. The discontinuous look of the high estimates is a result of the cut-off values in the soil moisture algorithm used (θ = 0.0 and 0.45 which give a $S_e$ of 0.1 and 0.35 respectively).

for the Niger data set for pixel-based estimates and from 1-3% for field-based estimates (Table 4). For Spain, the errors ranged from 5-10% for pixel-based estimates and from 3-8% for field-based estimates.

Currently, the INVIEM model uses — when available - both VV and HH polarisations to obtain soil moisture estimates. Usually, the ranges of soil moisture estimates from VV and HH overlap. When that happens, the minimum and maximum values of the overlap are taken as the boundaries of the soil moisture estimate range. A problem arises when there is no overlap and in that case the inversion is not possible. This was the case for field 3 for which no inversion was possible for a large number of pixels.

The soil moisture estimate results are plotted for each pixel for each case and averaged cumulatively as shown in Figure 4. After averaging over a certain number of pixels the average soil moisture estimate becomes almost constant. Each individual estimate can however deviate strongly from the reference or true value. When the estimates are averaged for all separate surface roughness cases or fields, the result generally becomes closer to the reference value (Figure 4, Table 5). In the case where all values are averaged over a number of pixels larger then 25 there is only a slight deviation in the soil moisture estimates from the reference values (Table 5). For the field-based estimates there are almost no differences between the cases with speckle and those without. For the pixel-based estimates, even after

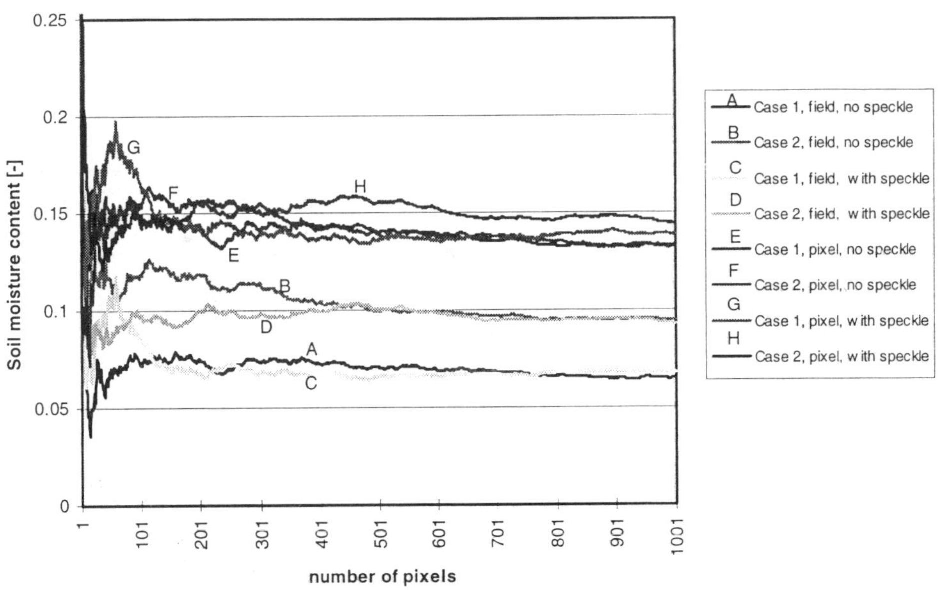


**Figure 4** The soil moisture estimates of the cumulatively averaged backscatter coefficient. In Table 5 some of the numerical results are listed.

**Table 5** The standard error of the soil moisture estimates as function of the number of pixels for a single realisation over which the ERS-1 and JPL-AIRSAR backscatter coefficients are averaged.

| Number of pixels | Niger Case 1 No Speckle | Speckle | Case 2 No Speckle | Speckle | Spain Case 1 No Speckle | Speckle | Case2 No Speckle | Speckle |
|---|---|---|---|---|---|---|---|---|
| 2 | 0.05 | 0.09 | 0.03 | 0.07 | 0.11 | 0.00 | 0.01 | 0.00 |
| 10 | 0.05 | 0.02 | 0.01 | 0.02 | 0.10 | 0.04 | 0.03 | 0.05 |
| 25 | 0.03 | 0.01 | 0.02 | 0.01 | 0.07 | 0.06 | 0.03 | 0.03 |
| 50 | 0.03 | 0.01 | 0.01 | 0.01 | 0.08 | 0.07 | 0.03 | 0.03 |
| 100 | 0.03 | 0.02 | 0.02 | 0.00 | 0.08 | 0.06 | 0.03 | 0.03 |
| 250 | 0.03 | 0.03 | 10.0 | 0.00 | 0.08 | 0.07 | 0.03 | 0.04 |
| 500 | 0.03 | 0.03 | 0.00 | 0.00 | 0.08 | 0.08 | 0.03 | 0.04 |
| 1000 | 0.04 | 0.03 | 0.00 | 0.01 | 0.08 | 0.07 | 0.03 | 0.03 |

averaging over a large number of pixels, there is a constant difference between the cases with and without speckle. Speckle thus introduces an off-set or additional constant error (Figure 4).

The results of the error analysis provide a basis on which to assess the quantitative value of the spatial information provided by detailed remote sensing data. The advantage of remote sensing data, especially if at high resolution, is that a number of samples (i.e. pixels) can be used to obtain estimates with smaller random errors. This can be done by averaging over a limited number of pixels, at the price of degrading the spatial resolution. However, to avoid mixing measurements relating to different physical objects, this averaging can be performed over stationary heterogeneous targets only.

## INFLUENCING FACTORS AND DIFFICULTIES IN SOIL MOISTURE ESTIMATION BY MEANS OF REMOTE SENSING

The first problem one encounters in estimation of soil moisture by means of remote sensing is the definition of soil moisture. The total volume of water per volume of soil or the total mass of water per mass of soil seem very clear definitions and not open to discussion. However, difficulties arise as soon as one has to measure this quantity. Conventional methods are usually point measurements, or more precisely measurements over a very small volume of soil. Soil water content measured in this way, even when averaged over a large number of points, does not necessarily represent the areal average correctly. Large areal estimates of soil moisture or indicators (e.g. Antecedent Precipitation Index, antecedent moisture index) often based upon precipitation measurements in combination with sparse soil moisture measurements, have to be considered carefully since they are influenced by numerous factors which are not accounted for. Some of these factors are: differences in soil, vegetation type and density, topography, spatial variability of precipitation and last, but not least, accuracy of ground measurements which can be difficult to assess in remote or over large areas. Most of these factors are related in a non-linear way and their influence on soil moisture estimation is therefore difficult to predict. At this point there is no way of dealing directly with this problem but the importance of differences between estimates of soil water content needs to be realized and thus defined at all times.

Remotely-sensed estimates of soil moisture concern areal averages of soil water content over a certain depth. The depth over which one can associate soil moisture retrieved by the model, the so-called sensing depth, poses a severe problem. The sensing depth varies per band and with the dielectric properties of the soil. In general, longer wavelengths correspond to higher penetration depths as compared to shorter wavelengths. On the other hand, wet soils - having a high dielectric constant - correspond to lower penetration depths as compared to drier soils, which have a low dielectric constant. An approximate formula to estimate the theoretical penetration depth $L_p$ is given by:

$$L_p \approx \frac{\lambda\sqrt{\varepsilon'}}{2\pi\varepsilon''} \tag{4}$$

An example of this relationship is given in Figure 5. The sensing depth is usually

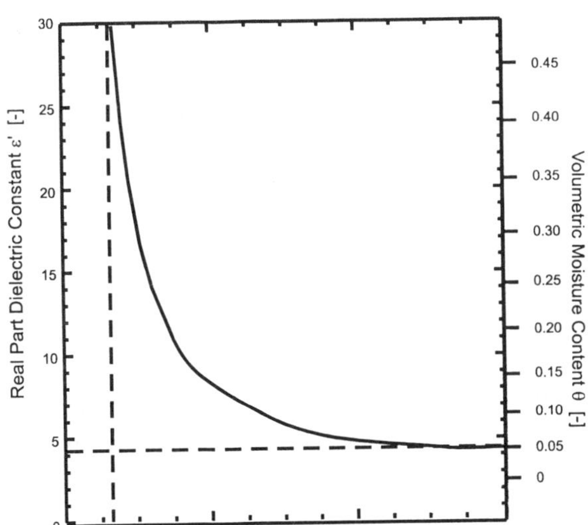


**Figure 5** The penetration depth $L_p$ as described by Eq. (4) as function of volumetric water content $\theta$ or versus the real part of the dielectric constant $\varepsilon'$. The soil type assumed is a loamy sand (sand fraction is 0.85 and clay fraction is 0.10) and the frequency used is 5.6 GHz (wavelength $\lambda$ = 5.3 cm).

much smaller than the theoretical penetration depth. Most models involving dielectric properties of soils assume these to be constant over depth as does Equation 4: this assumption of homogeneity usually does not hold for heterogeneous soils.

Remotely-sensed estimates of soil moisture are averages in space and depth. If these estimates are to be compared to ground-truth data or modelled data, the latter have to be averaged over the same depth. For C- and L-band this does not present a big problem since the sensing depth in these bands is such that soil moisture sampling using portable TDRs or gravimetric samples is feasible, even over 1-2 cm. Although TDR is difficult to use over such small depths, other methods such as the gravimetric method can be used. However, for P-band, the sensing depth becomes too large for routine gravimetric sampling or TDRs and therefore other options have to be considered, e.g. installing TDR probes at greater depths. Hence, the larger the area the more difficult it becomes to obtain reliable estimates.

The large variation of surface roughness makes it a difficult parameter to describe at different resolutions. The field measurements of surface roughness generally cover only a very small area which might be inadequate for a good description. Agricultural fields which are regularly tilled or ploughed have surface roughness characteristics that are particularly difficult to describe. Most models that describe the surface backscatter assume a certain distribution for the roughness characteristics and this has to be adapted for such field surfaces.

## CONCLUSIONS

The approach used in this study is to analyse the effect of spatial heterogeneity on remotely-sensed soil moisture estimates. The heterogeneity is treated as a random error source in the backscatter measurements. A specific case only was treated here, namely the use of INVIEM for given surface object parameters (e.g. surface roughness) and sensor parameters (e.g. frequency, polarization). Generalisation of the results is extremely difficult. A more general and systematic approach is needed to extend the validity of this type of error analysis. This will be executed in a follow up study that will also include the use of other microwave soil moisture retrieval algorithms, including those using passive microwave data. However, it is clear from this study that soil moisture estimates using radar data based upon a single pixel have to be treated with caution. The presence of speckle, inherent in radar data, makes this problem more complicated but does not pose a real problem when averaged over a number of pixels. The error in individual pixel estimates can become unacceptably high but averaging over a small number of pixels reduces the error significantly.

The soil moisture estimates using INVIEM requires a knowledge of the roughness characteristics of the surface and so far, available ground measurements have been used. Future developments will include generalisation of the INVIEM inversion algorithm to estimate surface roughness or surface soil moisture, or both, depending upon the information available. Furthermore, the model currently uses only HH- and

VV-polarisations: inclusion of multi-polarization data or employing polarization ratios can improve the performance of the model. To extend INVIEM to vegetated surfaces, vegetation correction algorithms will have to be developed.

## ACKNOWLEDGMENTS

The authors acknowledge all the people and institutions who have taken part in the EFEDA and HAPEX field experiments and made them a success. In particular we thank CNES, ORSTOM for their provision of data and their support. We thank ESTEC for the provision of the ERS-1 data and NASA-JPL for the AIRSAR data.

## REFERENCES

Bastiaanssen, W.M.G., Hoekman, D.H. and Roebeling, R.A. 1993. A methodology for the assessment of surface resistance and soil water storage variability at mesoscale based on remote sensing measurements. IAHS Special Publications, No. 2, IAHS Press, Institute of Hydrology Wallingford, UK: pp. 63.

Bastiaanssen, W.G.M., Menenti, M., van Oevelen, P.J. Hoekman, D.H. and Feddes, R.A. 1994. Spatial variability and accuracy of remote sensing estimates of evaporation and soil moisture within EFEDA'91. In: Proceedings of AGMET Group (Ireland) and the Agricultural group of the Royal Meteorological Society (UK) (Eds. T. Keane and E. Daly). Conference, Dublin September 7-9, 1994.

Bolle, H.J. and Streckenbach, B. 1992. The ECHIVAL Field Experiment in a desertification threatened area EFEDA. First Annual Report to EC, Published by the EFEDA-Secretariat, Free University of Berlin, Berlin, Germany.

Dobson, M.C., Ulaby, F.T., Hallikainen, M.T. and El-Rayes, M.A. 1985. Microwave dielectric behaviour of wet soil. Part II: dielectric mixing models. *IEEE Trans. geosci. Remote Sensing*, **GE-23**, 35-46

Dubois, P.C., van Zyl, J. and Engman, E.T. 1995. Measuring soil moisture with imaging radars. *IEEE, Trans. Geosci. Remote Sensing*, **30**, 915-926

Fung, A.K. 1994. *Microwave scattering and emission models and their applications.* Artech House, Boston-London.

Fung, A.K., Zongqian, L. and Chen, K.S. 1992. Backscattering from a randomly rough dielectric surface. *IEEE Trans. Geosci. Remote Sensing*, **30**, 356-369.

Goutorbe, J.P., Lebel, T., Tinga, A., Bessemoulin, P., Brouwer, J., Dolman, A.J., Engman, E.T., Gash, J.H.C., Hoepffner, M., Kabat, P., Kerr, Y.H., Monteny, B., Prince, S., Said, F., Sellers, P.J. and Wallace, J.S. 1994. Hapex-Sahel: a large scale study of land-atmosphere interactions in the semi-arid tropics. *Annales Geophysicae*, **12**, 53-64.

Hallikainen, M.T., Ulaby, F.T., Dobson, M.C., El-Rayes, M.A. and Wu, L. 1985. Microwave dielectric behaviour of wet soil. Part I: empirical models and experimental observations. *IEEE Trans. Geosci. Remote Sensing*, **GE-23**, 25-34.

Hoekman, D.H. 1991. Speckle ensemble statistics of logarithmically scaled data. *IEEE Trans. Geosci. Remote Sensing*, **29**, 180-182.

Oh, Y., Sarabandi, K. and Ulaby, F.T. 1992. An empirical model and an inversion technique for radar scattering from bare soil surfaces. *IEEE Trans. Geosci. Remote Sensing*, **30**, 370-381.

Ulaby, F.T., R.K. Moore and Fung, A.K. 1986. *Microwave Remote Sensing: Active and Passive.* Vol. III. From Theory to Applications. Artech House, Dedham, MA.

Van Oevelen, P.J., Hoekman, D.H. and Vissers, M.A.M. 1993. Soil moisture and surface roughness measurements during HAPEX-Sahel 1992. Ground data collection report. Report of the Department of Water Resources, Wageningen Agricultural University, Wageningen, The Netherlands.

Van Zyl, J.J. 1989. Unsupervised classification of scattering behavior using radar polarimetry data. *IEEE Trans. Geosci. Remote Sensing.*, **27**, 36-45.

Wang, J.R., and Schmugge, T.J. 1980. An empirical model for the complex dielectric permissivity of soils as a function of water content. *IEEE Trans. Geosci. Remote Sensing.*, **GE-18**, 288-295.

# 15 Radar and optical remote sensing to infer evapotranspiration and soil moisture

C. OTTLÉ[1], D. VIDAL-MADJAR[1], A.L. COGNARD[2],
C. LOUMAGNE[2] and M. NORMAND[2]
[1]*CETP/CNRS, Vélizy, France*
[2]*CEMAGREF-Hydrologie, Antony, France*

## INTRODUCTION

Remote sensing may be applied in different domains of hydrological modelling. The determination of surface characteristics such as land use, pedology and snow cover may be estimated by means of satellite data and are already used in hydrological models. Evapotranspiration and soil moisture content may also be estimated at the mesh scale of hydrological models and may be used to control the model simulations. In fact, most hydrological models are based on a mass balance between the input of the system (= precipitation) and the output (= stream flows). Generally, part of the input is stored in a surface reservoir from where it controls the distribution between runoff and infiltration. Thus, regular measurements of soil water content could be used to monitor the temporal evolution of the surface reservoir and which could then improve forecasts of runoff. The advantage of remote sensing as compared with ground-truth data is that it provides us with repetitive areal measurements that can be incorporated in regional models.

We present two studies where remotely-sensed data have been used in hydrological models. The first study has been performed within the framework of the HAPEX-MOBILHY experiment. Optical and thermal infrared data were used to infer land use, the annual cycle of the vegetation and soil moisture storage in the root zone for incorporation in a distributed hydrological model (Ottlé and Vidal-Madjar, 1994). The second study is still in progress and deals with the synergistic use of radar, optical and thermal infrared data for hydrological modelling of small agricultural watersheds in Brittany (France). About 30 basins have been selected at the intersection of ascending and descending tracks of the ERS1 satellite, permitting SAR images to be obtained at least once every three days during the orbiting phases B and D. Among these basins, the Naizin watershed is a test site for hydrological studies and has been selected to evaluate the SAR/ERS1 capacity to estimate soil

*Scaling up in hydrology using remote sensing*. Edited by J.B. Stewart, E.T. Engman, R.A. Feddes and Y. Kerr

moisture content (Cognard *et al.*, 1995). Ground-truth measurements of soil moisture content, vegetation cover, atmospheric and hydrological parameters have been made during two years 1992-1993: first results are presented here.

## METHODS FOR INFERRING HYDROLOGICAL SURFACE PARAMETERS FROM REMOTE SENSING DATA

### VEGETATION COVERAGE

Reflectances in the visible and near infrared domains are good indicators of the biomass. The reflectivity depends on the soil texture and on the vegetation cover. Because the chlorophyll responsible for plant photosynthesis reflects near infrared radiation more than visible radiation, comparison of the two signatures is related to the amount of vegetation. It is then possible to define vegetation indices which combine these two reflectances and take account of soil, atmospheric and geometric effects. The most simple and widely used is the Normalized Difference Vegetation Index (NDVI) (Deering *et al.*, 1975) which is the ratio of the difference to the sum of the reflectances in the visible and near infrared domains. This index is a good tool with which to estimate the vegetation density within the satellite pixel.

The NDVI, combined with ground-truth data, permit a land-use classification to be developed and the estimation of the vegetation fractional cover. This parameter is necessary to partition soil evaporation from transpiration. It can be derived from the value of the vegetation index as compared to its maximum and minimum values. In the Hapex study, all the cloudless NOAA/AVHRR images available over the two-year period have been processed and a rough classification of the vegetation has been made possible from study of the NDVI. Forested areas, summer and winter crop regions could be separated and the annual variation in this index permits us to identify the different phenological periods of the vegetation. As an illustration,

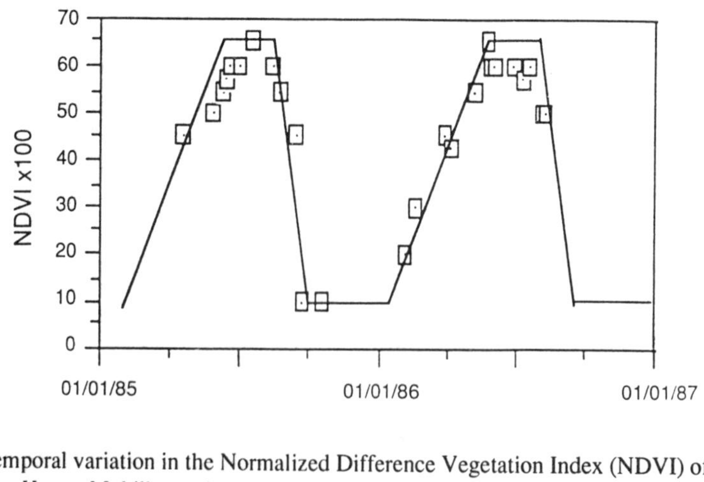


**Figure 1** Temporal variation in the Normalized Difference Vegetation Index (NDVI) of summer crops over the Hapex-Mobilhy region.

Figure 1 presents the time variation of NDVI of summer crops in the Hapex-Mobilhy region between January 1985 and January 1987. Hence, the bare soil period corresponding to minimum values of the NDVI can be clearly differentiated from the periods of vegetation growth when the NDVI rises to its maximum value, from the period of maturity as characterized by constant maximum values and finally from the period of senescence corresponding to a quick decrease in the index. From these plots, the state of maturity of the vegetation can then be estimated and also the fraction of vegetation in the pixel. The vegetation fractional cover $f_c$ can be derived from the value of NDVI as compared with its minimum and maximum value:

$$\frac{f_c}{f_{c\ max}} = \frac{NDVI - NDVI_{min}}{NDVI_{max} - NDVI_{min}} \qquad (1)$$

where $NDVI_{min}$ is the minimum and $NDVI_{max}$ the maximum values observed during the entire vegetative period.

The same approach has been followed over the Brittany watersheds and particularly over the Naizin river-basin where a precise vegetation survey was carried out in 1992. Figure 2 compares the computed NDVI from the NOAA/AVHRR images after correction of the atmospheric effects by means of the 5S model (Tanré *et al.*, 1990). On the same graph is plotted the fraction of vegetation $f_c$ calculated using Equation 1 and that measured from the ground-truth vegetation survey for the whole year 1992. A value of 0.7 has been assumed for $f_{c\ max}$ which corresponds to the averaged maximum cover for the region. Unfortunately, no cloudless images were available between Julian days 150 and 200, leaving an uncertainty in the exact

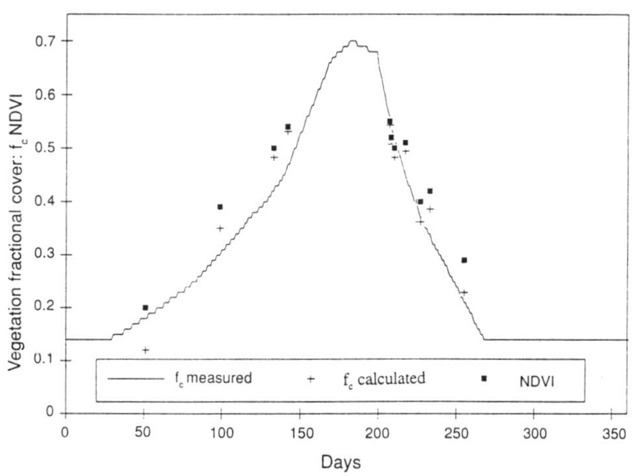


**Figure 2** Fraction of vegetation cover $f_c$ over the Naizin watershed as calculated from ground-truth measurements (full line) compared to the one deduced from the NDVI values (crosses). Also shown are the NDVI (solid squares).

location of the maximum. Nevertheless, the plot shows the good agreement between calculated and measured values of $f_c$, demonstrating the capacity of the NDVI to monitor the vegetation cover.

## SURFACE SOIL MOISTURE

One of the possible applications of satellite radar remote sensing is the estimation of surface soil moisture. The Synthetic Aperture Radar (SAR) aboard the experimental European Remote Sensing Satellites (ERS1/2) is equipped with a C band instrument suitable for the retrieval of soil moisture. The swath width is about 100 km and the resolution of the standard products is 20 x 20 $m^2$. The orbits of ERS1 were divided into two phases. The first one, known as the 'ice phase' (or phase B and D) has been set up with a repetition period of three days but no global coverage of the Earth surface. This is particularly interesting for hydrological studies since the characteristic time for surface soil moisture change is in the order of a few days. Phase B was active from the end of 1991 to the end of March 92 and phase D from the beginning of

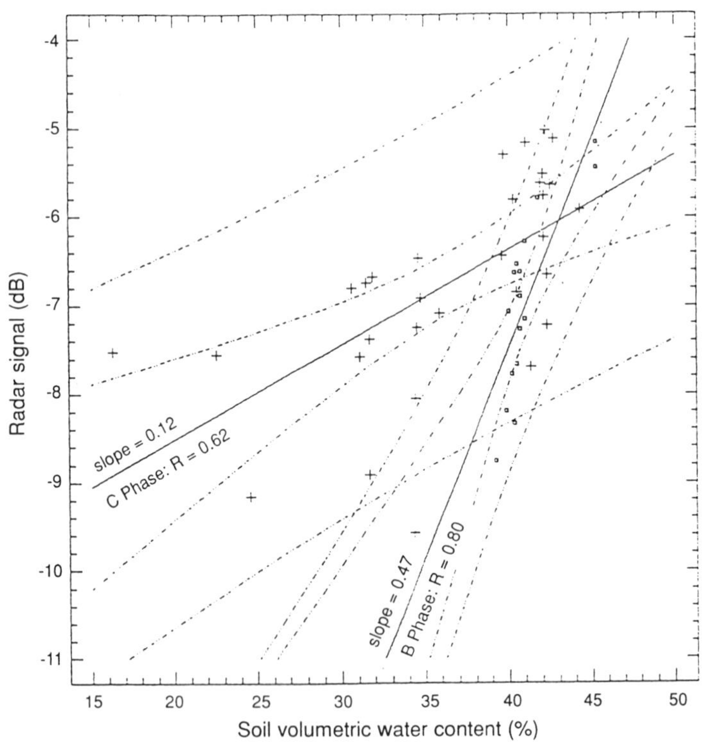


**Figure 3** Correlation between soil moisture measured by an automatic probe and mean SAR signal over the entire Naizin watershed during orbiting phases B (open circles) and C (pluses) of ERS1 satellite.

January 94 to the end of March 94. Between these two periods, the satellite was in phase C with global coverage but a repetition period of 35 days.

The Naizin watershed is a good site to evaluate the SAR/ERS1 capacity in this field. All possible SAR images were acquired together with two types of ground truth: intensive ground measurements during 14 field campaigns and point automatic measurements over the entire period. All the results are presented in Cognard *et al.* (1995). They have shown that at the field scale, the relation between the radar backscattering and the soil moisture depends on the nature of the surface, mainly on its roughness. At a larger scale (e.g. a few square kilometres) during the fall, winter and mid-spring periods (phase B and D), the correlation is high but is much poorer at the end of spring and in summer corresponding to the period of denser vegetation (phase C) as shown on Figure 3. It may then be possible to define a hydrological state index from ERS1/SAR images at least during the periods of low vegetation density. Research is under way to find a way to correct the C band radar signal for the vegetation attenuation (Taconet *et al.*, 1995).

## ROOT ZONE SOIL MOISTURE

Various studies have shown that surface temperature estimated from space can be used to infer surface fluxes such as the evapotranspiration and the soil moisture content by the use of a surface model describing the exchanges between the soil, the vegetation and the atmosphere, (e.g. Carlson *et al.*, 1981; Taconet *et al.*, 1986). Such a model has been developed by our team and applied in different field experiments.

This one-dimensional, soil-vegetation-atmosphere (SVAT) model is described in several papers, (Taconet *et al.*, 1986; BenMehrez *et al.*, 1992a, 1992b; Ottlé and Vidal-Madjar, 1994). The model calculates the surface fluxes, the surface temperature and the surface soil moisture by solving simultaneously the energy budget equation for the bare soil surface and above the canopy. An adequate partitioning of the incident energy fluxes between the vegetation and the soil is assumed as well as knowledge of the daily variation of the atmospheric forcing and the physical characteristics of the soil and of the vegetation. The soil is represented by a two-layer system derived from Deardorff (1978) as modified by Bernard *et al.* (1986).

The SVAT model can be used in a direct way (see Figure 4) to calculate the surface fluxes and the surface radiative temperature, if the soil and vegetation parameters, the soil moisture content and the atmospheric components are known. The input atmospheric data necessary to run the two-component model are the daily variation in the incoming solar radiation $R_g$, the air temperature $T_a$, moisture content $q_a$ and the windspeed $V_a$ at two metres above the surface.

The model can also be used in an inverse way. If the radiative surface temperature as well as all the functional parameters are known, the model may be inverted to retrieve one unknown variable which may be the soil moisture in one of the two distinguishable layers. In other terms, the soil moisture content may be adjusted in order to force the model simulated radiative surface temperature to be equal to the

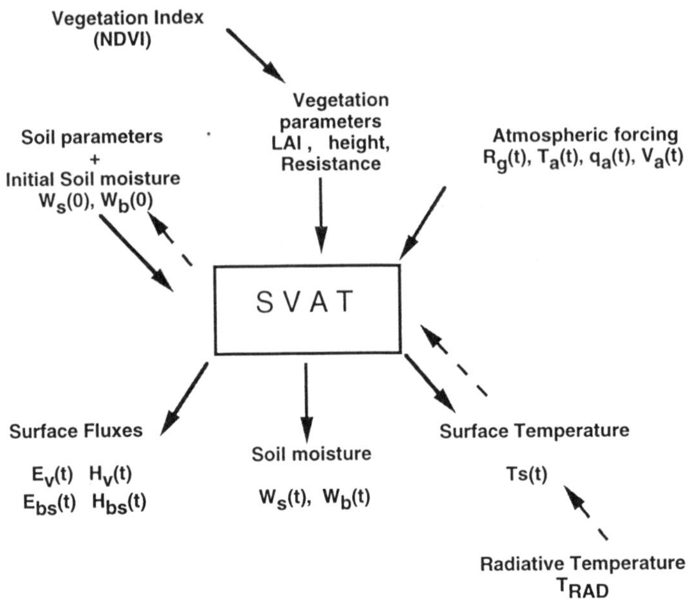


**Figure 4** Presentation of the SVAT model and the two ways of functioning (direct way: solid arrows; inverse way: dotted arrows). Subscript b refers to the bulk layer, s to the surface layer, v to vegetation and bs to bare soil.

measured one. Over bare soils, when the surface temperature is very sensitive to soil evaporation (which is mostly determined by soil moisture content in the surface layer $W_s$) the surface temperature allows the estimation of $W_s$, as shown by BenMehrez *et al.*, (1992b). In the same way, over dense vegetation covers, since evapotranspiration is strongly related to the soil moisture content in the root zone, the foliage temperature permits the estimation of the soil moisture of the bulk layer $W_b$ through estimation of stomatal resistance. In the case of partial covers, where the soil and the vegetation both contribute to the total evaporation, it could still be possible to estimate either of the two soil moistures: $W_s$ or $W_b$. But this case is still being considered (Taconet *et al.*, 1995) and needs further model development. This methodology has been applied to estimate the bulk soil moisture at a regional scale, using AVHRR thermal infrared data over the Adour river basin and over the Brittany basins.

## INCORPORATION OF REMOTELY-SENSED DATA IN HYDROLOGICAL MODELS

### METHODOLOGY

Such remotely-sensed variables are of great interest for physically-based hydrological models because their incorporation can improve the simulation of the water

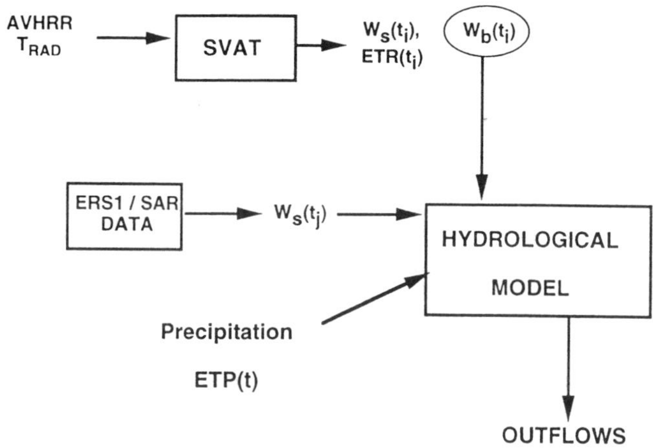


**Figure 5** Presentation of the methodology

exchange at the surface. We now want to show how soil moisture, land use and some vegetation characteristics inferred from space measurements can be used in hydrological models. From knowledge of the daily values of precipitation and potential evaporation, hydrological models simulate the daily flows at the outlet of the river basin. Every time a satellite image is available, the thermal and radar data permit us to monitor the soil moisture content in the surface layer and in the root zone. The optical data are used to adjust the vegetation parameters (Figure 5). This can be done for all the grid squares of the distributed hydrological model.

## APPLICATION TO THE ADOUR RIVER BASIN HYDROLOGICAL MODELLING

Vegetation parameters and root zone soil moisture inferred from AVHRR data have been used in a distributed hydrological model implemented on the Adour river basin within the Hapex-Mobilhy experiment. The land-use classification is used to assign to each grid square the dominant type of vegetation and its annual variation, namely the evolution of the fractional vegetation cover and the phenological stages. This information has been used in the hydrological model to improve the calculation of the total daily evapotranspiration. The soil moisture content inferred from the surface temperature measurements also has been introduced in the hydrological model to correct the simulated value for each grid square for each cloudless day that a satellite image was available.

### *Description of the hydrological model*

The Hapex-Mobilhy experiment (André *et al.*, 1988) took place in south-western France during a period of two years in 1985-1986. The chosen site is watered by the

Adour river system. The upper part of the region is covered by the Les Landes forest and the rest by mixed crops. Girard and Boukerma (1985) describe the simulation of the streamflows in the Adour river basin using the hydrological model of Girard (1974) and Ledoux (1980). This model, described in Ottlé and Vidal-Madjar (1994), is a deterministic model composed of two coupled domains, the surface and the underground. These two domains have been discretized over square grids ranging from 1.25 km to 5 km. The primary objective of the present work was first to improve the simulation of the water budget, especially the estimation of the actual evapotranspiration, and also to incorporate the soil water content of soil layers contributing to the evapotranspiration. For this purpose, it has been necessary to separate the surface reservoir into two layers: a superficial soil layer corresponding to the first ten centimetres of soil, and the root zone extending through approximately the first metre. Consequently, we have introduced into the hydrological model the surface parameterisation already used in our SVAT scheme (Taconet *et al.*, 1986) and the calculation of the evaporation has been modified by taking account of the presence of the vegetation.

To improve the calculation of the evapotranspiration, we have separated the evaporation from the bare soil from the transpiration, according to Deardorff (1978) and BenMehrez *et al.* (1992a; 1992b) who have both shown the importance of taking account of the nature of the surface and its vegetal covering. The total evapotranspiration can be calculated as the sum of the contributions of the vegetation and of the bare surface with respect to the fraction of vegetal covering. In the hydrological model, the total evapotranspiration from the surface layer is given by the following equation :

$$E_b = f_c\ E_v + (1 - f_c)\ E_{bs} \tag{2}$$

where $E_v$ is the daily evapotranspiration and $E_{bs}$ is the daily bare soil evaporation. The fraction of vegetation $f_c$ is estimated from the NDVI annual variation (Equation 1).

Since the input data of the hydrological model is the 10-day average potential evaporation, it has been necessary to parameterise $E_{bs}$ and $E_v$ in terms of the potential evaporation and of the soil moisture content in the respective soil layers. For the evaluation of $E_v$, we have also taken into account the state of maturity of the vegetation because it strongly influences the stomatal resistance. For more details, see Ottlé and Vidal-Madjar (1994).

*Use of the inferred soil moisture content in the hydrological model and results*

One of the objectives of this study was to use the estimation of soil moisture content deduced from the energy budget observed from space to correct the simulated content of the hydrological model's surface reservoir. For that purpose, we have kept only the images corresponding to dense covers with an LAI greater than two where Taconet *et al. (1986)* have shown that the radiative surface temperature is more sensitive to root zone moisture content than to superficial moisture content.

For each of the days for which all data and satellite images are available, the SVAT model can be run with the surface moisture content obtained from the hydrological model as an initial value at all the grid points. The soil and vegetation parameters of the interface model have been calibrated for each grid point and each day considered, according to the surface characteristics estimated from the combination of ground-truth measurements and NDVI data. The atmospheric parameters necessary to run our model have been obtained by interpolation of the PATAC meteorological network to the centre of each grid square. The soil moisture content $W_b$ is modified until the difference between the modelled surface radiative temperature and the observed value is less than 1.5 K. If the variation of the soil moisture has no effect on the surface simulated temperature, the model is not inverted.

The difference between the soil moisture inferred in this way and the initial value simulated by the hydrological model is then introduced in the hydrological model for correcting the modelled soil moisture content $W_b$. The model is then run until a new satellite image is available. This work has been done for 24 images acquired mostly during the growing period of the vegetation between May and September during the two years of the experiment.

It can be seen that when the actual soil moisture is obtained for each grid square by inversion of the interface model, the actual evapotranspiration is also retrieved

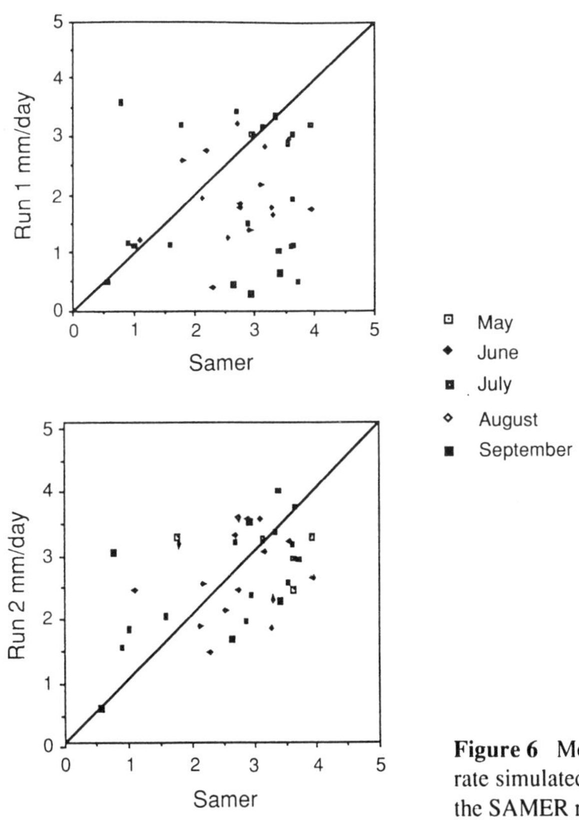


**Figure 6** Monthly averaged evapotranspiration rate simulated by Run 1 and Run 2 compared to the SAMER measurements.

and could be incorporated as well into the hydrological model. But, since it varies strongly in time, it is not very interesting to correct it for the few days that cloudless satellite images are available. As a result, it has not been introduced in the hydrological model. All the results have been presented in Ottlé and Vidal-Madjar (1994) and we will only present here those concerning the evapotranspiration.

Two simulations of the hydrological model for the two years 1985-1986 will be compared: the first one (Run 1) was run with the initial version of the hydrological model (Girard's (1974) surface parameterisation), the second (Run 2) was run with the second version of the model in which we have adjusted the vegetation parameters from the visible imagery and incorporated the soil moisture content inferred from the thermal infrared imagery.

An important point is the estimation of actual evapotranspiration. The SAMER stations give an estimate of evaporation at the field scale which is representative of a larger region. André *et al.* (1990) have shown that the SAMER measurements are well correlated with the airplane flux measurements integrated over a larger region.

We have compared evapotranspiration as measured by the SAMER stations during the five months of measurements of the Special Observing Period (SOP) in 1986 against the simulated values over the corresponding grid squares. Figure 6 presents monthly averaged values. The results show that Run 2 has a better agreement with the observed, especially for the months of July and August when Run 1 simulates very low and unrealistic values. The standard deviation is 1.1 mm for Run 1 against SAMER measurements and 0.7 mm for Run 2.

## APPLICATION OF HYDROLOGICAL MODELLING TO THE BRITTANY RIVER BASINS

The same methodology is under way for application to the 30 small river basins situated in central Brittany, where SAR/ERS1 and AVHRR/NOAA11 data have been acquired over a period of two years.

### *Description of the hydrological model*

The GRHUM model used in this study (Loumagne *et al., 1995),* is based on a daily conceptual rainfall-runoff model named GR4 (Edijatno and Michel, 1989) which has already been successfully applied to 120 French catchments with different soil, vegetation and climate characteristics (Edijatno, 1991).

The upper soil surface is modelled as a two-layer system: a surface layer representing approximately the first ten centimetres of the soil and a bulk layer representing the root zone (about the first metre of soil). This depth depends on the particular soil properties and vegetation type of the basin under study. The exchanges between the two layers depend on the soil moisture conditions (Chkir, 1994). Rainfall is divided into two terms: the first term feeds the soil reservoir and the second is routed directly to the outlet of the catchment through a transfer function. This partition depends on the moisture level in the bulk soil layer. The

NDVI index calculated from AVHRR data is used to estimate the fractional vegetation cover (as in the HAPEX study) to separate soil evaporation from transpiration.

A model validation has also been performed on the 36 basins of central Brittany. These basins have been chosen because streamflow and rainfall data are available for more than six years. The catchments range in area from 20 to 500 km² and most of them have a Brioverian schists substratum (loamy soils). This region is characterized by intensive agricultural exploitation.

One of them, the Naizin experimental watershed, has been chosen as a permanent test site for ground-truth measurements dedicated to the assessment of the ability of the ERS.1/SAR radar to evaluate the soil surface water content (Cognard *et al.*,1995). The accuracy of the model in reproducing discharges observed at the outlet of the catchment and its ability to simulate soil moisture in the first ten centimetres surface soil layer and in the first meter bulk layer has been checked.

*First results*

A preliminary study on the Naizin watershed has shown a good correlation between the backscattering coefficient averaged over the whole river basin and the soil moisture in the first ten centimetres as modelled by GRHUM during the periods of bare soils for the years 1992 and 1993. Figure 7 shows the correlation between the relative soil moisture content in the first ten centimetres simulated by GRHUM over

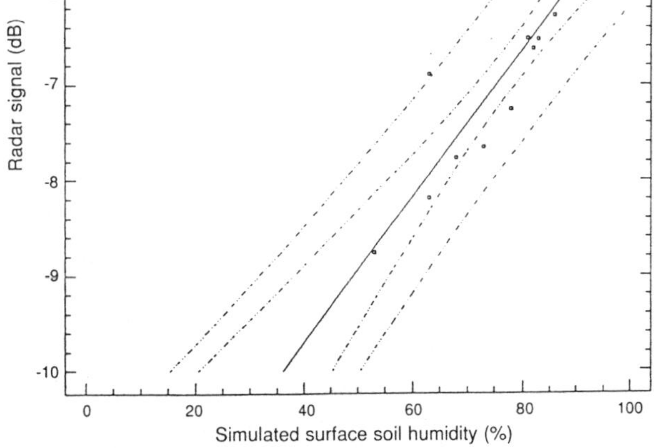


**Figure 7** Correlation between relative surface soil humidity simulated by the hydrological model and the mean radar signal over the entire Naizin watershed

the Naizin watershed and the mean radar signal (dB) over the basin. The measurements correspond to phase B where the Naizin basin is covered mostly by bare soils or permanent grass (fraction of vegetation $f_c$ is less than 0.2). The good correlation (R = 0.93) shows that it may be possible to use the radar signal to define a surface soil moisture index for bare soils.

The next step will be the treatment of the thermal infrared data for the estimation of root zone soil moisture and the definition of a second soil moisture index for the root zone in vegetated areas. The final step will be the introduction of these indices in the hydrological model.

## CONCLUSION

These two attempts to use optical, thermal infrared and radar remote sensing in hydrological models are very encouraging. They show particularly that the temporal evolution of soil moisture may be incorporated to monitor the surface soil moisture. It has been shown within the framework of the Hapex-Mobilhy experiment that a better simulation of the hydrological exchanges at the land-atmosphere interface leads to major improvements in the simulation of regional evapotranspiration and runoff. The synergy of radar with thermal infrared data opens new perspectives. The results make us confident that in the very near future, it will be possible to use remotely-sensed data to improve hydrological modelling of gauged and ungauged basins at a variety of scales.

## REFERENCES

André, J.C. and collaborators. 1988. HAPEX-MOBILHY: First results from the Special Observing Period. *Ann. Geophys.*, **6**, 477-492

André, J.C. , Bougeault, P. and Goutorbe, J P. 1990. Regional estimates of heat and evaporation fluxes over non-homogeneous terrain. Examples from the HAPEX-MOBILHY program. *Boundary-Layer Meteorol.*, **50**, 77-108

BenMehrez, M., Taconet, O., Vidal-Madjar, D. and Sucksdorff, Y. 1992a. Calibration of a fluxes model over bare soils during HAPEX-MOBILHY experiment. *Agric. For. Meteor.*, **58**, 257-283.

BenMehrez, M., Taconet, O., Vidal-Madjar, D. and Valencogne, C. 1992b. Estimation of canopy stomatal resistance and canopy evaporation during HAPEX-MOBILHY experiment. *Agric. For. Meteor.*, **58**, 285-313

Bernard, R., Soares, J.V. and Vidal-Madjar, D. 1986. Differential bare field drainage properties from airborne microwave observation. *Water Resour.Res.*, **22**, 869-875 .

Carlson, T.N., Dodd, J.K., Benjamin, S.G. and Cooper, J.N. 1981. Satellite estimation of the surface energy balance, moisture availability and thermal inertia. *J. Appl. Meteor.* **20**, 67-87.

Chkir, N., 1994. Mise au point d'un modéle hydrologique conceptuel intégrant l'état hydrique du sol dans la modélisation Pluie-Débit. Thése de L'Ecole Nationale des Ponts et Chaussées de Paris (France), "Sciences et Techniques de l'Environnement", Mars 1994

Cognard, A.L., Loumagne, C., Normand, M., Olivier, Ph., Ottlé, C., Vidal-Madjar, D., Louahala, S. and Vidal, A. 1995. Evaluation of the ERS1/synthetic aperture radar capacity to estimate surface soil moisture: two years results over the Naizin watershed. *Water Res. Res.*, **31**, 975-982.

Deering, D.W., Rouse, J.W., Haas, R.H. and Schall, J.A. 1975. Measuring forage production of grazing units from Landsat MSS data. Proc. 10th International Symposium on Remote Sensing of the Environment, Ann Arbor, Michigan, p 1169.

Deardorff, J.W. 1978. Efficient prediction of ground surface temperature and moisture, with inclusion of layer of vegetation. *J. Geophys. Res.*, **83**(C4), 1889-1903.

Edijatno, M.C. 1989. Un modéle pluie-débit journalier trois paramètres. *La Houille Blanche* n°2, 113-121.

Edijatno, M.C. 1991. Mise au point d'un modéle élémentaire pluie-débit au pas de temps journalier, Thèse de doctorat, Université de Strasbourg, CEMAGREF, 625 pp.

Girard, G., 1974. Modèle global ORSTOM. Première application du modèle journalier discretisation spatiale sur le bassin versant de la crique Grégoire en Guyane in "Atelier hydrologique sur les modèles mathématiques", ORSTOM, PARIS.

Girard, G. and Boukerma, B. 1985. Projet HAPEX-MOBILHY, Calage du Modèle Hydrologique. Rapport LHM/RD/85/110, (available from CIG, 35 rue Saint-Honoré, 77305 Fontainebleau Cédex, France)

Ledoux, E., 1980. Modélisation intégrée des écoulements de surface et des écoulements souterrains sur un bassin hydrologique. Thèse de Docteur-Ingénieur, ENSMP et Université P.M. Curie, Paris, France.

Loumagne, C., Chkir, N., Normand, M., Ottlé, C. and Vidal-Madjar, D. 1995. Introduction of the soil-vegetation-atmosphere continuum in a rainfall-runoff model for remote sensing data incorporation. Submitted to IAHS review.

Ottlé, C. and Vidal-Madjar, D. 1994. Assimilation of soil moisture inferred from infrared remote sensing in a hydrological model over the HAPEX-MOBILHY region. *J. Hydrol.*, **158**, 241-264.

Taconet O., Vidal-Madjar, D., Emblanch, Ch. and Normand M. 1995. Taking into account vegetation effects to estimate soil moisture from C-Band radar measurements. *Remote Sens Environ*. In press.

Taconet, O., Bernard, R. and Vidal-Madjar, D. 1986. Evapotranspiration over an agricultural region using a surface flux / temperature model based on NOAA-AVHRR data. *J.Climate Appl. Met.*, **25**, 284-307.

Tanré, D., Deroo, C., Duhaut, P., Herman, M., Morcrette, J.J., Perbos, J. and Deschamps, P.Y. 1990. Description of a computer code to simulate the satellite signal in the solar spectrum: the 5S code, *Int. J. Remote Sensing*, **11**, 659-668.

# 16 Daytime evaporation from conservation of surface flux ratios

R.D. CRAGO
*University of Illinois at Chicago, USA*

## INTRODUCTION

Although they would be useful for studies of hydrology, climate, and climate change, continuous land-based measurements representative of all regions worldwide are not likely to be available in the near future. Polar orbiting satellites help solve the problem of spatial coverage but temporal coverage is still problematic, since an orbit pattern may provide coverage of a region only once per day. Therefore, methods to extend approximately instantaneous measurements to daily or longer periods are needed to make full use of satellite data.

The approach taken here is to exploit general patterns of diurnal variability and to predict the daytime partition of the surface energy budget based on only a few instantaneous measurements during the day. An objection to such a study is that models are already available which can assimilate newly available data into a model run, and can realistically extrapolate from instantaneous to daily values. However useful the modeling techniques may be, the approach adopted in this paper has the important advantage of simplicity. If enough can be learned by observation and modelling to make such simple methods reliable, a valuable tool will be available that requires minimal data. An introduction to the method follows.

Each term of the surface energy budget,

$$R_n - G = H + LE \qquad (1)$$

typically varies considerably during the diurnal cycle. In Equation 1, $R_n$ is the net incoming radiation at the surface, G is the heat flux into the ground, H is the sensible heat flux into the atmosphere, and LE is the latent heat flux into the atmosphere. Jackson *et al. (1983)*, Shuttleworth *et al.* (1989), Gurney & Hsu (1990), Sugita & Brutsaert (1991), Brutsaert & Sugita (1992) and others have shown that various ratios of the energy budget components often vary little during the daytime. For example, Jackson et al. (1983) used an assumed conservation of the ratio $LE/S_d$, where $S_d$ is the downwelling solar radiation, and Shuttleworth *et al.* (1989) showed that the Evaporative Fraction, EF, is often conserved. EF is the ratio of latent heat flux to the available energy A i.e.

*The scaling issue in hydrology*. Edited by J.B. Stewart, E.T. Engman, R.A. Feddes and Y. Kerr.

$$EF = \frac{LE}{R_n - G} \tag{2}$$

None of the works cited claim the flux ratios are conserved at night, and indeed, there may be singularities in the ratios at night or near the dawn and dusk transition periods. Sugita & Brutsaert (1991) found that night time evaporation was about 7% of the daily total during FIFE, the First ISLSCP (International Satellite Land Surface Climatology Project) Field Experiment.

Conservation of flux ratios allows the extrapolation of a single measurement of a flux to a daytime total or average flux. For example, a single estimate of LE may be extrapolated to a daytime average value of LE if EF is conserved and if the daytime progression of A is available.

This paper will explore the extent to which flux ratios are conserved, reasons for the conservation and some ways in which the conservation may be exploited to provide daytime average sensible and latent heat fluxes into the atmosphere. As more attention has been given to the evaporative fraction than to other flux ratios, this paper will also concentrate on it.

## CONSERVATION OF FLUX RATIOS

*Evidence for conservation*

Figure 1 shows the daytime progression of the net radiation and the evaporative fraction for one mostly sunny day (15 August 1987) during FIFE. The experiment took place mostly in a natural prairie in Kansas, USA. The curves represent the areal average half-hourly values from at least eight surface flux stations distributed across the 15 by 15 km experiment site. The net radiation has a smooth, nearly sinusoidal progression, while the evaporative fraction varies only slightly from about 0900 to

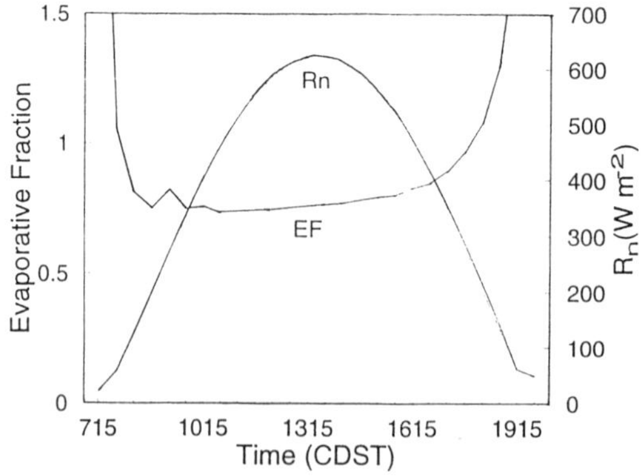


**Figure 1** Daytime progression of net radiation $R_n$ and evaporative fraction EF on August 15, 1987 at FIFE. [After Crago (1996a)]

1700 Central Daylight Saving Time (CDST). This behaviour is typical of sunny days at FIFE (e.g. Crago, 1996a).

Sugita & Brutsaert (1991), Brutsaert & Sugita (1992), Shuttleworth et al. (1989), Gurney & Hsu (1990), and Crago (1996a) found that EF displayed a remarkable tendency towards conservation under a wide variety of conditions during FIFE. Sugita & Brutsaert (1991) noted that the EF progression is typically not actually flat during the middle of the day, but rather slightly concave up. This results in slightly lower EF values near noon than the daytime average value. Nichols & Cuenca (1993) applied statistical tests to data from the HAPEX–MOBILHY (Hydrologic Atmospheric Pilot Experiment–MOdlisation du BILan HYdrique) experiment in an agricultural and forested region of France. They rejected (at the 0.05 significance level) the hypothesis that the daytime value of EF was equal to the midday value, even for the partial data set of mostly clear days. They did find that midday and daytime EF were well-correlated with a coefficient of correlation $r = 0.95$. Crago (1996b) reached similar conclusions from the FIFE data set. The difference between midday and daytime values of EF and the high correlations between them indicate that there is a fairly regular and systematic amount of concavity for each day within these experiments. Although there are days which are exceptions, the consistency of the pattern is striking.

Both HAPEX–MOBILHY and FIFE took place under mostly humid conditions in well-vegetated areas. In contrast, Kustas et al. (1993) used data from the MONSOON 90 experiment at the partially-vegetated and semi-arid Walnut Gulch watershed in Arizona, USA. Midday values of EF ranged from less than 0.1 to about 0.8, indicating a wide range of moisture availabilities. Nevertheless, the correlation coefficient for midday and daytime average EF was $r = 0.98$. Scatter about the linear regression line was similar for very moist and very dry conditions, indicating similar temporal variability in the daytime progression of EF under a wide range of conditions.

Other flux ratios have been considered as well. For example, Brutsaert & Sugita (1992) suggested that the ratios $LE/R_n$, $LE/S_d$ and $LE/L_u$ may be fairly well conserved, where $S_d$ is the downwelling shortwave radiation, and $L_u$ is the upwelling longwave radiation from the surface. Reasonable estimates of daytime evaporation were obtained assuming a conserved ratio $LE/S_d$, both when $S_d$ was measured directly and when it was parameterised as a function of extraterrestrial solar radiation and cloud cover [Brutsaert & Sugita (1992)]. While Brutsaert & Sugita (1992) only gave a one-day example of the progression of $LE/L_u$, it deserves further consideration, especially due to the availability of radiometric measurements of $L_u$. Nevertheless, since the available energy, A, is the primary driver of evaporation (e.g., Brutsaert & Sugita, 1992), $EF = LE/(A)$ is intuitively appealing and has received by far the most attention.

*Reasons for conservation*

It is likely that the conservation of EF during daytime hours is the result of the interaction of a number of variables (e.g., Crago, 1996a). However, the role of the

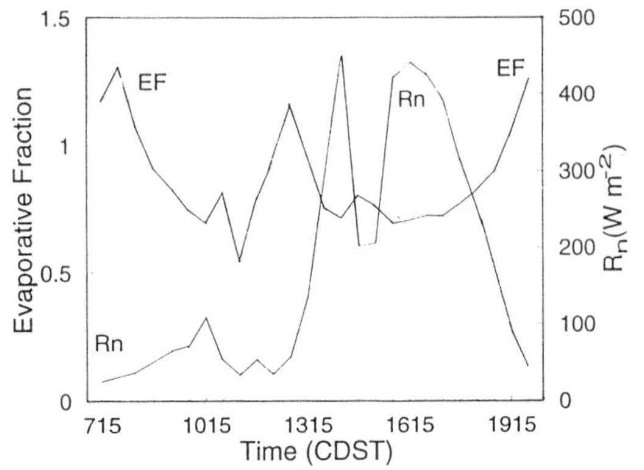


**Figure 2** Daytime progression of net radiation $R_n$ and evaporative fraction EF on July 4, 1987. [After Crago (1996a)]

available energy as the driving force behind evaporation is probably crucial. Thus Sugita & Brutsaert (1991) noted that sporadic increases in EF corresponded to sudden decreases in LE as clouds passed over. Under nearly cloud-free and steady weather conditions, well-conserved EF appears to be the norm (Crago, 1996a). Cloudiness causes variability in solar radiation, net radiation, and available energy. Figure 2 illustrates the diurnal variation of EF and $R_n$ for 4 July 1987 at FIFE, an unsteady, variably-cloudy day. Both the net radiation and evaporative fraction are much more variable and chaotic in Figure 2 than in Figure 1.

The role of available energy in determining evaporative fraction was discussed by Rowntree (1991) and Nichols & Cuenca (1993). Its role in the daytime EF progression was studied in detail by Crago (1996a). Using a Penman-Monteith formulation for LE, these studies showed that EF decreases with increasing $R_n$–G, with increasing surface resistance $r_s$, and with decreasing humidity deficit of the air. Rowntree (1991) noted that the resulting formulation for EF is most sensitive to small changes in A when A is small. Since available energy changes most rapidly in the morning and evening and is large near noon, variations in EF tend to be small during the middle of the day, and EF itself tends to be smallest near midday, especially on mostly clear days. The resulting concave up shape of the temporal EF progression may be moderated by the temporal progression of the air temperature, surface resistance and humidity deficit cycles (c.f. Nichols & Cuenca, 1993). Temperature and humidity deficit typically peak near midday, and $r_s$ is typically at a minimum near midday. These trends reduce but generally do not reverse the curvature of the concave up progression of EF.

Significant variations from the smooth concave up shape are often associated with significant changes in net radiation or in energy and moisture advection (Crago, 1996a). Passage of clouds lowers the solar radiation reaching the earth and usually also the available energy, causing EF to increase slightly (e.g. Sugita & Brutsaert, 1991). Crago (1996a) suggested that a 20% decrease in A, caused for instance by an increase in low level clouds from 0-50% coverage, might typically result in a 5%

increase in EF. This may not always be significant or discernable if there is other noise in the EF progression.

In a variation on the work of Rowntree (1991) and Nichols & Cuenca (1993), Crago (1996a) followed Betts *et al.* (1993) in investigating the role of surface temperature ($T_s$) in EF variability. Changes in surface heating are closely associated with changes in available energy, and directly affect the surface sensible and latent heat fluxes. Crago (1996a) used the following equations:

$$H = \rho c_p [T_s - T_a]/r_a \tag{3}$$

$$LE = (\rho c_p \gamma)[e_s^* - e_a]/(r_s + r_a) \tag{4}$$

and

$$EF = LE/(H + LE) \tag{5}$$

where $\rho$ and $c_p$ are the density and specific heat at constant pressure of air, respectively, $T_a$ and $e_a$ are the temperature and water vapour pressure at a given height, $r_a$ is the aerodynamic resistance found using a formulation by Choudhury *et al.* (1986), $\gamma$ is the psychrometric constant, and $e_s^*$ is the saturation vapour pressure at $T_s$ (see Crago, 1996a for details). Following Crago (1996a) and using typical values for the variables, Figure 3 shows sample values of LE (Figure 3 A) and EF (Figure 3 B) as functions of $T_s$-$T_a$ computed from Equations (3) (4) and (5) for three different conditions: moist soil, little dry advection (curve a); moist soil, dry advection (curve b); and dry soil, little dry advection (curve c). Specific parameter values are given in the caption. Curve b extends to the end of the realistic $T_s$–$T_a$ range; beyond the range, implied A values are much greater than 1000 W $m^{-2}$. Figure 3 shows that LE increases and EF decreases with increasing $T_s$–$T_a$. Changes in $T_s$–$T_a$ affect EF more when $T_s$-$T_a$ is small. Dry advection causes the EF curve to flatten out slowly with increasing $T_s$-$T_a$, while high surface resistance causes greater sensitivity for small values of $T_s$-$T_a$. Recognizing the relationship between $T_s$–$T_a$ and A, these results are similar to those mentioned above, obtained from the Penman-Monteith equation.

As discussed previously, EF may be conserved under very moist and very dry conditions. The Priestley-Taylor parameter may be defined by $\alpha = EF(\Delta+\gamma)/\Delta$, where $\Delta$ is the slope of the saturation vapour pressure versus temperature curve. Thus, $\alpha$ behaves like EF except for a fairly weak dependence on temperature and a very weak dependence on atmospheric pressure. Crago (1996a) calculated the daytime average and the daytime standard deviation of half-hourly values of $\alpha$ from a single FIFE Bowen ratio flux station. Daytime was defined as the period for which $R_n$>0. A positive correlation between the average value of $\alpha$ and soil moisture was apparent. However, there was no discernable relationship between the daytime standard deviation of half-hourly $\alpha$ values and the soil moisture. This indicates that the amount of variability in $\alpha$ (and presumably in EF) during the daytime is

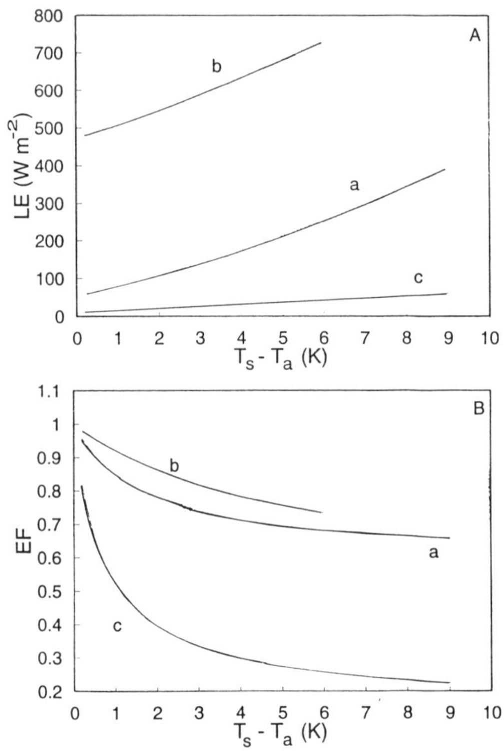


**Figure 3** The variability of LE and EF with $T_s$-$T_a$, using Eqs. (3), (4) and (5). $T_a$, $e_a$ and wind speed correspond to values at a height of 2m. Variable values for curve a are: roughness length for momentum is 8 cm, and for vapor is 8 mm, $T_a$ is 295 K, wind speed is 1.5 m $s^{-1}$, $r_s$ is 50 s $m^{-1}$, $e_a$ is 2.2 kPa, and atmospheric pressure is 97.5 kPa. Curve b is the same except dry advection is increased, so wind speed is 5 m $s^{-1}$ and $e_a$ is 0.6 kPa. Curve c is the same as curve a except $r_s$ is increased to 600 s $m^{-1}$. [Figure 3 B after Crago (1996a)]

independent of soil moisture.

The role of vegetation and topography on daytime average EF may at times be significant (e.g. Smith *et al.* (1992), but their effects on the daytime variability of the energy budget are still not completely clear (e.g., Crago, 1996a). One effect may be that if soil moisture is limited, stomata may close when evaporation demand becomes too high, contributing to a decrease in EF at midday (see Sellers *et al.*, 1992, for discussion of surface conductance).

Atmospheric effects on the temporal variability of the energy budget may at times be important. Pereira & Villa Nova (1992) suggested that the diurnal variations in $\alpha$ are actually governed by H, and found a nearly linear relationship between the two for a range of conditions ranging from near potential to arid and oasis situations. However, this finding does not actually explain the partitioning of available energy between H and LE.

Varying advective temperature or humidity at times may dominate the available energy in determining the evaporative fraction. This usually occurs near sharp discontinuities in the surface properties, such as irrigated fields in arid regions (e.g., Cooper *et al.*, 1992), or the shores of a large water body. Advection is most likely to change the partition of the energy budget significantly when strong advection occurs along with small available energy (Crago, 1996a).

To end this section on reasons for flux ratio conservation, the work of de Bruin (1983), (c.f., McNaughton & Spriggs, 1986; 1989) will be discussed. De Bruin (1983) indicated that conservation of the Priestley-Taylor parameter $\alpha = EF(\Delta+\gamma)/\Delta$ is a natural result of the evolution of the Atmospheric Boundary Layer (ABL). De Bruin (1983) used a mixed layer slab model of the ABL in which advection was assumed negligible, and surface sensible and latent heat fluxes were determined using the Penman-Monteith equation. The surface and aerodynamic resistances were considered to be constant throughout the daytime. In this model, the surface sensible and latent heat fluxes, the ABL temperature and humidity, and the ABL depth and growth rate were all interdependent. The resulting progression of $\alpha$ displayed little variability during the daytime other than a slight concave up shape, when the model was driven by solar radiation typical of a clear summer day in the Netherlands. This shape may thus result from the typical interactions between the surface and the ABL. Note that $\alpha$ tends to be slightly more concave up than EF, because of the temperature dependence of $\Delta$ and $\gamma$.

## DAYTIME AVERAGE LATENT HEAT FLUXES

An evaporative flux ratio may be defined as

$$R = \frac{LE}{F} \tag{6}$$

where F is any flux relevant to the partition of the energy budget [Brutsaert & Sugita (1992)]. When R is $\beta^{-1}$, where $\beta$=H/LE is the Bowen ratio, then F is taken as the sensible heat flux H. When R is EF, F is the available energy A. If R is assumed to be conserved during the day, the daytime average value of LE may be estimated from

$$LE_{Rd} = \frac{R_m}{(t_2 - t_1)} \int_{t_1}^{t_2} F\, dt \tag{7}$$

where $LE_{Rd}$ is the daytime average of LE based on conservation of R, $R_m$ is the midday value of R, measured in the few hours around noon, and the period from $t_1$ to $t_2$ corresponds to the daytime period, usually defined as the period in which A>0. Thus when R=EF, the midday value of EF is multiplied by the average available energy during the daytime to give the average daytime latent heat flux, $LE_{EFd}$. If R=$\beta^{-1}$, the midday value of $\beta^{-1}$ is multiplied by the average value of H to give

$LE_{\beta d}$ (e.g., Crago & Brutsaert, 1996).

Equation 7 has been applied with several versions of R and with several methods of determining the midday value of R. Brutsaert & Sugita (1992) looked at R=EF and R=LE/$S_d$, where the midday values of LE to use in $R_m$ were determined by the ABL bulk similarity approach with the FIFE data set. Both versions gave estimates of $LE_d$ well correlated with measured values ($r \geq 0.9$). However, the estimates consistently underestimated the measured values by 2-5 %, due to the concave-up progression during the day of both versions of R. Crago (1996b) also used the FIFE data set and used all the available surface Bowen ratio station data for days with at least three of the stations reporting throughout the daytime period, defined in this case by $R_n > 15$ W m$^{-2}$ and $A > 15$ W m$^{-2}$. A total of 77 days were included. This study considered R given by EF and also a variation of Equation 7, in which R was replaced by the Priestley-Taylor $\alpha$, and F was replaced by the equilibrium evaporation equation. Midday measurements were taken to be the half-hour surface station averages centred at 1215 CDST. Both versions of Equation 7 produced strong correlations between $LE_d$ from Equation 7 and from measurements ($r > 0.98$), but with an underestimation of about 8%.

Application of Equation 7 can give counter-intuitive results at times. For example, the Bowen ratio, $\beta = H/LE$, has a one-to-one relationship with the evaporative fraction, $EF = 1/(1+\beta)$. The application of Equation 7 to EF and to $\beta^{-1}$ has been described above. Surprisingly, for the range of conditions commonly encountered in temperate climates, EF tends to give more precise estimates of $LE_d$ in Equation 7 than $\beta^{-1}$ does. Crago & Brutsaert (1996) showed this by propagating uncertainties through the equations and demonstrating that the uncertainty in the daytime average evaporation is likely to be smaller when Equation 7 is used with EF than with $\beta^{-1}$. Data from the FIFE surface stations also showed this. Intuitively, LE is often already well-correlated with the available energy, while H is not. Thus, while $\beta^{-1}$ is more sensitive to small flux variations than EF is, it typically must play a larger role in Equation 7.

A series of papers by Jackson *et al.* (1977), Itier & Riou (1982), Seguin & Itier (1983), Lagouarde & McAneney (1992), Seguin (1993), and others explored a different method of extrapolating from a single measurement to daily fluxes. Although the formulation does not use a conserved flux ratio, the method does depend on regular daytime progressions from day to day, and Seguin & Itier (1983) based an analysis of the method on the conservation of the flux ratio $H/R_n$. As presented in a review by Seguin (1993), the daily total evaporation $LE_t$ takes the form:

$$LE_t = R_{nt} + a - b(T_s - T_a) \tag{8}$$

where $R_{nt}$ is the daily total net radiation, $T_s$ is the "instantaneous" surface temperature, usually measured by means of sensors on satellites operating in the thermal infrared band, $T_a$ is the "instantaneous" air temperature at a specified height, and "a" and "b" are empirical parameters. Equation 8 represents the energy balance

(Equation 1) solved for LE and applied to daily totals. The term $b(T_s - T_a)$ represents the sensible heat flux, and 'a' absorbs any bias. Seguin & Itier (1983) suggested that universal values of 'a' and 'b' are unlikely to be acceptable, although they suggested a constant value for 'b' corresponding to stable conditions and a different value for unstable conditions. Seguin & Itier (1983) suggested that the method was accurate to within 10-20% and was most appropriately applied to total evaporation over a period of at least several weeks.

## CONCLUSIONS

The progression of the daytime energy budget is a complex process involving the interactions between the atmosphere, vegetation and soil. Nevertheless, it has been shown that certain ratios of fluxes found in the budget are surprisingly invariant during the daytime period, and that techniques are available to exploit their invariance. Such methods may be useful in the extrapolation of 'instantaneous' flux measurements derived from satellite measurements to longer period total fluxes.

## REFERENCES

Betts, A.K., Ball, J.H. and Beljaars, A.C.M. 1993. Comparison between the land surface response of the ECMWF model and the FIFE-1987 data. *Quart. J. R. Met. Soc.*, **119,** 975-1001.

Brutsaert, W. and Sugita, M. 1992. Application of self-preservation in the diurnal evolution of the surface energy budget to determine daily evaporation. *J. Geophys. Res.*, **97,** 18,377-18,382.

Choudhury, B.J., Reginato, R.J. and Idso, S.B. 1986. An analysis of infrared temperature observations over wheat and calculation of latent heat flux. *Agric. For. Meteorol.*, **37,** 75-88.

Crago, R. 1996a. Conservation and variability of the evaporative fraction during the daytime. *J. Hydrol.* In press.

Crago, R. 1996b. A comparison of the evaporative fraction and the Priestley-Taylor $\alpha$ for parameterizing daytime evaporation. *Water Resources Res.* In press.

Crago, R. and Brutsaert, W. 1996. Daytime evaporation and the self-preservation of the evaporative fraction and the Bowen ratio. *J. Hydrol.* in press.

Cooper, D.I., Eichinger, W.E., Holtkamp, D.B., Karl, R.R. Jr., Quick, C.R., Dugas, W. and Hipps, L. 1992. Spatial variability of water vapor turbulent transfer within the boundary layer. *Boundary-Layer Meteorol.*, **61,** 389-405.

De Bruin, H.A.R. 1983. A model for the Priestley-Taylor parameter $\alpha$. *J. Climate and Appl. Meteorol.*, **22,** 572-578.

Gurney, R.J. and Hsu, A.Y. 1990. Relating evaporative fraction to remotely sensed data at the FIFE site. In: Hall, F.G. and Sellers, P.J. (eds). Symposium on FIFE, 112-116.

Itier, B. and Riou, Ch. 1982. Une nouvelle mthode de dtermination de l'vapotranspiration relle par thermographie infrarouge. *J. Rech. Atmos.*, **16,** 113-125.

Jackson, R.D., Reginato, R.J. and Idso, S.B. 1977. Wheat canopy temperature: A practical tool for evaluating water requirements. *Water Resour. Res.*, **13,** 651-656.

Jackson, R.D., Hatfield, J.L., Reginato, R.J., Idso, S.B. and Pinter, P.J. Jr. 1983. Estimation of daily evapotranspiration from one time of day measurements. *Agric. Water Manage.*, **7,** 351-362.

Kustas, W.P., Schmugge, T.J., Humes, K.S., Jackson, T.J., Parry, R. Weltz, M.A. and Moran, M.S. 1993. Relationships between evaporative fraction and remotely sensed vegetation index and microwave brightness temperature for semiarid rangelands. *J. Appl. Meteorol.*, **32,** 1781-1790.

Lagouarde, J.-P. and McAneney, K.J. 1992. Daily sensible heat flux estimation from a single measurement of surface temperature and maximum air temperature. *Bound.-Layer Meteorol.*, **59,** 341-362.

McNaughton, K.G. and Spriggs, T.W. 1986. A mixed-layer model for regional evaporation. *Bound.-Layer Meteorol.*, **34,** 243-262.

McNaughton, K.G. and Spriggs, T.W. 1989. An evaluation of the Priestley-Taylor equation and the complementary relationship using results from a mixed-layer model of the convective boundary layer. IAHS Publ. no. **177,** 89-104.

Nichols, W.E. and Cuenca, R.H. 1993. Evaluation of the evaporative fraction for parameterization of the surface energy balance. *Water Resour. Res.*, **29,** 3681-3690.

Pereira, A.R. and Villa Nova, N.A. 1992. Analysis of the Priestley-Taylor parameter. *Agric. Forest Meteorol.*, **61,** 1-9.

Rowntree, P.R. 1991. Atmospheric parameterization schemes for evaporation over land: Basic concepts and climate modeling aspects. In: Schmugge, T.J. and Andre, J.C. (ed). *Land Surface Evaporation*, 5-29.

Seguin, B. 1993. Areal evaporation from satellite thermal infrared data. Presented at NATO Advanced Research Workshop, "Global environmental change and land surface processes in hydrology: The trials and tribulations of modeling and measuring", Tucson, AZ, May, 1993, 29pp..

Seguin, B. and Itier, B. 1983. Using midday surface temperature to estimate daily evaporation from satellite thermal IR data. *Int. J. Remote Sensing*, **4,** 371-383.

Sellers, P.J., Heiser, M.D. and Hall, F.G. 1992. Relations between surface conductance and spectral vegetation indices at intermediate (100 $m^2$ to 15 $km^2$) length scales. *J. Geophys. Res.*, **97,** 19033-19059.

Shuttleworth, W.J., Gurney, R.J., Hsu, A.Y. and Ormsby, J.P. 1989. FIFE: The variation in energy partition at surface flux sites. *IAHS Publ. No.* **186,** 67-74.

Smith, E.A., Hsu, A.Y., Crosson, W.L., Field, R.T., Fritschen, L.J., Gurney, R.J., Kanemasu, E.T., Kustas, W.P., Nie, D., Shuttleworth, W.J., Stewart, J.B., Verma, S.B., Weaver, H.L. and Wesely, M.L. 1992. Area-averaged surface fluxes and their time-space variability over the FIFE experimental domain. *J. Geophys. Res.*, **97,** 18,599-18622.

Sugita, M. and Brutsaert, W. 1991. Daily evaporation over a region from lower boundary layer profiles measured with radiosondes. *Water Resour. Res.*, **27,** 747-752.

# 17 Extrapolation of evaporation at time of satellite overpass to daily totals

**J.B. STEWART**
*Institute of Hydrology, Wallingford, UK*

## INTRODUCTION

Satellite sensors measure the radiation reflected or emitted from a pixel in a micro second. From this instantaneous acquisition of data, fluxes of energy, such as of incoming shortwave radiation and sensible heat at the Earth's surface, can be estimated. These estimates can be used to validate and improve numerical weather prediction models (NWPM) and hydrological models for estimating regional evaporation from soil and vegetation. However, fluxes at the earth's surface vary significantly over all time scales, from minutes to hours, days, seasons and years. So linear extrapolation of the instantaneous satellite estimates of the fluxes to longer periods will not necessarily give the correct fluxes over longer periods.

With geostationary satellites, instantaneous estimates of the fluxes can be obtained hourly or even half-hourly and linear extrapolation should be adequate. It is usually accepted that over a day or longer, the effect of having frequent instantaneous rather than continuous measurements only introduces random rather than systematic errors into the estimated fluxes.

Though sensors on geostationary satellites provide adequate temporal resolution, their spatial resolution is coarse even at the equator and becomes ever coarser away from the equator. They do not cover high latitudes at all. With polar orbiting satellites, the spatial resolution is higher - in some cases orders of magnitude higher - however the temporal frequency is much lower. Overpasses may be available twice or four times daily e.g. NOAA satellites, or much less frequently e.g. every 16 days for Landsat satellite.

Ideally, to test relationships between instantaneous estimates of surface fluxes and their daily totals, a long series of measurements is required so that a range of conditions are encountered. Having to combine satellite and ground-based data greatly reduces the amount of data available and introduces the complication of the unknown accuracy of the satellite estimates. To avoid these problems the following analysis uses only ground-based data to investigate relationships between hourly and daily total measurements of evaporation. The ground-based data were hourly measurements made over two types of natural vegetation in the Republic of Niger.

*Scaling up in hydrology using remote sensing.* Edited by J.B. Stewart, E.T. Engman, R.A. Feddes and Y. Kerr

Most previous studies to determine daytime totals from instantaneous data have concentrated on estimating daily evaporation. This chapter also reports on results for evaporation, even though sensible heat flux can be determined more directly from satellite data. Determination of evaporation requires estimation of the other components of the energy budget involving additional information and empiricism.

## THEORY

To determine the daytime evaporation total from satellite measurements at a single time of day, it is necessary to know the diurnal variation of evaporation. The latent heat flux, $\lambda E$, i.e. the energy associated with evaporation, can be described by the Penman-Monteith equation i.e.

$$\lambda E = \frac{\Delta A + \rho c_p D/r_a}{\Delta + \gamma (1 + r_s/r_a)} \qquad (1)$$

where $\Delta$ is slope of the curve of vapour pressure against temperature, A is the available energy (the sum of net radiation, $R_n$, and soil heat flux, G), $\rho$ and $c_p$ are the density and specific heat respectively of dry air, D is the vapour pressure deficit, $r_a$ is the aerodynamic resistance to the transfer of water vapour from the surface to the reference height, $\gamma$ is the psychometric constant and $r_s$ is the surface resistance. If the diurnal variation of these variables is known then the diurnal variation in evaporation can be determined.

Alternatively, there are a number of empirical relationships that can be used to extrapolate from instantaneous values to daily totals. Only the simple assumption of constant evaporative fraction is used here, since this method can be applied to satellite data without the need for any other source of data.

The assumption that the evaporative fraction, the ratio of latent heat flux to available energy or net all-wave radiation, is conservative leads to:

$$\int_{day} \lambda E dt = E_F \int_{day} (R_n - G) dt \qquad (2)$$

$$\text{where } E_F = \lambda E/(R_n - G) \qquad (3)$$

where $E_F$, the instantaneous evaporation fraction at the time of the overpass, is computed from the satellite data; $R_n$, the net all-wave radiation, is computed from geostationary satellite or ground based data and G, the soil heat flux, is either neglected or derived as an empirical fraction of $R_n$.

## SITES AND DATA

The sites were within or nearby the ICRISAT Sahelian Center (ISC) (latitude 13° 15'N, longitude 2° 17'E, altitude 240 m) about 45 km south of Niamey, Republic of Niger. The long-term mean precipitation is 564 mm with 95% occurring in the months of May to September. In 1989/1990 measurements were made at a site within ISC which had been used for agricultural crops and then remained fallow for over seven years allowing regeneration of savannah vegetation. The vegetation consisted primarily of annual grasses and woody shrubs with an average height of 2-3 m. Details of the vegetation cover are given in Table 1. Another site 6 km south-west of ISC was within an area of open natural forest, known locally as tiger bush. This vegetation is confined to dense bands about 10-30 m wide by 100-300 m long separated by a completely bare laterite soil.

Hourly measurements of the sensible and latent heat fluxes were made with an Institute of Hydrology Mk2 Hydra, an eddy correlation system using a vertical sonic anemometer, a fine-wire thermocouple and an infrared hygrometer (Shuttleworth *et al.*, 1988). Table 2 indicates the distribution of observation days throughout the measurement period.

**Table 1** Proportions of the components of the surface cover at sites near Niamey in Republic of Niger

| | | Surface Cover (%) | |
|---|---|---|---|
| Surface | Bare soil | Grasses/Forbes | Shrubs/Trees |
| Savannah | 3 | 78 | 19 |
| Open forest | 67 | | 33 |

## RESULTS

Between October 1989 and September 1990, there were more than 100 days with measurements from 0800 to 1700 over each type of vegetation. Since a number of polar orbiting satellites, e.g. Landsat, SPOT and ERS - 1/2, pass over the equator between 0900 and 1000 local solar time, the current analysis used the evaporative fraction calculated for this hour to estimate the daytime total latent heat flux.

Figure 1 shows the estimated total latent heat fluxes plotted against the measured values for (a) savannah and (b) open natural forest. These figures show that there is considerable scatter. The agreement between estimated and measured latent heat flux for open natural forest is poorer than for savannah: for the mean value of the measured daytime total latent heat flux, the 95% prediction limits for open natural forest are ±49% whereas for savannah it is ±42%. Regressions are given in Table 2.

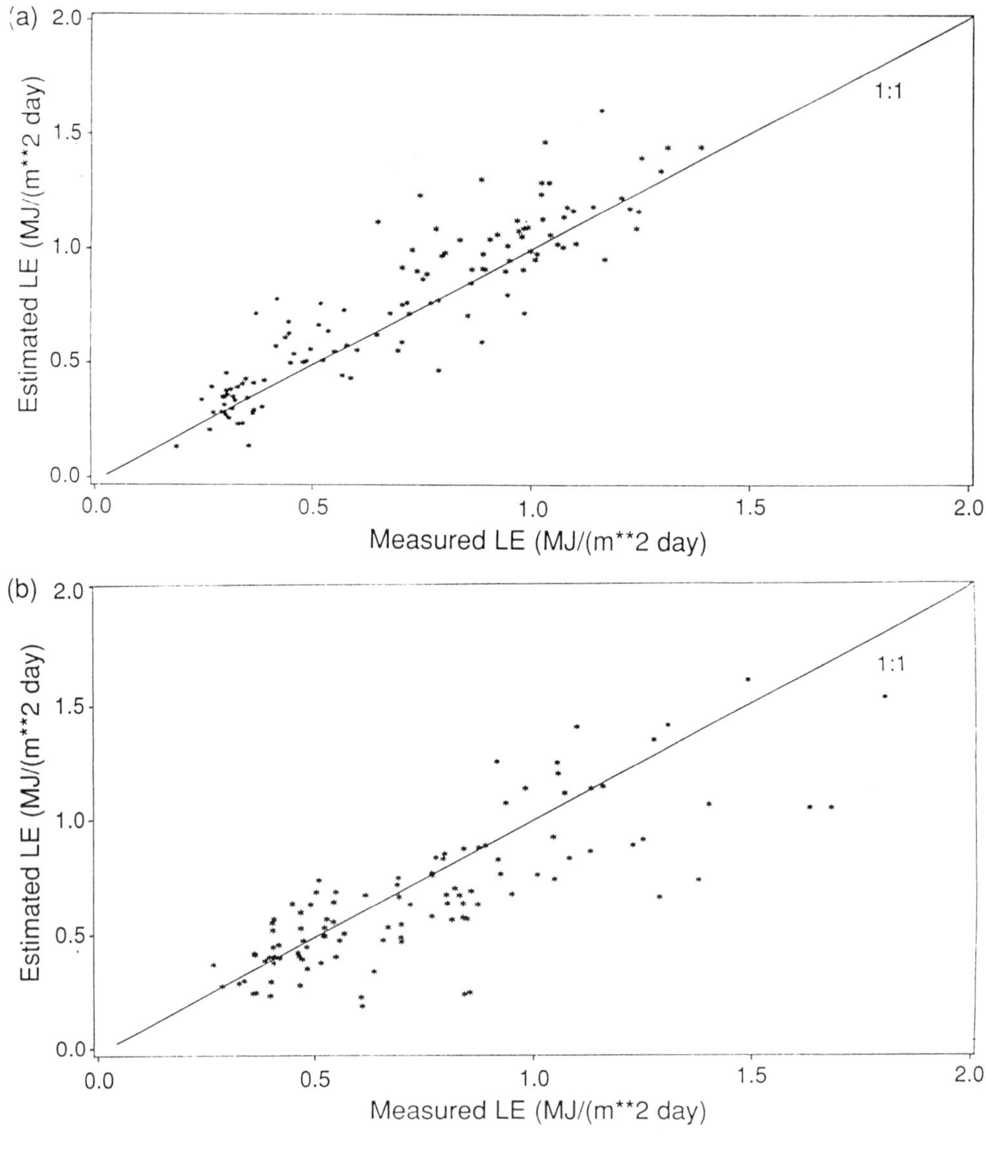


**Figure 1** Estimated against measured daytime total latent heat flux a) for savannah and b) for natural open forest. Daytime total latent heat flux was estimated from the daytime total net radiation and the ratio of latent heat flux to net radiation for the hour starting at 0900.

Though the overall bias is less than 15%, there is considerable scatter. This is demonstrated more explicitly in Figure 2 which shows the ratio of estimated to measured daytime total latent heat flux plotted against date. The agreement for open forest is poorer than that for savannah: for the mean value of the measured daytime total latent heat flux the 95% prediction limits for open forest are ±49% whereas for

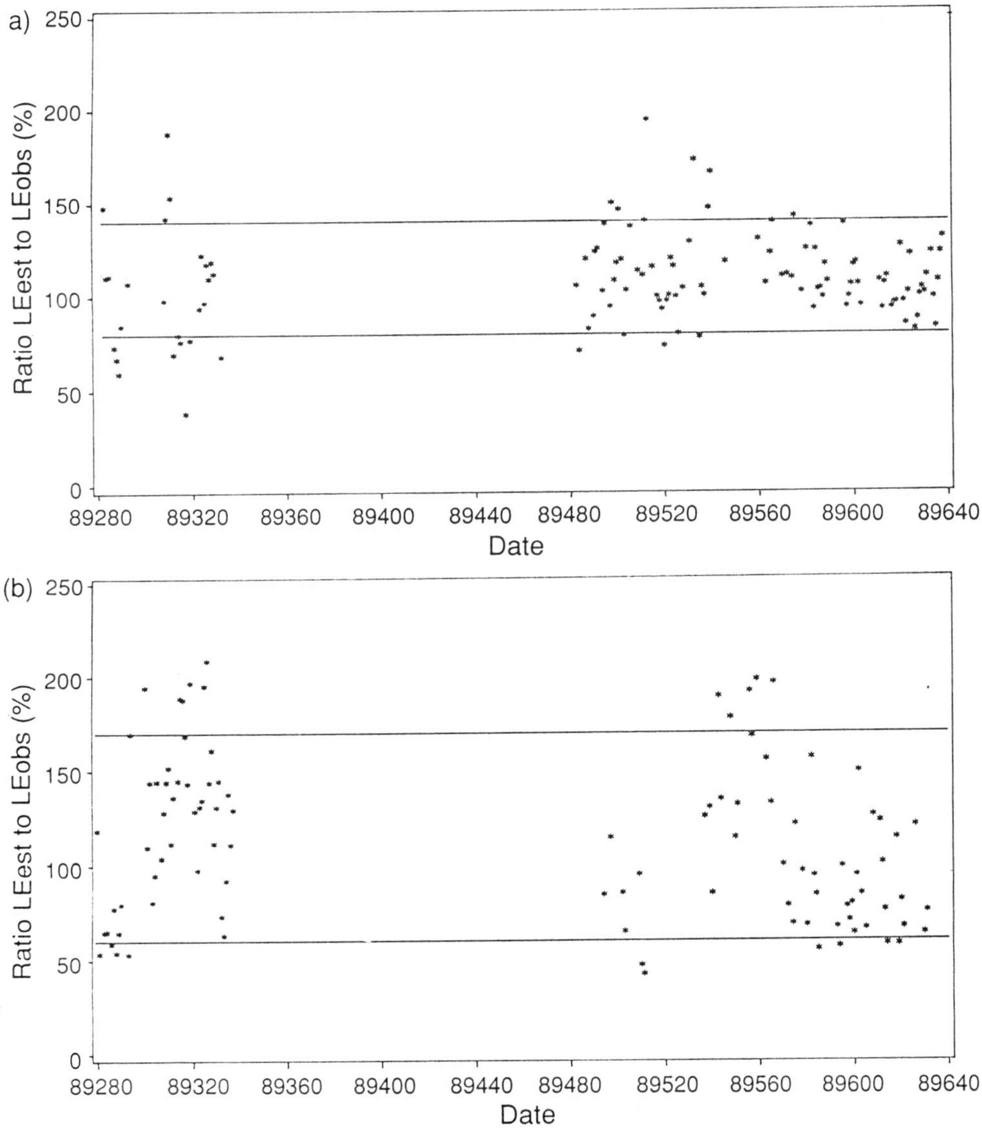


**Figure 2** Variation of the ratio of estimated to measured daytime total latent heat flux a) for savannah and b) for natural open forest for the period October 1989 to September 1990.

savannah it is ±42%.

The measured ratio of latent heat flux to net radiation and the net radiation for the hour 0900 to 1000, the daytime total measured and estimated latent heat fluxes and the ratio of estimated to measured fluxes for the whole period and individual months for each vegetation type is presented in Table 2. For savannah for the whole period,

**Table 2** Linear regression with an intercept and forced through the origin of calculated against measured daytime (0800 to 1700) total fluxes for two types of natural evegetation in Republic of Niger

| | Fallow / Savannah | | | | | | Tiger Bush | | | | | |
|---|---|---|---|---|---|---|---|---|---|---|---|---|
| Period | No. of Days | Intercept | Slope | $r^2$ (%) | Slope | $r^2$ | No. of days | Intercept | Slope | $r^2$ | Slope | $r^2$ |
| Oct 1989 | 11 | −0.305 ±0.356 | 1.306 ±0.364 | 54.3 | 1.000 ±0.069 | 55.5 | 22 | 0.045 ±0.095 | 0.846 ±0.151 | 59.2 | 0.914 ±0.045 | 60.7 |
| Nov 1989 | 17 | −0.496 ±0.205 | 2.498 0.587 | 51.6 | 1.092 ±0.096 | 37.0 | 27 | −0.198 ±0.126 | 1.528 ±0.284 | 51.8 | 1.089 ±0.042 | 49.1 |
| Dec 1989 | 0 | - | - | - | - | - | 3 | −0.091 ±0.017 | 1.171 ±0.045 | 99.7 | 0.934 ±0.98 | 95.7 |
| April 1990 | 2 | −0.171 | 1.634 | - | 0.961 ±0.155 | 94.9 | 0 | - | - | - | - | - |
| May 1990 | 21 | 0.014 ±0.060 | 1.158 ±0.152 | 74.2 | 1.192 ±0.051 | 75.4 | 7 | 0.290 ±0.208 | −0.006 ±0.305 | 0 | 0.412 ±0.062 | 86.2 |
| June 1990 | 19 | −0.108 ±0.224 | 1.287 ±0.362 | 39.3 | 1.116 ±0.065 | 41.9 | 5 | −0.083 ±0.353 | 0.905 ±0.354 | 58.0 | 0.824 ±0.054 | 68.0 |
| July 1990 | 8 | 0.047 ±0.235 | 1.158 ±0.274 | 70.7 | 1.211 ±0.053 | 74.7 | 14 | 0.497 ±0.291 | 0.400 ±0.224 | 14.3 | 0.774 ±0.052 | 1.7 |
| Aug 1990 | 18 | 0.111 ±0.245 | 0.982 ±0.257 | 44.3 | 1.098 ±0.034 | 46.9 | 18 | −0.169 ±0.142 | 1.159 ±0.163 | 74.4 | 0.971 ±0.043 | 73.8 |
| Sept 1990 | 24 | 0.228 ±0.140 | 0.805 ±0.135 | 59.9 | 1.021 ±0.026 | 57.0 | 11 | −0.135 ±0.296 | 1.104 ±0.308 | 54.2 | 0.966 ±0.052 | 57.8 |
| Oct 1989 to Sept 1990 | 120 | 0.043 ±0.033 | 1.019 ±0.043 | 82.4 | 1.071 ±0.018 | 96.9 | 107 | 0.105 ±0.043 | 0.751 ±0.054 | 64.8 | 0.870 ±0.022 | 93.4 |

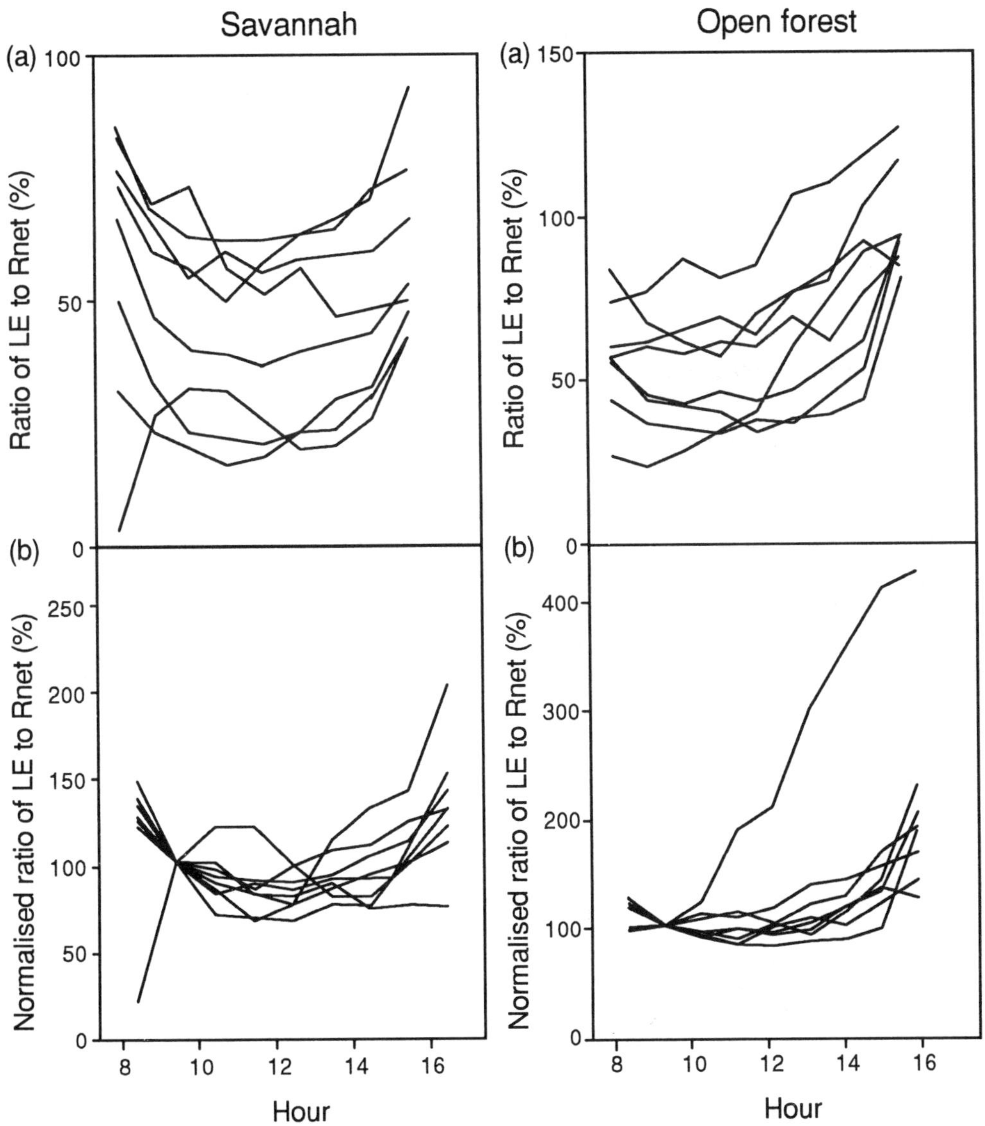


**Figure 3** (a) Diurnal variation of ratio of latent heat flux to net radiation for each month for savannah and natural open forest. (b) Diurnal variation of the same ratio after normalising to the value for the hour starting at 0900.

the ratio of estimated to measured latent heat flux is 108%, whereas for open natural forest it is 89%. So the departure of the mean estimated value is only just over 10%. However, the departure from the measured values can be very large for individual days. Figure 2 also presents the results for the individual days in the form of the ratio of estimated to measured daytime total latent heat flux plotted against the date. It shows that the highest and lowest deciles are concentrated around particular dates; but that these dates differ for the two types of vegetation.

Table 3 presents the linear regressions of estimated against measured daytime total latent flux for individual months for the two vegetation types. Linear regressions forced through the origin show that the slopes are greater than unity in 6 out of 8 months for savannah, whereas for open natural forest it is only 1 out of 8 months. Figure 3a presents the average monthly diurnal variation of the ratio of measured latent heat flux to the net radiation for savannah and open natural forest. In Figure 3b, the ratios have been normalised with respect to the ratio for the hour starting at 0900 for each month. It is then more obvious that more of the values lie below the 100% line for savannah than for open natural forest. The results for the open natural forest also show a larger variation from month to month than for savannah.

Theoretically, it would be expected that using either the available energy, i.e. the net radiation minus soil heat flux, or excluding days with variable cloud would both reduce the scatter in the relationship between estimated and measured daytime total latent heat flux. In tests with this data set there was no improvement using either of these modifications to the analysis. In the case of using measurements of soil heat flux to determine the available energy, it is thought that the uncertainty in these measurements introduced additional noise into the relationship. Excluding days with variable cloud considerably reduced the number of days available for the analysis and reduced $r^2$ by about half a per cent.

**Table 3** Linear regressions with an intercept and forced through the origin, of the estimated against measured daytime (0800 to 1700) total latent heat fluxes for two types of natural vegetation in Republic of Niger

| Dates | | 0900 to 1000 | | 0800 to 1700 | Estimated | |
|---|---|---|---|---|---|---|
| | No of days | $\lambda E/R_n$ (%) | $R_n$ (W m$^{-2}$) | $\lambda E$ (W m$^{-2}$) | $\lambda E$ (W m$^{-2}$) | Est/obs(%) |
| *Savannah* | | | | | | |
| 1 - 31 Oct 89 | 14 | 53.5 | 428 | 236 | 229 | 97.0 |
| 1 - 30 Nov 89 | 17 | 28.0 | 360 | 95 | 101 | 105.2 |
| 1 - 31 May 90 | 23 | 31.8 | 367 | 100 | 117 | 115.8 |
| 1 - 30 June 90 | 19 | 46.8 | 396 | 169 | 185 | 109.5 |
| 1 - 31 July 90 | 8 | 67.4 | 418 | 234 | 282 | 120.5 |
| 1 - 31 Aug 90 | 18 | 64.7 | 444 | 262 | 287 | 109.5 |
| 1 - 30 Sep 90 | 24 | 65.9 | 440 | 283 | 290 | 102.5 |
| *Open forest* | | | | | | |
| 1 - 31 Oct 89 | 23 | 44.9 | 325 | 166 | 146 | 88.0 |
| 1 - 30 Nov 89 | 27 | 43.8 | 302 | 122 | 132 | 108.2 |
| 1 - 31 May 90 | 7 | 20.7 | 373 | 186 | 77 | 41.4 |
| 1 - 30 June 90 | 5 | 60.6 | 366 | 273 | 222 | 81.3 |
| 1 - 31 July 90 | 14 | 72.5 | 372 | 351 | 270 | 76.9 |
| 1 - 31 Aug 90 | 18 | 58.8 | 372 | 234 | 219 | 93.6 |
| 1 - 30 Sep 90 | 11 | 67.3 | 373 | 263 | 251 | 95.4 |

# DISCUSSION

The results for the sites in the Republic of Niger show much more variation than the earlier results presented by Shuttleworth *et al.* (1989). Shuttleworth *et al.* (1989) showed that the evaporative fraction for a number of FIFE sites was nearly constant over the central part of the day, when most of the transfer occurs. Crago (1996a and 1996b) also found the evaporative fraction tended to be conserved during much of the daytime under a surprisingly wide range of conditions.

Earlier work (Jackson *et al.*, 1977; Seguin & Itier, 1983; Riou *et al.*, 1988; Lagouarde & McAneney, 1992) used mid-day surface temperature measurement combined with near-surface air temperature measurements to estimate daytime evaporation empirically. There is considerable scatter in the reported relationships caused by the variation in the parameter which is related to the transfer of sensible heat from the surface. This scatter obscures the variation in the relationships between instantaneous and daytime values.

Assessment of conditions necessary for the evaporative fraction to be constant can be obtained by rearranging the Penman-Monteith equation:

$$\frac{\lambda E}{A} = \frac{\Delta}{\Delta + \gamma(1 + r_s/r_a)} + \frac{\rho c_p D}{A(\Delta r_a + \gamma(r_a + r_s))} \quad (4)$$

of these quantities $c_p$ and $\lambda$ are constants, while $\rho$ and $\Delta$ only change slowly with temperature. $r_a$ depends on wind speed and shows no consistent change during daytime hours. A shows a strong sinusoidal variation with, on average, its maximum value at mid-day. The increase in temperature during the day depends on the magnitude of A and the interaction between the convective boundary layer and the surface fluxes. D depends on absolute humidity and on the change in temperature and generally peaks an hour or so after mid-day. The absolute humidity near the surface depends on the interaction of the convective boundary layer and the evaporation from the surface, therefore the diurnal variation of D depends on a complex interaction between a number of variables. Similarly, $r_s$ depends on a number of biological and environmental variables, the result is that $r_s$ usually increases during the day.

Since many of the variables controlling the evaporative fraction show pronounced diurnal variations, it could be ,expected that it is not conservative; in fact it is remarkable that the evaporative fraction can ever be considered as a conservative quantity. Further research studying the interaction of these variables, their interaction with each other, and with the development of the convective boundary layer, will be necessary to define the conditions when the evaporative fraction can be treated as conservative.

# CONCLUSIONS

Using more than 100 days of data, spread over a year, the total latent heat flux for the period 0800 to 1700 was estimated from the ratio of latent heat flux to net radiation for the 0900 to 1000 period and the total net radiation for the period 0800 to 1700. From the regression analysis of estimated against measured daytime evaporation, it can be concluded that averaged over more than 100 days this method estimates the daytime evaporation to within ±15% for two types of natural vegetation.

However, for individual days the estimated daytime evaporation could be as little as 40% or as much as 200% of the measured value. For the measurements over savannah 80% of the estimated values were between 80 and 140% of the measured values, while for open forest 80% lay between 60 and 170%.

Plots of diurnal variation of the ratio of latent heat flux to net radiation averaged over each month show considerable changes during the year particularly for the open forest. This variation could be due to a number of causes; such as diurnal variation in surface conductance (described in Gash *et al.*, 1991), the relative contributions to evaporation from vegetation and bare soil and the interaction of the surface fluxes with the convective planetary boundary layer.

The frequency of data from satellites is strictly limited, owing to the combination of revisit frequency and the occurrence of obscuring clouds. This analysis suggests that simple methods of extrapolating from satellite estimates to daytime totals of the fluxes are likely to introduce unacceptably large errors. Combining measurements from morning and afternoon overpasses of the NOAA satellites may improve the estimates of daytime total latent heat flux.

## ACKNOWLEDGEMENTS

This work was funded by the UK Overseas Development Administration and the UK Natural Environment Research Council, under a collaborative link between the Institute of Hydrology (IH) and the ICRISAT Sahelian Centre (ISC). The author thanks all the staff of IH and ISC who contributed to this successful set of measurements.

## REFERENCES

Crago, R.D. 1996a. Conservation and variability of the evaporative fraction during the daytime. *J. Hydrol.* In press.

Crago, R.D. 1996b. Daytime evaporation from conservation of surface flux ratios. This volume.

Gash, J.H.C., Wallace, J.S., Lloyd, C.R., Dolman, A.J., Sivakumar, M.V.K. and Renard, C. 1991. Measurements of evaporation from fallow Sahelian savannah at the start of the dry season. *Q.J.R. Met. Soc.*, **117**, 747-760.

Jackson, R.D., Reginato, R.J. & Idso, S.B. 1977. Wheat canopy temperature: a practical tool for evaluating water requirements. *Water Resour. Res.*, **13**, 651-656.

Lagouarde, J-P. & McAneney, K.J. 1992. Daily sensible heat flux estimation from a single measurement of surface temperature and maximum air temperature. *Boundary-Layer Meteorol.*, **59**, 341-362.

Riou, C., Itier, B. & Seguin, B. 1988. The influence of surface roughness on the simplified relationship between daily evaporation and surface temperature. *Int. J. Remote Sensing.*, **9**, 1529-1533.

Seguin, B. & Itier, B. 1983. Using mid-day surface temperature to estimate daily evaporation from satellite thermal IR data. *Int. J. Remote Sensing*, **4**, 371-383.

Shuttleworth, W.J., Gash, J.H.C., Lloyd, C.R., McNeil, D.D., Moore, C.J. & Wallace, J.S. 1988. An integrated micrometeorological system for evaporation measurement. *Agric. For. Meteorol.*, **43**, 295-317.

Shuttleworth, W.J., Gurney, R.J., Hsu, A.Y. & Ormsby, J.P. 1989. FIFE: the variation in energy partition at surface flux sites. Proc. Remote sensing and large scale global processes (Ed. A. Rango) *IAHS Publ. No. 186*, 67-74.